Lecture Notes in Electrical Engineering 233

For further volumes:
http://www.springer.com/series/7818

Mourad Fakhfakh, Esteban Tlelo-Cuautle, and
Rafael Castro-López (Eds.)

Analog/RF and Mixed-Signal Circuit Systematic Design

Springer

Editors
Mourad Fakhfakh
University of Sfax
Sfax
Tunisia

Esteban Tlelo-Cuautle
INAOE
Department of Electronics
Puebla
Mexico

Rafael Castro-López
Instituto de Microelectrónica de Sevilla
IMSE-CNM, CSIC. c/ Américo
Vespucio, s/n
Seville
Spain

ISSN 1876-1100
e-ISSN 1876-1119
ISBN 978-3-642-36328-3
e-ISBN 978-3-642-36329-0
DOI 10.1007/978-3-642-36329-0
Springer Heidelberg New York Dordrecht London

Library of Congress Control Number: 2012956292

Printed on acid-free paper

Springer is part of Springer Science+Business Media (www.springer.com)

Foreword

This book introduces contributions to the topic of systematic design of analog, RF and mixed signal circuits. In my view, the material represents a significant step towards advancing the state-of-the-art in the field of robust analog and mixed signal design automation. Although The topic has been researched for many years primarily for pure analog circuits, it is quite evident that extending the research effort to include mixed signal and RFIC design is very timely and relevant in light of the ever increasing complexity of complete Systems-on-Chip (SoCs) which include analog, RF and mixed signal on the same die with Ultra large scale digital. Such SoCs are also widely recognized as the "More-than-Moore" scaling extending the end of the silicon roadmap for many years to come. The field of systematic circuit design has long been a mature science area for many years for digital circuits but has represented a formidable challenge to analog circuits. However, the analysis and synthesis techniques and flows presented in this book provide practical solutions to meet the challenge. Not only does the material address some of the traditional bottlenecks of analog and RF design automation such as device sizing and layout generation, but also incorporates tools and methodologies to deal with worst case corners and random process variations. This will indeed result in automated and robust design solutions that lend themselves naturally to implementations in deep nanometer process nodes and with enhanced yields. Automation of these More-than-Moore SoCs will also help with meeting narrow market windows and with reducing development costs of complex nano-scale chip sets.

The material is organized in two parts and presented in 16 chapters. It strikes a good balance between theory and practice and includes case studies or design examples to reinforce understanding of basic concepts. The book is highly recommended for mixed signal, RF and SoC design engineers and practitioners in the semiconductor industry as well as researchers and graduate students in electrical and computer engineering with a major in circuit design and design automation.

November 2012 Mohammed Ismail

Mohammed Ismail
Ohio State University, Columbus, USA
Currently with Khalifa University of Science, Technology and Reserach (KUSTAR), UAE

Preface

Advances in electronics technologies have led to a kind of a 'boom' in a very wide range of fields, such as, informatics, bioengineering, communications, electronic gadgets, to name a few.

Despites the fact that in the digital domain, designers can take full benefits of IPs and design automation tools to synthesize and design very complex systems, the analog designers' task is still considered as a 'handcraft', cumbersome and very time consuming process. This is mainly due to the lack of support by computer-aided design programs, which has led to a so-called 'productivity gap' (difference between what technology can offer and what can be manufactured). Thus, tremendous efforts are being deployed by researchers, R/D engineers, etc. to develop new design methodologies in the analog/RF and mixed-signal domains.

Actually, the analog/RF and mixed signal fields rely on three major areas, namely Synthesis, Design and Optimization. These domains form a trilogy in this realm of analog/RF and mixed-signal circuit and system design. Endeavors are being made to develop new synthesis techniques (building novel active circuits, for instance), design methodologies (proposing new circuits) and sizing/optimization techniques (offering more complex functionalities with advanced performances, higher frequency operating ranges, less power consumption, etc.).

On this basis, this book collects in sixteen Chapters, recent theories, synthesis techniques and design methodologies, as well as new sizing approaches. It highlights their application to the design of high performance analog/RF and mixed-signal circuits and systems. This book is intended to researchers and R/D engineers, as well. The book encompasses two parts: *Methodologies* and *Techniques*.

The first part, *Methodologies*, is composed of seven Chapters, very briefly introduced in the following:

Chapter 1, entitled 'Towards Automatic Structural Analysis of Mixed-Signal Circuits', is proposed by *M. Eick and H. Graeb*. It presents a new method for the automatic structural and functional analysis of analog, digital and mixed-signal circuits.

Chapter 2, 'Efficient Synthesis Methods for mm-wave Frequency Passive Components and Amplifiers', authored by *B. Liu and G. Gielen*, deals with an efficient high-frequency synthesis methods for integrated passive components as well as for the synthesis of mm-wave-frequency linear amplifiers, using the memetic machine learning-based differential evolution method and the efficient machine learning-based differential evolution method, respectively.

Chapter 3, entitled 'Self-Healing Circuits Using Statistical Element Selection' and proposed by *V. H.-C. Chen, G. Keskin, and L. T. Pileggi*, analyzes the statistical element selection methodology for the implementation of low-power self-healing circuits and systems.

Chapter 4, 'Improving Design Feature Reuse in Analog Circuit Design through Topological-Symbolic Comparison and Entropy-based Classification', authored by *C. Ferent and A. Doboli* introduces a novel circuit synthesis methodology based on concept comparison, combination, learning, and re-use.

Chapter 5 that is entitled 'Graph-based symbolic and symbolic sensitivity analysis of analog integrated circuits' and proposed by *S. Rodriguez-Chavez, A.A. Palma-Rodriguez, E. Tlelo-Cuautle, and S.X.-D. Tan*, describes a graph-based technique for the solution of a system of equations for analog ICs formulated by applying symbolic NA and for symbolic sensitivity analysis.

Chapter 6 titled 'A Designer Centric Analog Synthesis Flow', which is authored by *F. Javid, S. Youssef, R. Iskander, and M.-M. Louërat*, presents a designer centric analog synthesis flow that is fully controlled by the designer and offers an intuitive design approach that is composed of a sizing tool and a layout generation tool.

Chapter 7; 'Analog Circuit Design based on Robust POFs using an Enhanced MOEA with SVM Models' by *N. Lourenço, R. Martins, M. Barros, and N. Horta* highlights a multi-objective design methodology for automatic analog integrated circuits synthesis, which enhances the robustness of the solution by varying technological and environmental parameters, and by the inclusion of corner cases.

The second part of the book, *Techniques*, encompasses the nine following Chapters:

Chapter 8; 'Applications of symbolic analysis in the design of analog circuits' by *F. Grasso, A. Luchetta, and M. C. Piccirilli*, describes the use of symbolic techniques in the realization of efficient automatic tools for designing analog circuits. In particular three phases of the design cycle of an integrated circuit are considered: the simulation phase, the design centering phase and the fault diagnosis phase.

Chapter 9, titled 'Synthesis of Electronically-Controllable Signal Processing/Signal Generation Circuits using Modern Active Building Blocks', is authored by *R. Senani, D. R. Bhaskar, A. K. Singh, and V. K. Singh* focuses on the synthesis of various electronically-controllable signal processing/signal generation circuits. The coverage includes the basics and hardware implementation of various building blocks mentioned above and includes some elegant representative applications using them.

Chapter 10, entitled 'Synthesis of Generalized Impedance Converter and Inverter Circuits Using NAM Expansion' by *A. M Soliman* proposes the use of the nodal admittance matrix expansion technique to generate all possible voltage generalized impedance converter and the current generalized impedance converter circuits, and the realizations of two types of the generalized impedance inverter circuits.

Chapter 11; 'Fractional Step Analog Filter Design', by *T. Freeborn, B. Maundy, and A. Elwakil* outlines the process to design, analyze, and implement continuous-time fractional-step filters, and presents new methods and design equations for the physical realization of these filters using fractional capacitors, SABs, FPAA hardware, and FDNR topologies.

Chapter 12, entitled 'The Flipped Voltage Follower: Theory and applications' and that is authored by *J. Ramirez-Angulo, M. R. Valero-Bernal, A. Lopez-Martin, R. G. Carvajal, A. Torralba, S. Celma-Pueyo, and N. Medrano-Marqués,* exposes and summarizes in a tutorial way, the most relevant information published to date on the FVF, and presents several improved FVF cells and structures and gives a comparison of their performances and characteristics.

Chapter 13, titled 'Synthesis of Analog Circuits using only Voltage and Current Followers as Active Elements', by *R. Senani, D.R. Bhaskar, A.K. Singh, and R.K. Sharma,* presents a brief account of some prominent works done on the analog circuit design using VFs and CFs as active elements, together with the design of VFs and CFs themselves.

Chapter 14; 'Design of Setable Active Lossy Inductors', proposed by *M. Pierzchala, and M. Fakhfakh* is concerned with transformation of passive LC filters into active RC-circuits using signal-flow graphs in the two-graph by using exclusively RC-elements and the newly introduced 'active switches'. The Chapter also deals with the reduction of the complexity of the constructed active circuits.

Chapter 15, entitled 'MIDAS: Microwave Inductor Design Automation on Silicon' by *L. Aluigi, F. Alimenti, L. Roselli, D. Pepe, and D. Zito* emphasizes a methodology to automate the design of microwave inductor on silicon and presents the implementation of an auxiliary CAD tool for Microwave Inductor Design Automation on Silicon.

Chapter 16; 'LC-VCO Design Challenges in the Nano-Era' authored by *P. Pereira, H. Fino, M. Fakhfakh, F. Coito, and M. Ventim-Neves* exposes an optimization based methodology for the design of LC-VCOs whose efficiency is granted by the use of analytical models to characterize the behavior of active and passive elements.

Finally, we want to use this opportunity to thank all the authors for their high quality contributions, and the reviewers for their valuable help. We are also thankful to Prof. Mohamed Ismail (Ohio State University, Columbus, USA. Currently with Khalifa University of Science, Technology and Reserach (KUSTAR), UAE) for writing the foreword of the book. Our thanks go also to the SPRINGER team for his support and assistance.

Mourad Fakhfakh
Esteban Tlelo-Cuautle
Rafael Castro-López

Contents

Part I: Methodologies

Part II: Techniques

Part I

Methodologies

Chapter 1
Towards Automatic Structural Analysis of Mixed-Signal Circuits

Michael Eick and Helmut Graeb

Abstract. A new approach for the structural analysis of integrated circuits is presented in this chapter. As a unique feature this approach can handle circuits that contain analog and digital components at the same time. Such a situation occurs, e.g., in mixed–signal circuits. First, the approach analyzes the circuit for basic analog and digital building blocks. Next, a structural signal flow analysis partitions the circuit into an analog and digital part. It is also used to determine true pass–gate directions and break feedback loops. Finally, the logic functions of the building blocks as well as the complete digital circuit part are extracted. The chapter presents application examples for digital standard cell libraries and mixed–signal circuits. For industrial grade standard cell libraries more than 95% of the contained cells are analyzed correctly. The mixed–signal examples include a charge pump as well as voltage–controlled ring oscillator.

1.1 Introduction

Mixed–signal circuits play an important role in most modern integrated circuits. Typical examples are analog–to–digital and digital–to–analog converters, voltage–controlled ring oscillators and charge pumps. Like pure analog circuits, mixed–signal circuits are subject to several constraints, e.g., certain MOSFET transistors must work in saturation region and special layout styles must be applied to some devices to achieve good matching. The availability of such constraints in machine–readable form is an indispensable prerequisite for the automation of design steps such as sizing and layout synthesis. Usually such a machine–readable documentation is not available, which requires algorithms to extract these constraints from the schematic.

Michael Eick · Helmut Graeb
Institute for Electronic Design Automation, Technische Universität München, Munich, Germany
e-mail: {eick,graeb}@tum.de

M. Fakhfakh et al. (Eds.): *Analog/RF & Mixed-Signal Circuit Sys. Design*, LNEE 233, pp. 3–25.
DOI: 10.1007/978-3-642-36329-0_1

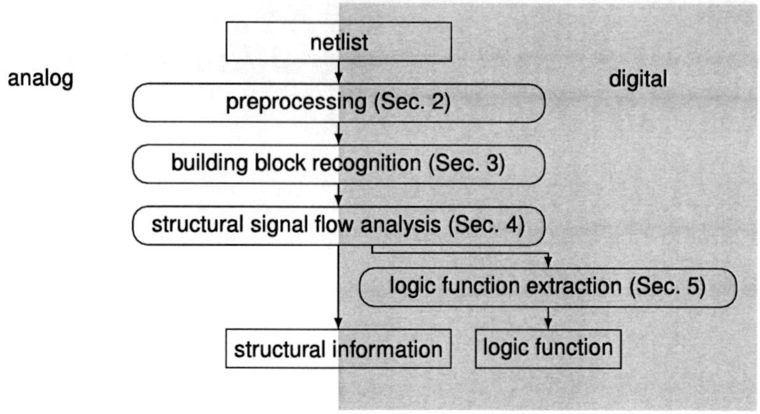

Fig. 1.1 Overall structural analysis flow

Previous work has shown that such constraints can be generated automatically for analog circuits [6, 7, 14]. The authors of [14] use building block recognition to identify analog blocks such as current mirrors. The available building blocks are defined through a library, which can contain CMOS and bipolar structures. Ambiguities are resolved using a dominance graph. The authors of [6, 7] compute symmetry in analog circuits using the recognized building blocks. Based on detected building blocks and symmetries, constraints for sizing and placement are generated.

These methods cannot be applied to mixed–signal circuits. This is because mixed–signal circuits consist of common analog components, such as current mirrors, common digital components, such as inverters and logic gates, as well as pass–gates and pass–transistors. In addition, continuous time and signal values can be assumed for analog circuits, for mixed-signal circuits time and signal values can be discrete.

Current approaches for the structural analysis of digital circuits can be divided into two classes. The first class assumes a CMOS structure and analyzes the parallel and serial connections of the transistors using special algorithms, e.g., [4, 5, 9, 11, 20]. These approaches can handle nearly all digital CMOS circuits but are limited to this type of circuit, which makes them infeasible for mixed–signal circuits. Some approaches can generate a graph representing the circuit structure, e.g., [11, 20]. The second class compares a netlist to a given library using subgraph isomorphism algorithms, e.g., [13, 16, 21, 22]. They are applicable for a wide range of circuit types but are limited to the provided library. Both approaches can yield a logic function for each identified subcircuit, which in turn allows to compute the overall logic function.

In this book chapter, we will present a new method enhancing the approaches of [7, 14] to handle mixed–signal circuits. The overall analysis flow is shown in Fig. 1.1. First, a netlist is read and some preprocessing is performed. After that, a building block recognition algorithm is executed. Compared to the state of the art, it provides the following new features,

- a versatile building block library for analog, digital and mixed–signal circuits,
- a corresponding dominance relation,
- a new recognition algorithm that can handle this library.

The approach uses a hierarchical library combining the benefits of library based approaches and algorithmic approaches. Next, a structural signal flow analysis is performed. It enhances the analysis presented in [6, 7] for analog circuits to handle digital and mixed–signal circuits. Algorithms to assign pass–gate directions and to break feedback loops are added. Finally, the logic function of the digital circuit part is extracted.

Preprocessing, enhanced building block recognition and structural signal flow analysis are discussed in Sections 1.2, 1.3 and 1.4, respectively. Section 1.5 introduces the logic function extraction algorithm. Application examples are shown in Section 1.6. Section 1.7 concludes the chapter.

1.2 Preprocessing

The netlist can contain parasitic resistors and capacitors which inhibit a correct building block recognition. Therefore parasitic devices are replaced by short–circuits and open–circuits as appropriate.

In addition, the source and drain assignment of MOSFETs in the netlist does not always match the actual assignment during operation. The actual assignment is required for correct building block recognition. It is determined by traversing the netlist from *Vdd*– to *Vss*–nets nets and vice versa.

1.3 Building Block Recognition

In the following, an algorithm is presented that recognizes basic building blocks, e.g., simple current mirrors (in analog circuits) and inverters (in digital circuits). This is done by comparing the circuit netlist to a given library of building blocks. A library for analog, digital and mixed–signal circuits is presented after some formal definitions. Next, a dominance relation is presented, which is used to resolve recognition ambiguities. Finally, the recognition algorithm is discussed.

A circuit consists of several devices such as MOSFETs. The set of all devices is $\mathscr{D}$. Each device $d \in \mathscr{D}$ has several attributes associated with it. We denote these attributes using a pseudo object–oriented notation, e.g., $d.a$ is attribute a of device d. A device d has the following attributes:

type t — The type $d.t \in T_{\mathscr{D}} = \{\text{trans}, \text{res}, \text{cap}, \ldots\}$ describes whether the device is a transistor (trans), a resistor (res), a capacitor (cap), etc.

subtype s — The subtype $d.s \in \{\text{none}, \text{nmos}, \text{pmos}\}$ is used to distinguish between NMOS and PMOS transistors.

Fig. 1.2 Stack chain consisting of three stacks

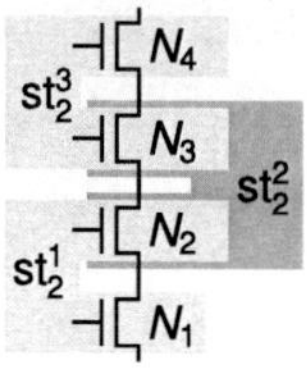

pins p — Pins are used to connect the device to nets. The set $d.p$ lists the available pins, e.g., $d.p = \{\mathrm{gt}, \mathrm{dn}, \mathrm{sc}\}$ for a mosfet-transistor with gate (gt), drain (dn) and source (sc).

Definition 1 (building block). A building block $b \in \mathscr{B}$ consists of several devices or other building blocks, where $\mathscr{B}$ is the set of all building blocks. It has the following associated attributes:

children c — A tuple $b.c \in (\mathscr{D} \cup \mathscr{B})^{n_{c,s}}$ listing the included building blocks or devices, where $n_{c,s}$ is the number of children.
type t — A type $b.t \in T_{\mathscr{B}}$ similar to the type defined for devices. The available building block types depend on the used library.
subtype s — A subtype $b.s$ similar to the type defined for devices.
pins p — A set of pins $b.p$ similar to the type defined for devices.

Devices and building blocks connect to the nets $n \in \mathscr{N}$ of the circuit using their pins.

Definition 2 (Connectivity function η)
The connectivity function $\eta(x,p) \in \mathscr{N}, x \in (\mathscr{D} \cup \mathscr{B}), p \in x.p$ describes the connectivity of a circuit. A device or building block x connects to a net n by pin p iff $n = \eta(x,p)$.

1.3.1 Analog, Mixed-Signal and Digital Building Block Library

The recognition algorithm is based on the building block library shown in Fig. 1.3. The unshaded part covers analog building blocks, the gray shaded part covers digital building blocks and the gray striped part covers building blocks used in analog and digital circuits. The figure does not show the complete library of analog building blocks, which can be found in [14].

The library is based on three different generic building blocks.

pair — A pair consists of two building blocks or devices, e.g., a simple current mirror or a stack.
array — An array consists of n building blocks or devices connected in parallel, e.g., a diode transistor array.
chain — A chain consists of pairs y_1, y_2 to y_n, where two pairs $y_i, y_{i+1}, i = 1 \ldots (n-1)$ share one child. Figure 1.2 illustrates this for a stack chain consisting of stacks st_2^1 to st_2^3. Stacks st_2^1 and st_2^2 share N_2, stacks st_2^2 and st_2^3 share N_3. A chain can have a single child only.

For a part of the building blocks, all children have the same subtype, i.e., they are either all nmos or all pmos. For these building blocks only the nmos variant is shown in Fig. 1.3. Examples are simple current mirror and stack. Other building blocks consist of children with nmos and pmos type. Examples are logic gate and pass–gate.

The library is organized into different hierarchy levels. Building blocks from one hierarchy level are built out of building blocks from lower hierarchy levels. The lowest hierarchy level, hierarchy level 0, is formed by the transistors from the netlist. The overall number of hierarchy levels n_L depends on the circuit and is automatically determined by the recognition algorithm.

Hierarchy level 1 contains building blocks that group parallel transistors together. For example, a diode transistor array, consists of parallel diode connected transistors.

Hierarchy level 2 contains the analog building blocks simple current mirror (scm), voltage reference II (vrII), differential pair (dp) and level-shifter (ls). Simple current mirror and level–shifter consist of a diode transistor array connected to a normal transistor array. The other building blocks consist of normal transistor arrays only. Stack, pass–gate and cross–coupled pair can be used for analog as well as digital circuits. Logic gate, logic array and stack chain are useful for digital circuits only. The current hierarchy level is used as index for stack, logic gate, logic array and stack chain because they are repeated on higher hierarchy levels. The gate pins of the logic gate can be connected to an inverter or independently controlled.

Hierarchy Level 3 contains a stack chain which is needed for digital circuits only. It is constructed from multiple stack building blocks that overlap at one transistor.

The analog part of *hierarchy level 4* contains the cascode current mirror, which is formed from a simple current mirror and a level–shifter as well as the wide–swing current mirror, which is formed from a voltage reference II and a stack from level 2. For digital circuits the tristate base block is defined. It consists of a pass–gate and a logic array.

For all *even hierarchy levels starting from 4 up to* n_L digital building blocks are defined recursively. A logic array on hierarchy level $k = 4, 6, \ldots$ can be formed by stack chains from lower hierarchy levels as well as normal transistor arrays. At least one of the stack chains must be from hierarchy level $k-1$. The same principle applies to stacks which are formed out of logic arrays and normal transistor arrays. A logic gate combines a logic array, stack chain or normal transistor array with PMOS–subtype and a logic array, stack chain or normal transistor array with NMOS–subtype.

All *odd hierarchy levels starting from 5 up to* $n_L - 1$ contain a stack chain which is formed from stacks from the hierarchy level before.

The analog part of *hierarchy level 6* defines the differential stage, consisting of a current mirror and a differential pair. In addition to the recursively defined building blocks, the digital part of hierarchy level 6 contains the tristate control block, which consists of two tristate base blocks. It is needed to handle one type of tristate buffers correctly.

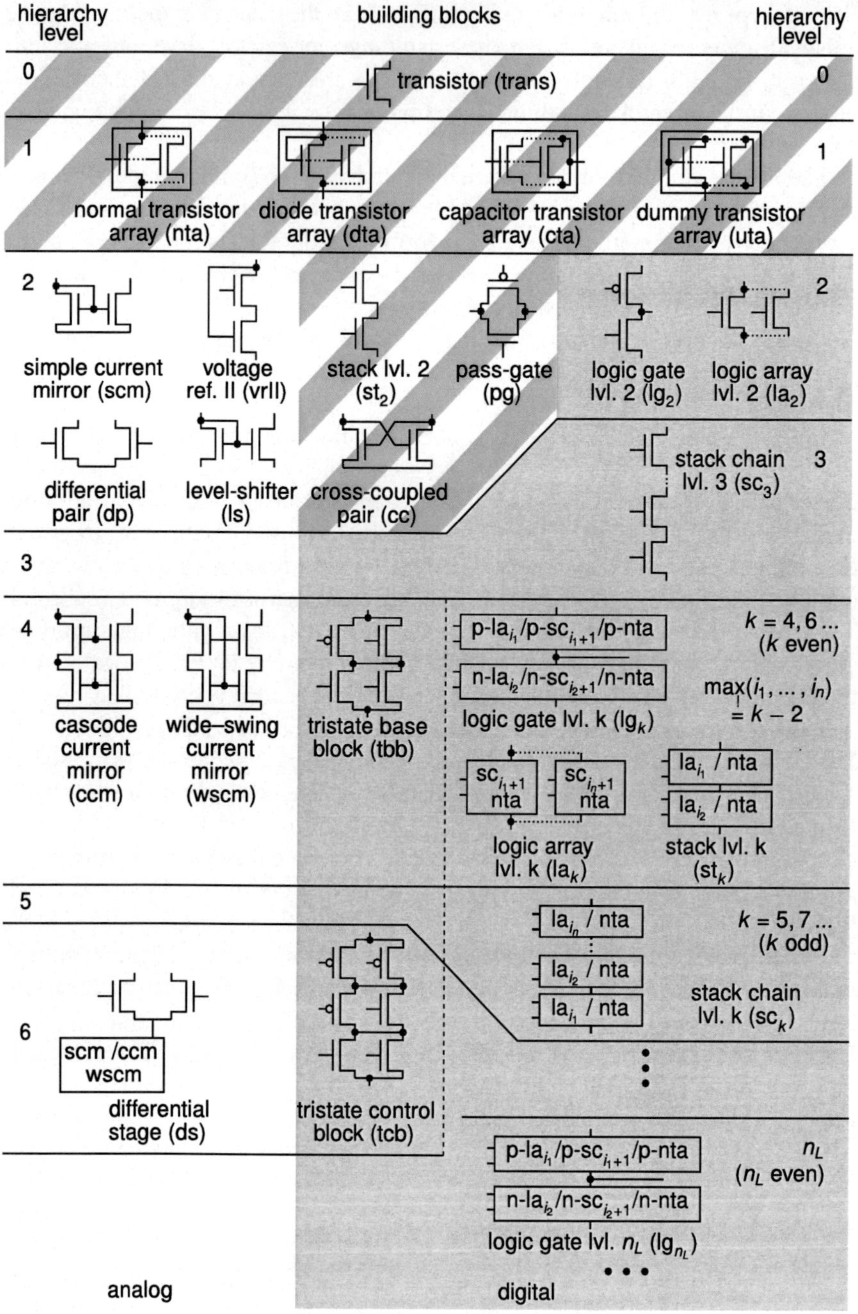

Fig. 1.3 Library for building block recognition of analog, mixed-signal and digital circuits. The analog part shows a subset from the library presented in [14].

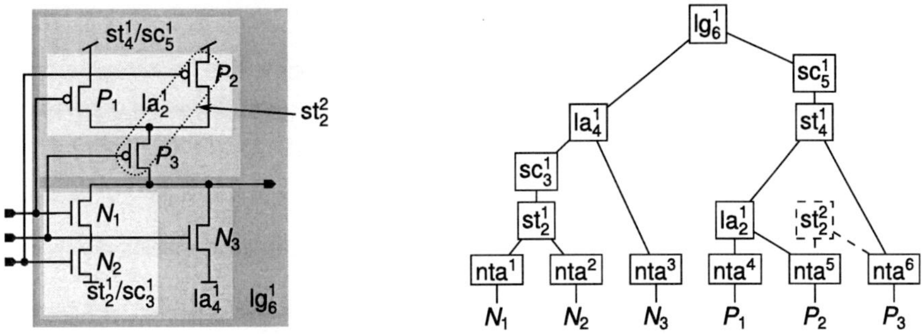

Fig. 1.4 And-nor gate [18] with recognized building blocks

Figure 1.4 illustrates how this library can be used to recognize the building blocks of the and-nor gate from [18]. First, normal transistor arrays nta^1 to nta^6 are recognized for every transistor in the circuit. After that, a stack st_2^1 covering N_1, N_2 and a logic array la_2^1 covering P_1, P_2 are found. For the third hierarchy level a stack chain (sc_3^1) is formed out of stack st_2^1. In hierarchy level four a logic array (la_4^1) and a stack (st_4^1) are recognized. Stack st_4^1 becomes part of a stack chain sc_5^1 on hierarchy level five. Finally, logic gate lg_6^1 is recognized on hierarchy level six.

Comparing the netlist to the library does not unambiguously yield this result. Additional building blocks can be recognized, e.g., the stack st_2^2. Normal transistor array nta^5 would be part of la_2^1 and st_2^2 at the same time. In the following, we will show how such ambiguities can be resolved by determining a dominating building block, i.e., one building block is kept and one is removed.

1.3.2 *Recognition Conflicts and Their Resolution*

For pairs used in analog circuits an ambiguity resolution concept was presented by [14]. An enhanced version, capable of handling chains and arrays as well, is described in the following.

For ambiguity resolution two building blocks are considered together with their transitive children. The set of transitive children $C_\star(x)$ of a building block x contains the children $x.c$ of x as well as all elements of their sets of transitive children, i.e.,

$$C_\star(x) = \begin{cases} \bigcup_{y \in x.c} \left(\{y\} \cup C_\star(y)\right) & x \in \mathscr{B} \\ \emptyset & x \in \mathscr{D} \end{cases} . \tag{1.1}$$

Set $C_i(x)$ is the set of transitive children limited to the i-th child $x.c_i$ of x, i.e.,

$$C_i(x) = \{x.c_i\} \cup C_\star(x.c_i) \tag{1.2}$$

The ambiguity resolution is based on a dominance graph (Fig. 1.5).

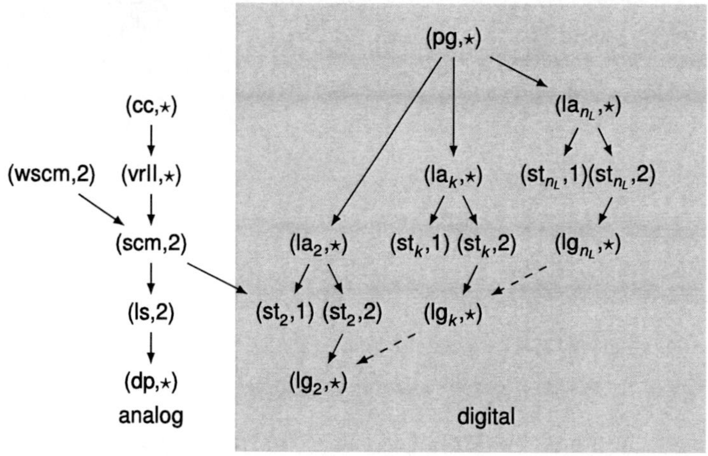

Fig. 1.5 Dominance graph for the library shown in Fig. 1.3. The analog part is based on [14].

Definition 3 (Dominance Graph). A dominance graph G_D is a directed graph $G_D = (N_{G_D}, E_{G_D})$. The nodes are pairs $(t,i) \in N_{G_D} = T_{\mathscr{B}} \times \{1,2,\star\}$, where t is a building block type and i refers to one of the set of transitive children defined above. The edges are pairs of nodes, i.e., $E_{G_D} = N_{G_D}^2$.

Definition 4 (Dominance). A building block x_1 dominates a building block x_2 iff

$$\exists_{(i,j)\in\{1,2,\star\}^2} \left(C_i(x_1) \cap C_j(x_2) \neq \emptyset\right) \wedge \mathrm{reachable}_{G_D}((x_2.t, j), (x_1.t, i)) . \quad (1.3)$$

The first part checks if there is a common transitive child using one of the sets C_1, C_2 and $C_\star$. The second part checks if the node in the dominance graph corresponding to x_2 is reachable from the node corresponding to x_1. Function $\mathrm{reachable}_{G_D}(\mu, \nu)$ is true if node μ is reachable in G_D from node ν. This definition is based on [14].

The dominance graph for the building block library for analog, digital and mixed–signal circuits is shown in Fig. 1.5. The left, non–shaded part handles conflicts between analog building blocks. It is based on the graph presented in [14]. The right, gray shaded part handles conflicts between digital building blocks. It has to consider the recursive nature of the library. Inside each hierarchy level the following holds: Transistors that are part of a logic array must not be part of a stack. The upper transistor of a stack must not be part of a logic gate. In the example (Fig. 1.4) this prevents recognition of a false logic gate consisting of N_1 and P_3. A logic gate from a higher hierarchy level will always dominate logic gates from a lower hierarchy level. This means, in case that multiple logic gates are detected only the largest one is kept. This includes transitive relations, e.g., a stack from level 4 will dominate a logic gate from level 2. In case a transistor is part of a pass–gate it must not be part of another building block.

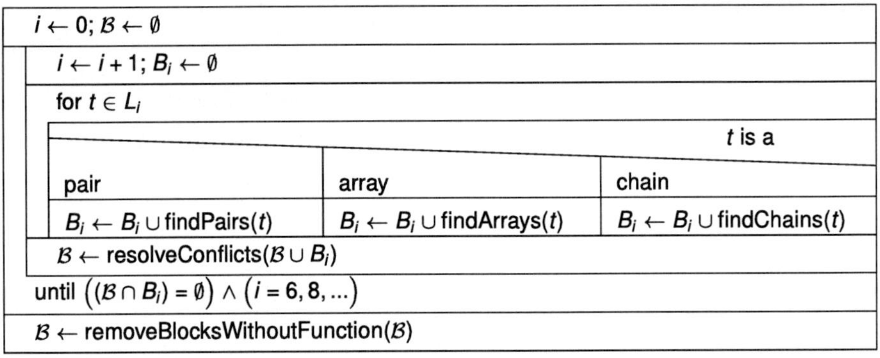

Fig. 1.6 Building block recognition algorithm

This allows to resolve the conflict from the example (Fig. 1.4). Building block nta^5 is a child of la_2^1 and second child of st_2^2, i.e., $C_\star(\mathrm{la}_2^1) \cap C_2(\mathrm{st}_2^2) = \{P_2, \mathrm{nta}^5\}$. Since node $(\mathrm{st}_2, 2)$ is reachable from $(\mathrm{la}_2, \star)$, logic array la_2^1 dominates stack st_2^2.

1.3.3 Recognition Algorithm

The recognition algorithm for analog, digital and mixed-signal circuits is shown in Fig. 1.6. It is based on the algorithm presented in [14]. It was enhanced to handle the recursive library, recognize arrays and chains as well as recognize pairs faster.

The algorithm iterates over all hierarchy levels $L_i \subseteq T_{\mathscr{B}}, i = 1, 2 \ldots n_L$ of the library. In each iteration, pairs, arrays and chains are found by calling functions findPairs, findArrays and findChains, respectively. These functions are discussed below. All building blocks recognized for the current hierarchy level are collected in set B_i. Conflicts are resolved in each hierarchy level, leading to an update of the overall set of recognized building blocks $\mathscr{B}$. In contrast to a conflict resolution at the very end as suggested by [14], this has the benefit that the overall number of building blocks is kept low. Consequently, less components must be considered during subsequent steps. According to Definition 4, it is sufficient to check for each new building block $x_1 \in B_i$,

- if it is dominated by some other building block $x_2 \in B_i \cup \mathscr{B}$, or,
- if it dominates some other building block $x_2 \in B_i \cup \mathscr{B}$.

The outer loop ends if the following two conditions are met,

- no new building blocks were found in this iteration or all found building blocks were dominated, and,
- the current hierarchy level number is even and greater or equal to six.

Finally, building blocks are removed that do not have a function if they are not part of a bigger building block. For example, voltage references II, which are not part of a wide–swing cascode current mirror, are removed.

Fig. 1.7 Function findPairs

```
findPairs(t)
B ← ∅
X ← candidatePairs(t)
for (c1, c2) ∈ {(c1, c2) ∈ X | r_t(c1, c2)}
    B ← B ∪ {newPair(t, c1, c2)}
return B
```

1.3.3.1 Finding Pairs

Function findPairs is shown in Fig. 1.7. First a set of candidate pairs $X \subseteq (\mathscr{B} \cup \mathscr{D})^2$ is determined. Below, this will be described in more detail. Next, a rule function r_t is evaluated for each of these candidate pairs. In case the function is true for a pair, a new pair is created and added to the set of found pairs, which is returned in the end. The rule function r_t is specific for each pair type t. It can contain conditions about type, subtype, required and forbidden connections as well as existence of parents. For example, the rule function $r_{\mathrm{st}_k}(x_1, x_2)$ for stack type st_k on level k contains the following conditions,

$$\begin{array}{lcl} & (x_1.t, x_2.t) \in \left\{ \begin{array}{l} (\mathrm{la}_{k-2}, \mathrm{nta}),\ (\mathrm{la}_{k-2}, \mathrm{la}_2),\ \cdots) \\ (\mathrm{nta}, \mathrm{la}_{k-2}),\ (\mathrm{la}_2, \mathrm{la}_{k-2}),\ \cdots) \end{array} \right\} & \text{(type)} \\ \wedge & x_1.s = x_2.s & \text{(same subtype)} \\ \wedge & \eta(x_1, \mathrm{dn}) = \eta(x_2, \mathrm{sc}) & \text{(required connection)} \\ \wedge & \eta(x_1, \mathrm{sc}) \neq \eta(x_2, \mathrm{dn}) & \text{(forbidden connections)} \\ \wedge & \mathrm{parents}(x_1) = \mathrm{parents}(x_2) = \emptyset & \text{(no parents)} \end{array} \tag{1.4}$$

The *type* condition requires one component to be a logic array from hierarchy level $k-2$. The other component can either be a normal transistor array or another logic array from any hierarchy level. Both components must have the *same subtype*. The *required connection* condition requires the drain of the first building block to connect to the source of the second building block. The *forbidden connection* condition forbids a connection between source of the first building block and drain of the second building block. Both components must have *no parents*.

Runtime of findPairs is dependent on the number of candidate pairs X. This number can be kept low by including some of above conditions in the candidate pair computation. The authors of [14] use all pairs of devices and building blocks that are of correct types. The authors of [8] use all pairs that are at least connected by one net. We combine both methods. For the stack at level k candidate set X is computed as follows,

$$X_1(n) = \{x_1 \in \mathscr{D} \cup \mathscr{B} | x_1.t \in \{\mathrm{nta}, \mathrm{la}_2, \mathrm{la}_4, \ldots\}) \wedge \eta(x_1, \mathrm{dn}) = n\} \tag{1.5a}$$

$$X_2(n) = \{x_2 \in \mathscr{D} \cup \mathscr{B} | x_2.t \in \{\mathrm{nta}, \mathrm{la}_2, \mathrm{la}_4, \ldots\}) \wedge \eta(x_2, \mathrm{sc}) = n\} \tag{1.5b}$$

$$X = \bigcup_{n \in \mathscr{N}} X_1(n) \times X_2(n) \tag{1.5c}$$

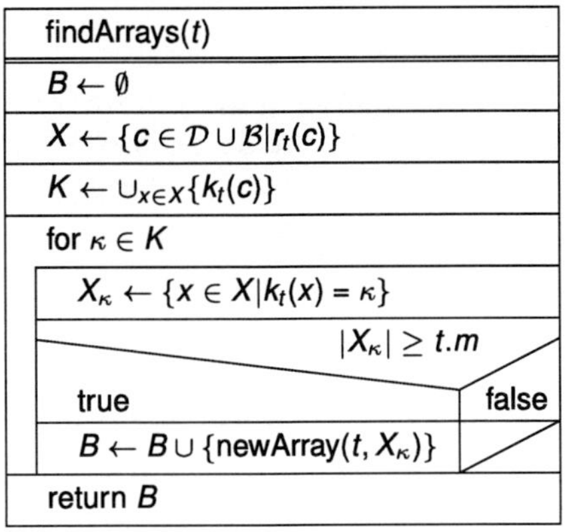

Fig. 1.8 Function findArrays

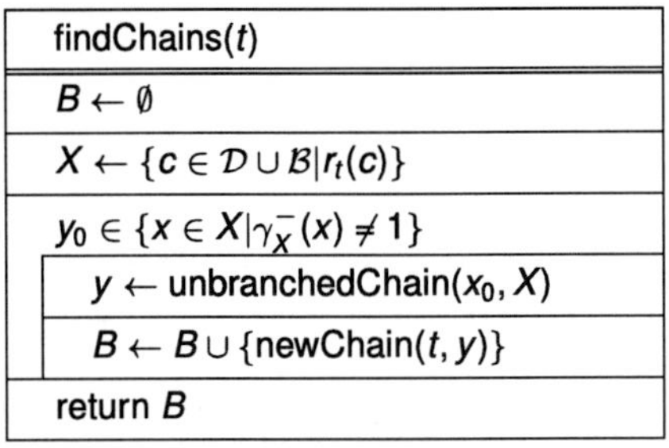

Fig. 1.9 Function findChains

Functions $X_1(n)$ and $X_2(n)$ return candidates for the first and second component for a specific net $n \in \mathscr{N}$, respectively. Pairs are then computed for each net $n \in \mathscr{N}$ by evaluating $X_1(n)$ and $X_2(n)$. The resulting set X only contains pairs where the connection condition and parts of the type condition are fulfilled.

1.3.3.2 Finding Arrays

The algorithm to find arrays is depicted in Fig. 1.8. First, the algorithm creates a set X of candidate children by evaluating a rule function r_t, which is specific for each array type t. It can consist of conditions about type, subtype, connectivity and existance of parents. The rule function $r_{\text{dta}}(x)$ for a diode transistor array contains the following conditions:

$$\underbrace{x.t = \text{trans}}_{\text{(type)}} \wedge \underbrace{\eta(x,\text{gt}) = \eta(x,\text{dn})}_{\text{(required connection)}} \wedge \underbrace{\eta(x,\text{dn}) \neq \eta(x,\text{sc})}_{\text{(forbidden connection)}} \tag{1.6}$$

It enforces type transistor and a gate drain connection. It forbids a connection between drain and source. The key function k_t maps each component in X to a tuple of nets, such that components connected in parallel get the same key. The key function k_{dta} for a diode transistor array is,

$$k_{\text{dta}}(x) = (\eta(x,\text{dn}), \eta(x,\text{sc})). \tag{1.7}$$

This means, for a diode transistor array all transistors are grouped together that connect to the same net at their drain pins and their source pins. If more than a minimum number $t.m$ building blocks are connected in parallel then a new array is created. For the diode transistor array dta.$m = 1$, i.e., an array is always created. Finally, the set B of new arrays is returned.

1.3.3.3 Finding Chains

Function findChains is shown in Fig. 1.9. First, the algorithm computes a set X of candidate children using a rule function r_t, which is specific for each chain type t. It can use the conditions described for arrays. All candidate children must be pairs. The rule function $r_{\text{sc}_k}(x)$ for a stack chain on level k contains the following conditions:

$$x.t = \text{st}_{k-1} \,. \tag{1.8}$$

It requires x to be a stack on level $k-1$.

Next, all tuples $y = (y_0, y_1, \dots y_{\text{last}})$ with the following properties are found:

- $\gamma^-(y_0) = |\{x \in X | x.c_2 = y_0.c_1\}| \neq 1$, i.e., more than one or no candidate in X share the second child with y_0.
- $y_i.c_2 = y_{i+1}.c_1$ for $y_i \neq y_{\text{last}}$, i.e., the second child of each building block y_i is the first child of the next building block y_{i+1}.
- $|\{x \in X | x.c_1 = y_{\text{last}}.c_2\}| \neq 1$, i.e., the chain can not be continued beyond y_{last}.

Finally, a new chain is created for each y and returned in B.

1.3.3.4 Discussion

The analog building block recognition described in [14] is included in this algorithm. It corresponds to the analog part of the library in Fig. 1.3 and the dominance graph in Fig. 1.5. The algorithm corresponds to the algorithm of Fig. 1.6 when all building blocks are pairs. Consequently, the results obtained for the algorithm of [14] can be transfered to the new algorithm.

The authors of [1] suggested to recognize simple current mirrors and level-shifters by recognizing diode connected transistors first. Application of the principle from [1] to the library from [14] resulted in the new hierarchy level 1 shown in Fig. 1.3. This has the advantage of faster recognition of pairs because less rules must be evaluated.

1.4 Structural Signal Flow Analysis

After applying the building block algorithm from the previous section to a circuit, basic building blocks such as pass-gates or simple current mirrors are known. Figure 1.10 illustrates this for a latch from [19]. It consists of logic gates lg_2^1 to lg_2^3 as well as pass–gates pg_2^1 and pg_2^2. This information is now used to generate the Enhanced Structural Signal Flow Graph [6] (ESFG) of the circuit, which combines qualitative behavioral and structural information. This graph is then used to assign a direction to each pass–gate, partition the circuit into an analog and digital part and to identify feedback loops.

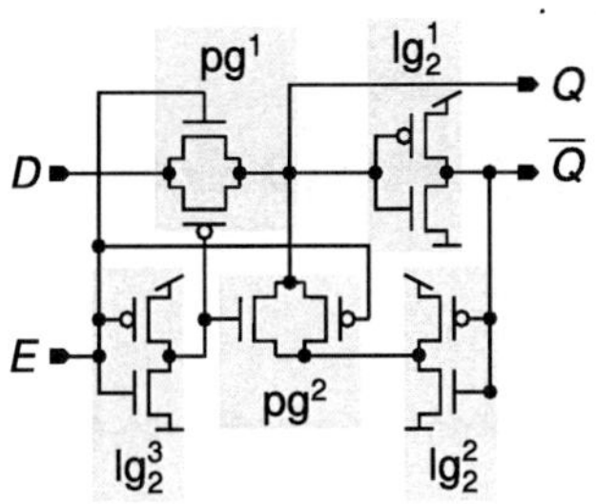

Fig. 1.10 Latch [19] with recognized building blocks

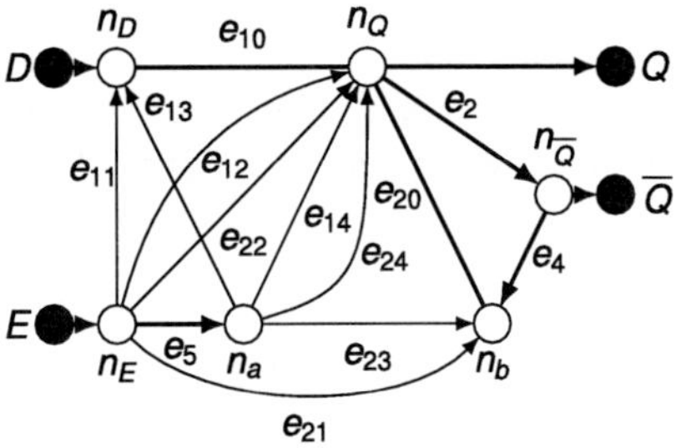

Fig. 1.11 Generated ESFG

1.4.1 Generation

An ESFG [6] is a directed graph. The nodes of the graph are formed by the nets of the circuit. An edge models a qualitative influence from one net to another. An edge from net n_i to n_j means that a change of a branch current or voltage of n_i causes a change of a branch current or voltage of n_j. The relation between edges and the recognized building blocks is modeled by some edge attributes. Only top–level building blocks without parents are considered. The ESFG is generated as follows.

- For a logic gate edges from each input to the output are generated.
- A pass–gate generates an undirected edge from drain to source. Directed edges are generated from both gates to drain and source. These edges are called control edges.
- For analog building blocks the generation is described in [6, 7]. For example, for current mirrors one edge from the input to the output is generated.
- For each port of the circuit a port node is generated and connected to the corresponding net.

For the latch example this is illustrated by Fig. 1.11. Logic gates lg_2^1 to lg_2^3 are represented by edges e_1 to e_3. Pass–gate pg^1 is represented by undirected edge e_{10} and control edges e_{11} to e_{14}. Pass–gate pg^2 is represented by edges e_{20} to e_{24}. Circuit ports E, D, Q, $\overline{Q}$ are represented by port nodes.

1.4.2 Assignment of Pass–Gate Directions

After the generation step, pass–gates are represented as undirected edges, e.g., edge e_{10} in Fig. 1.11. In reality pass–gates are only used in one direction. The problem is related to the problem of determining the signal flow direction of transistors in switch–level simulation [2]. However, only a small part of the ESFG edges is undirected in our case, which allows to used a different approach which is described in the following.

Assume e is an undirected edge connecting nodes ν and μ. It is replaced by a directed edge from μ to ν, if the following conditions hold simultaneously.

- An output node is reachable from ν without traversing e, and,
- no edge representing a logic gate ends at ν.

Simultaneously with the assignment, edges pointing from the control inputs of the pass–gate to μ are removed.

For the undirected edge e_{10}, connecting n_D and n_Q in the example of Fig. 1.11, output node $n_{\overline{Q}}$ is reachable from n_Q but not from n_D without traversing edge e_{10}. Consequently the edge points from n_D to n_Q. The undirected edge e_{20} between n_2 and n_Q is found to point from n_2 to n_Q. The control edges are removed accordingly (Fig. 1.12).

In some cases, it is not possible to assign directions to all pass–gates at once. In these cases above conditions are repeatedly evaluated for all pass–gates without assigned direction. In each iteration at least one pass–gate direction is assigned. The algorithm needs n_{pg} iterations at maximum, where n_{pg} is the number of pass–gates.

The computation of the logic functions (see Section 1.5.1) for building blocks can be done before this step. In this case, it can be checked whether the output of a logic gate can be in high impedance state.

1.4.3 Analog / Digital Partitioning

For further processing, the ESFG must be partitioned into an analog and digital part. Therefore a signal type is assigned to each node. This signal type can be either *unknown*, *analog* or *digital*. The signal type for each node is determined based on the edges of the graph and the building blocks they represent. For each building block type a specific set of conditions for the connected nodes exists. For example, input and output of a current mirror must be of type analog. In addition, the user can specify the signal type of inputs and outputs of the circuit. Overall, we get a set of conditions forming a constraint satisfaction problem which is solved by a constraint programming method, e.g., [17].

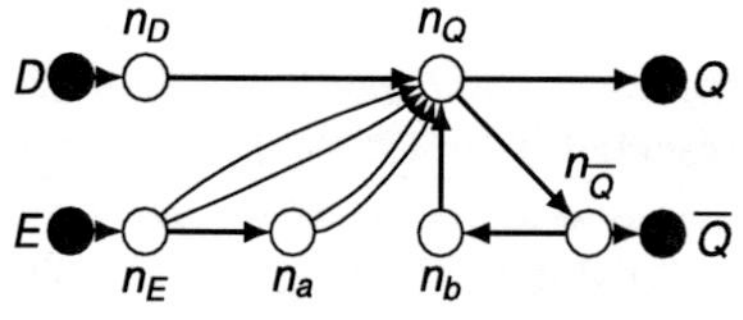

Fig. 1.12 ESFG after assignment of pass-gate directions

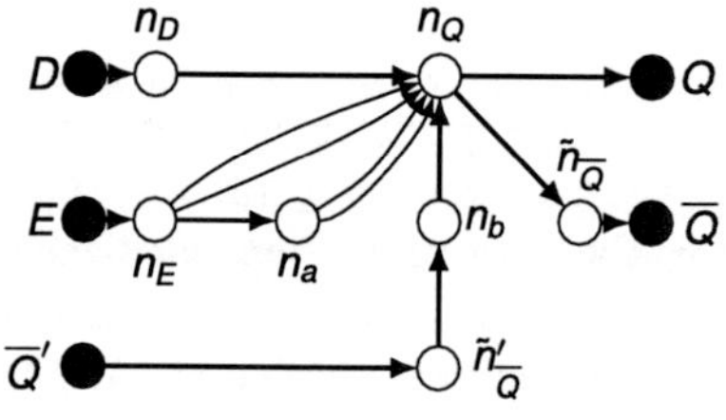

Fig. 1.13 Temporal ESFG

In some cases, this leads to conflicting requirements for a node, i.e., it must be analog and digital at the same time. This happens in case an analog building block was wrongfully recognized in the digital part or vice versa. Such conflicts are resolved by back–annotating the signal type to the nets of the circuit. Next, the building block recognition is rerun using additional rules for the signal types at the pins of a building block.

For a pure analog or digital circuit this step has no effect. Therefore the ESFG in Fig. 1.12 does not change.

1.4.4 Transformation to Temporal ESFG

In case of sequential circuits such as latches, the ESFG contains feedback loops. In order to compute the logic function of such circuits, a temporal ESFG is introduced, which is an acyclic ESFG and adds a time concept.

Definition 5 (Temporal ESFG). A temporal ESFG is an acyclic ESFG. It refers to a virtual normalized clock with clock period 1 that is at least twice the real clock frequency, i.e., the real clock can be sampled. Each node gets an additional clock cycle attribute, indicating if the node belongs to the current or a previous clock cycle.

The transformation from the ESFG to the temporal ESFG is described in the following. All loops of the ESFG are computed by finding strongly connected components in the graph. All nodes, where an edge to a node outside the loop starts, are called output nodes of the loop. All nodes, where an edge from a node outside the loop ends, are called input nodes of the loop. The feedback path of a loop is the path from an output node to an input node that does not contain any other output or input node of the loop. Some node n_s of this path, which is not an input node of the loop, is then selected as node to represent the state of the loop. This node is split up into two nodes $\tilde{n}_s$ and $\tilde{n}'_s$, which represent the state at the current and previous time step, respectively. All edges going from n_s to a node inside the loop are assigned to $\tilde{n}'_s$. All other edges are assigned to $\tilde{n}_s$. In addition, an output port node corresponding to $\tilde{n}_s$ and an input port node corresponding to $\tilde{n}'_s$ is created.

Figure 1.13 illustrates this for the example. The ESFG (Fig. 1.12) contains a loop consisting of n_Q, $n_{\overline{Q}}$ and n_b. Node n_Q is an input node of the loop and nodes n_Q, $n_{\overline{Q}}$ are output nodes. The feedback path is $n_{\overline{Q}}, n_b, n_Q$. Node $n_{\overline{Q}}$ is split up into node $\tilde{n}_{\overline{Q}}$ and $\tilde{n}'_{\overline{Q}}$. Input port $\overline{Q}'$ is created. The resulting temporal ESFG is shown in Fig. 1.13.

1.5 Logic Function Extraction

Based on the temporal ESFG the logic function of the circuit can now be computed. This is done in two steps. First, the logic function for all recognized building blocks is computed. Afterwards, the logic function for the complete circuit is determined.

Unless denoted otherwise, a four valued logic [4] with $0, 1, Z$ (high impedance), U (unknown) is used in the following. All logic functions are represented using ROBDDs [3].

1.5.1 Computation of Logic Function for Building Blocks

In this step, the logic function of single logic gates is determined. For CMOS circuits this requires in general to evaluate the serial and parallel connections of the pull–up and pull–down network [19]. Algorithmic implementations can be found in, e.g., [4, 5, 9, 11, 20]. Our approach builds on the hierarchical recognition result computed by the algorithm presented in Sec. 1.3.

Table 1.1 lists the logic function associated with each building block from the library for digital circuits. It uses the operators $\oplus$ and $\diamond$, which are defined as follows.

$$a \oplus b :\Leftrightarrow \begin{cases} a & (a=b) \vee (b=Z) \\ b & a=Z \\ U & \text{otherwise} \end{cases} \qquad a \diamond b :\Leftrightarrow \begin{cases} a & b=U \\ b & a=U \\ a \oplus b & \text{otherwise} \end{cases} \tag{1.9}$$

The operator $\oplus$ is the "merge" operator from [4]. The result is a defined logic state, i.e., zero or one, if a and b have the same value or one is high impedance. If both are high impedance the result is "Z", otherwise the result state is undefined. The operator $\diamond$ considers in addition, that undefined states can be canceled out in case of parallel connections, i.e., the result is a defined logic state in case a or b is 0 or 1 and the other one is undefined.

An NMOS transistor for example shows a logic "0" at the drain pin if the gate is at logic "1" (i.e., vdd) and source is at logic "0". The drain pin is at high impedance state if the transistor is off. This is the case for "0" at the gate or "Z" at the source pin. In all other cases the output is unknown (Table 1.1). The logic function for a PMOS transistor is found analogously.

The logic function at the drain pin of a stack chain is formed out of the logic functions f_1 to f_n of its children. The gate inputs of these children are described by vectors $\mathbf{g}_1$ to $\mathbf{g}_n$. The overall logic function is the logic function f_n with f_{n-1} substituted for the source variable. These substitutions are continued until f_1 is reached.

The logic function at the output o of a logic gate are the logic functions of the p– and the n–block combined by the $\oplus$ operator. This includes that the output becomes high–impedance in case no block is on or unknown in case both blocks are on at the same time.

The logic function of a logic array is the logic functions f_1 to f_n of the children combined by the $\diamond$ operator. The logic function at the output o of a pass–gate is equal to the input i in case the pass–gate is on, otherwise it is "Z".

Table 1.1 Logic functions for digital building blocks

n–transistor

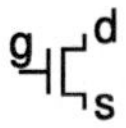

$$d = f_{\text{n-t}}(g,s) = \begin{cases} 0 \; (g=1)\wedge(s=0) \\ Z \; (g=0)\vee(s=Z) \\ U \text{ otherwise} \end{cases}$$

p–transistor

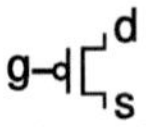

$$d = f_{\text{p-t}}(g,s) = \begin{cases} 1 \; (g=0)\wedge(s=1) \\ Z \; (g=1)\vee(s=Z) \\ U \text{ otherwise} \end{cases}$$

stack chain

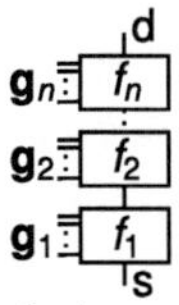

$$d = f_{\text{sc}}([\mathbf{g}_1\mathbf{g}_2\cdots\mathbf{g}_n],s) = f_n(\mathbf{g}_n, f_{n-1}(\mathbf{g}_{n-1},\cdots f_1(\mathbf{g}_1,s)\cdots))$$

logic gate

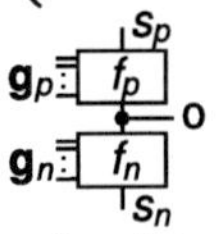

$$o = f([\mathbf{g}_p\mathbf{g}_n],s_p,s_n) = f_n(\mathbf{g}_n,s_n)\oplus f_p(\mathbf{g}_p,s_p)$$

logic array

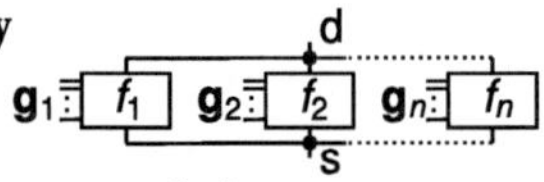

$$d = f_{\text{la}}([\mathbf{g}_1\mathbf{g}_2\cdots\mathbf{g}_n],s) = f_1(\mathbf{g}_1,s)\diamond f_2(\mathbf{g}_2,s)\diamond\cdots\diamond f_n(\mathbf{g}_n,s)$$

pass–gate

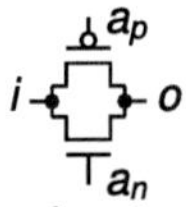

$$o = f(i,a_p,a_n) = \begin{cases} i \; (a_n=1)\wedge(a_p=0) \\ Z \; (a_n=0)\wedge(a_p=1) \\ U \text{ otherwise} \end{cases}$$

This is illustrated by the example shown in Fig. 1.14. The logic function for the complete gate,

$$f_{\text{NAND}}([a\;b]) = \underbrace{f_{N_2}(a, f_{N_1}(b,0))}_{f_N([a,b],0)} \oplus \underbrace{(f_{P_1}(a,1)\diamond f_{P_2}(b,1))}_{f_P([a,b],1)} \tag{1.10}$$

is formed by the logic function f_P of the logic array consisting of P_1 and P_2 as well as the logic function f_N of the stack chain consisting of N_1 and N_2. Logic function f_P is formed by the logic functions of P_1 and P_2 and logic function f_N is formed by the logic functions of N_1 and N_2, yielding

$$f_N([a\;b],0) = \begin{cases} 0 \; (a=1)\wedge(b=1) \\ Z \; (a=0)\vee(b=0) \\ U \text{ otherwise} \end{cases} \quad f_P([a\;b],1) = \begin{cases} 1 \; (a=0)\vee(b=0) \\ Z \; (a=1)\wedge(b=1) \\ U \text{ otherwise} \end{cases}. \tag{1.11}$$

Logic function f_N represents the output of the pull–down network, which is the drain of N_1. It is vss ("0") in case both inputs are one, if both inputs are zero it is high-impedance. In case one input is high–impedance or unknown f_N is unknown. Logic function f_P represents the output of the pull–up network at point x_1 in Fig. 1.14.

This results in the following logic function for the complete gate,

$$f_{\text{NAND}}([a\;b]) = \begin{cases} 0 \; (a=1)\wedge(b=1) \\ 1 \; (a=0)\vee(b=0) \\ U \text{ otherwise} \end{cases}, \tag{1.12}$$

which is the logic function of a NAND gate. The unknown case occurs if one of the inputs is in unknown or high–impedance state. Since the NAND gate is no tristate gate, the overall logic function does not include a high–impedance case.

1.5.2 Computation of Overall Logic Function

The overall logic function is computed by assigning logic variables to each node. We use a temporal logic, i.e., the logic variables refer to different time steps. Next, the temporal ESFG is traversed in topological order, i.e., each node in the graph is visited after all nodes it depends on. During this traversal, the logic functions are substituted into each other. In case two building blocks (e.g., pass gates) have outputs o_1, o_2 on the same node, the logic function for the node is calculated as $o_1 \oplus o_2$. It is assumed, that the inputs of the circuit are in a defined logic state, i.e., they are not "U" or "Z".

For the example circuit from Fig. 1.10 and the temporal ESFG from Fig. 1.13 the assigned logic variables are shown in Fig. 1.15. Logic variables $a(t), b(t), D(t)$, $E(t)$, $Q(t)$ and $\overline{Q}(t)$ refer to the current time step. Logic variable $\overline{Q}(t-1)$ refers to the previous time step. It holds.

$$a(t) = \begin{cases} 0 & E(t) = 1 \\ 1 & E(t) = 0 \end{cases} \qquad b(t) = \begin{cases} 0 & \overline{Q}(t-1) = 1 \\ 1 & \overline{Q}(t-1) = 0 \\ U & \text{otherwise} \end{cases} \tag{1.13}$$

$$Q(t) = \begin{cases} 0 & \left[(E(t)=1) \wedge (D(t)=0)\right] \vee \left[(E(t)=0) \wedge (\overline{Q}(t-1)=1)\right] \\ 1 & \left[(E(t)=1) \wedge (D(t)=1)\right] \vee \left[(E(t)=0) \wedge (\overline{Q}(t-1)=0)\right] \\ U & \text{otherwise} \end{cases} \tag{1.14}$$

$$\overline{Q}(t) = \begin{cases} 0 & \left[(E(t)=1) \wedge (D(t)=1)\right] \vee \left[(E(t)=0) \wedge (\overline{Q}(t-1)=0)\right] \\ 1 & \left[(E(t)=1) \wedge (D(t)=0)\right] \vee \left[(E(t)=0) \wedge (\overline{Q}(t-1)=1)\right] \\ U & \text{otherwise} \end{cases} \tag{1.15}$$

Logic function $a(t)$ can only become "0" or "1" because input $E(t)$ is assumed to be in a defined logic state. No such assumption is made for $\overline{Q}(t-1)$. Consequently, $b(t)$ can become unknown in case $\overline{Q}(t-1)$ is unknown or high-impedance. Overall

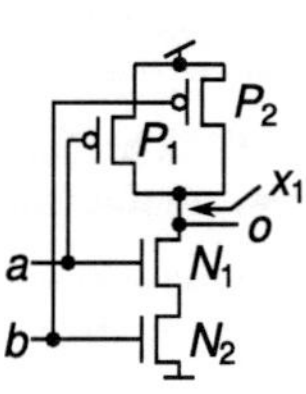

Fig. 1.14 NAND gate

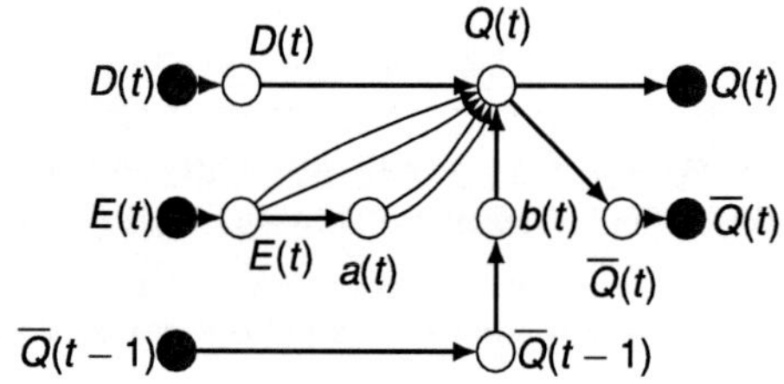

Fig. 1.15 ESFG with assigned logic variables

Table 1.2 Recognition results for different standard cell libraries

Library	No. Cells	Analysis Time	Coverage
Lib 1	32	1 sec.	100.0%
Lib 2	134	4 sec.	100.0%
Lib 3 – Tech 1	~ 600	18 sec.	97.6%
Lib 3 – Tech 2	~ 550	13 sec.	99.6%
Lib 3 – Tech 3	~ 700	27 sec.	99.1%
Lib 4	~ 850	37 sec.	95.2%

logic function $Q(t)$ is input $D(t)$ in case $E(t)$ is set otherwise it is the inversion of $\overline{Q}(t-1)$. Logic function $\overline{Q}(t)$ is the inversion of $Q(t)$. This corresponds to a latch.

1.6 Application Examples

In the following application to digital standard cell libraries and mixed-signal circuits is discussed including experimental results.

1.6.1 Description Generation for Digital Standard Cell Libraries

The approach is used to automatically generate a library description for digital standard cell libraries. The description includes a decomposition into pass–gates and logic gates, the ESFG, the logic function of the standard cell and a table listing possible single input switching events together with the possible values at the other inputs and the resulting output behavior. These events are a necessary input for automatic timing characterization of digital standard cell libraries. The decomposition into logic gates and pass–gates corresponds to a decomposition of multi–stage gates into single–stage gates. This is a required input for the current–source modeling approach of [10] and the aging analysis approach of [12].

In the experiment, the building block recognition with the digital part of the library is used as well as the structural signal flow analysis and the logic function extraction. Additional post–processing is used to generate the table of all possible single input switching events.

We performed this analysis for 4 different standard cell libraries (Table 1.2). Library 1 is the standard cell library included in the FreePDK presented in [18]. Library 2 is the Nangate open cell library. Library 3 is an industrial standard cell library which was available for three different technology nodes. Library 4 is an industrial standard cell library, too.

Table 1.2 shows that these libraries contained between 30 and 850 cells. In all cases the analysis for the complete library took less than 1 minute. All runtimes were normalized to an Intel® Xeon® 2.33 GHz computer with 4 GB RAM running Ubuntu and using 4 of 8 cores in parallel.

Column four of Table 1.2 gives the recognition coverage of the presented method. For libraries 1 and 2 all cells were recognized correctly. For libraries 3 and 4, the building block analysis was not able to fully decompose all cells into pass–gates and logic gates. Typically, these cells were not designed according to standard CMOS principles. However, these cells can be included by extending the library accordingly. Overall, more than 95% of all cells were correctly recognized for the industrial libraries.

1.6.2 Structural Analysis of Mixed-Signal Circuits

The new mixed–signal capabilities of the structural analysis were evaluated using a voltage–controlled ring oscillator (Fig. 1.16) and a charge–pump (Fig. 1.18).

The voltage–controlled ring oscillator generates a digital clock signal. The frequency of the clock signal can be adjusted by the analog control voltage applied at input *c*. The building block recognition computed 4 NMOS simple current mirrors, 3 PMOS simple current mirrors and 5 logic gates on level 2, i.e., inverter. It is not possible to get the correct recognition result by computing analog and digital building blocks independently: A logic gate on level 4 consisting of N_3, N_4, P_3, P_4 would be found, which would contradict the current mirrors formed by N_1,N_4 and P_1,P_2.

Fig. 1.17 shows the corresponding ESFG of the voltage–controlled ring oscillator. The partitioning into analog and digital part is symbolized by the node shape. The analog control circuitry as well as the digital feedback loop are clearly visible.

The charge pump shown in Fig. 1.18 is based on [15]. The output is usually connected to the loop filter of a PLL. Digital inputs *D* and *U* control the direction of the output current. The building block recognition computed 2 NMOS simple current mirrors, a PMOS simple current mirror and two logic gates. Transistors N_6 and N_7 as well as P_5 and P_6 would match a differential pair. These differential pairs were dropped because they connect to digital inputs *D* and *U*. Transistors N_7 and P_6 would form a logic gate, which was dropped because output *o* is specified as analog. The corresponding ESFG is shown in Fig. 1.19.

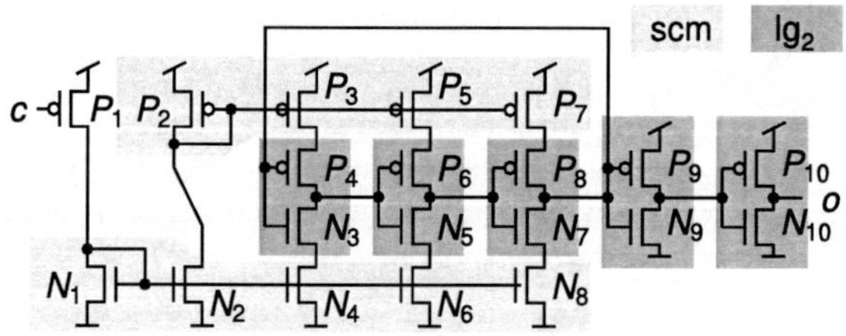

Fig. 1.16 Voltage–controlled ring oscillator with recognized building blocks

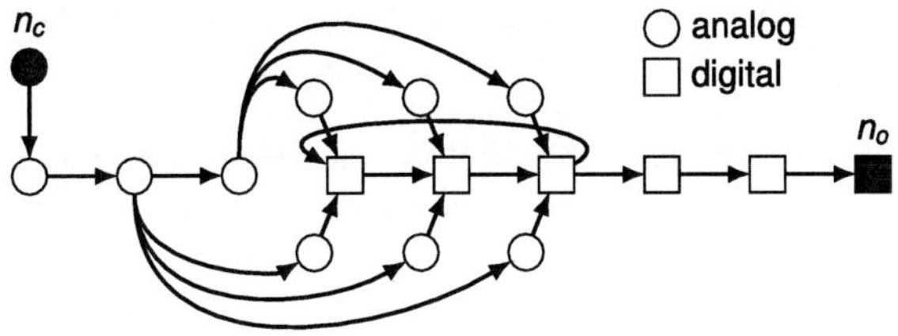

Fig. 1.17 ESFG of voltage–controlled ring oscillator

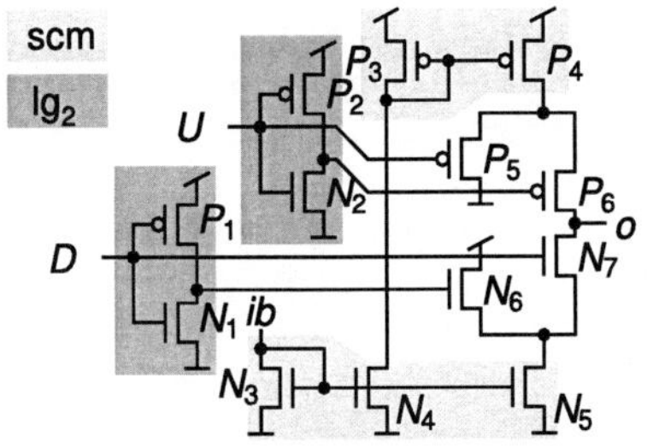

Fig. 1.18 Charge pump with recognized building blocks

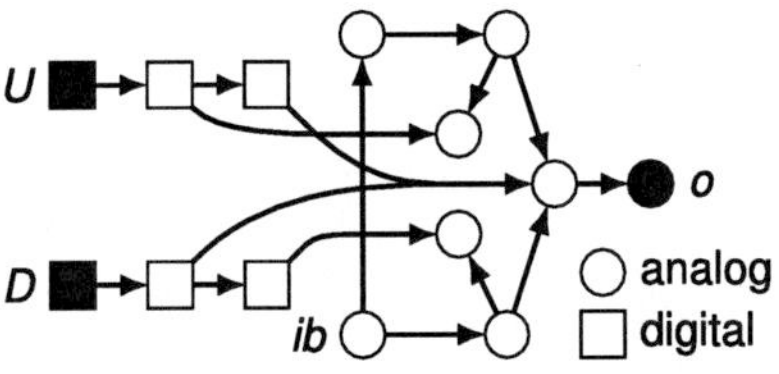

Fig. 1.19 ESFG of the charge pump

1.7 Conclusion

This chapter presented a new method for the automatic structural and functional analysis of analog, digital and mixed-signal circuits. Its first step is the recognition of building blocks such as simple current mirrors and logic gates. These results are then used to generate an Enhanced Structural Signal Flow Graph (ESFG). Based on that, true pass-gate directions are computed and feedback paths are broken up. Finally, the logic function is determined for the digital circuit parts.

Experimental results show successful application of the algorithm to several digital standard cell libraries with more than 95% of correctly recognized cells. Structural analysis of mixed-signal circuits was demonstrated using a voltage-controlled ring oscillator and a charge pump.

Acknowledgements. We would like to thank Jens Bargfrede from Infineon Technologies AG for all his help and encouragement. We would also like to remember Michael Mirbeth († 2009) from Infineon Technologies AG, who gave us great support in this work.

References

[1] Arsintescu, B.G.: A Method for Analog Circuits Visualization. In: IEEE International Conference on Computer Design (ICCD), pp. 454–459 (1996)

[2] Blaauw, D.T., Saab, D.G., Long, J., Abraham, J.A.: Derivation of signal flow for switch-level simulation. In: ACM/IEEE Design Automation Conference (DAC), pp. 301–305 (1990)

[3] Bryant, R.E.: Graph-Based Algorithms for Boolean Function Manipulation. IEEE Transactions on Computers 35(8), 677–691 (1986)
[4] Bryant, R.E.: Extraction of gate level models from transistor circuits by four-valued symbolic analysis. In: IEEE International Conference on Computer-Aided Design, ICCAD 1991. Digest of Technical Papers, pp. 350–353 (1991)
[5] Dagenais, M.R.: Efficient algorithmic decomposition of transistor groups into series, bridge, and parallel combinations. IEEE Transactions on Computer-Aided Design of Integrated Circuits and Systems 38(6), 569–581 (1991)
[6] Eick, M., Graeb, H.: MARS: Matching-driven Analog Sizing. IEEE Transactions on Computer-Aided Design of Integrated Circuits and Systems (2012)
[7] Eick, M., Strasser, M., Lu, K., Schlichtmann, U., Graeb, H.: Comprehensive Generation of Hierarchical Placement Rules for Analog Integrated Circuits. IEEE Transactions on Computer-Aided Design of Integrated Circuits and Systems 30(2), 180–193 (2011)
[8] Graeb, H., Zizala, S., Eckmueller, J., Antreich, K.: The Sizing Rules Method for Analog Integrated Circuit Design. In: IEEE/ACM International Conference on Computer-Aided Design (ICCAD), pp. 343–349 (2001)
[9] Kim, W., Shin, H.: Hierarchical LVS based on hierarchy rebuilding, pp. 379–384 (1998)
[10] Knoth, C., Kleeberger, V.B., Nordholz, P., Schlichtmann, U.: Fast and Waveform Independent Characterization of Current Source Models. In: IEEE/VIUF International Workshop on Behavioral Modeling and Simulation (BMAS), pp. 90–95 (2009)
[11] Lester, A., Sabet, P.B., Greiner, A.: YAGLE, a second generation functional abstractor for CMOS VLSI circuits. In: Proceedings of the Tenth International Conference on Microelectronics, ICM 1998, pp. 265–268 (1998)
[12] Lorenz, D., Barke, M., Schlichtmann, U.: Aging analysis at gate and macro cell level. In: IEEE/ACM International Conference on Computer-Aided Design (ICCAD), pp. 77–84 (2010)
[13] Lullau, F., Hoepken, T., Barke, E.: A Technology Independent Block Extraction Algorithm. In: ACM/IEEE Design Automation Conference (DAC), pp. 610–615 (1984)
[14] Massier, T., Graeb, H., Schlichtmann, U.: The Sizing Rules Method for CMOS and Bipolar Analog Integrated Circuit Synthesis. IEEE Transactions on Computer-Aided Design of Integrated Circuits and Systems 27(12), 2209–2222 (2008)
[15] Rhee, W.: Design of high-performance CMOS charge pumps in phase-locked loops. In: IEEE International Symposium on Circuits and Systems (ISCAS), vol. 2, pp. 545–548 (1999)
[16] Rubanov, N.: A High-Performance Subcircuit Recognition Method Based on the Nonlinear Graph Optimization. IEEE Transactions on Computer-Aided Design of Integrated Circuits and Systems 25(11), 2353–2363 (2006)
[17] Schulte, C., Tack, G., Lagerkvist, M.Z.: Modeling and Programming with Gecode (2010), `http://www.gecode.org/doc/3.4.0/MPG.pdf`
[18] Stine, J., Castellanos, I., Wood, M., Henson, J., Love, F., Davis, W., Franzon, P., Bucher, M., Basavarajaiah, S., Oh, J., Jenkal, R.: FreePDK: An Open-Source Variation-Aware Design Kit. In: IEEE International Conference on Microelectronic Systems Education, MSE 2007, pp. 173–174 (2007)
[19] Weste, N.H.E., Harris, D.: CMOS VLSI Design - A Circuits and Systems Perspective. Pearson Education, Inc. (2005)

[20] Yang, L., Shi, C.J.R.: FROSTY: A Fast Hierarchy Extractor for Industrial CMOS Circuits. In: IEEE/ACM International Conference on Computer-Aided Design (ICCAD), pp. 741–746 (2003)
[21] Zhang, N., Wunsch II, D.C.: Speeding up VLSI Layout Verification Using Fuzzy Attributed Graphs Approach. IEEE Transactions on Fuzzy Systems 14(6), 728–737 (2006)
[22] Zhang, N., Wunsch II, D.C., Harary, F.: The subcircuit extraction problem. In: Proceeding of IEEE International Behavioral Modeling and Simulation Workshop 2005, vol. 22(3), pp. 22–25 (2003)

Chapter 2
Efficient Synthesis Methods for High-Frequency Integrated Passive Components and Amplifiers

Bo Liu and Georges Gielen

Abstract. Existing design automation methods for RF ICs and microwave passive components often rely on parasitic-aware lumped equivalent circuit models. That framework is difficult to apply to synthesis tasks at high frequencies (e.g. 40GHz and above) due to the distributed effect. When directly embedding the computationally expensive electromagnetic (EM) simulations in the optimization loop, a too long synthesis time results. This chapter presents a new method for high-frequency integrated passive component synthesis, called Memetic Machine Learning-based Differential Evolution (MMLDE), and the first method for mm-wave integrated circuit synthesis, called Efficient Machine Learning-based Differential Evolution (EMLDE), both addressing the problem of obtaining highly optimized design solutions in a very practical time. The common idea of these two methods is the on-line surrogate model assisted evolutionary algorithm (SAEA), where a computationally cheap surrogate model is constructed adaptively in the optimization process to replace expensive EM simulations. The differences between the two algorithms are that a memetic SAEA is built to enhance the optimization ability and efficiency in MMLDE, while a decomposition method is used to address the "curse of dimensionality" of SAEA in EMLDE. Experimental results show the effectiveness and the high efficiency obtainable with MMLDE and EMLDE.

2.1 Introduction

In recent years, design methodologies for high-frequency and mm-wave circuits have attracted a lot of attention. In particular, research and applications on RF building blocks for 40 GHz to 120 GHz and beyond are increasing drastically [1]. Existing RF IC synthesis methodologies, however, focus on low-GHz cases [2,3]. Even till now, the synthesis methodologies for mm-wave frequencies are still

Bo Liu · Georges Gielen
ESAT-MICAS, KU Leuven, Kasteelpark Arenberg 10, B3001, Leuven, Belgium
e-mail: {Bo.Liu,Georges.Gielen}@esat.kuleuven.be,
liubo168@gmail.com

M. Fakhfakh et al. (Eds.): *Analog/RF & Mixed-Signal Circuit Sys. Design*, LNEE 233, pp. 27–52.
DOI: 10.1007/978-3-642-36329-0_2

lacking. Designers rely on experience and simulation verifications when designing these circuits. Due to the high-performance and tightening time-to-market requirements, this "experience and trial" method or local optimization is often not good enough.

The reason why existing synthesis methods cannot be extended to mm-wave frequencies is that they all rely on parasitic-aware equivalent circuit models for passive components [2,3,4]. Due to the distributed effects, however, an accurate equivalent circuit model is difficult to find at mm-wave frequencies. The solution is to include electromagnetic (EM) simulation based on the actual layout structure in the optimization loop. However, EM simulation is computationally very expensive. When combining it directly with techniques like evolutionary computation (EC) [5], like at low frequencies, high-quality solutions can be obtained, but the time consumption is extremely large. For example, the synthesis of a transformer typically needs more than 20 hours, and the synthesis of a linear amplifier needs about 10 days. This is not practical for real-world applications.

In this chapter, efficient synthesis method for mm-wave-frequency passive components and linear amplifiers will be introduced. The Memetic Machine Learning-based Differential Evolution (MMLDE) method [6] for the synthesis of integrated passive components will briefly be introduced first. The key idea of MMLDE is the on-line surrogate model-based memetic evolutionary optimization mechanism, whose training data are generated adaptively in the optimization process. By using the Gaussian Process with the expected improvement prescreening method and an artificial neural network with the prediction value in the proposed search mechanism, surrogate models are constructed on-line to predict the performances. Hence, the computationally expensive EM simulations are only used in the necessary part of the design space, which is guided by the prediction and prescreening methods. Compared with directly using EC algorithms, MMLDE can obtain comparable results, and has approximately a tenfold improvement in computational efficiency. The Efficient Machine Learning-based Differential Evolution (EMLDE) method [7] for the synthesis of mm-wave linear amplifiers will then be elaborated next. A decomposition method is used, which separates the design variables that require expensive EM simulations and the variables that only need cheap S-parameter circuit simulations. Hence, a low-dimensional but more complex expensive optimization problem is generated. By the proposed core algorithm integrating adaptive population generation, naive Bayes classification, Gaussian process and differential evolution, the generated low-dimensional expensive optimization problem can be solved efficiently (thanks to the on-line surrogate model), and global search can be achieved (thanks to the evolutionary computation algorithm). A 100GHz three-stage differential amplifier in a 90nm CMOS technology is shown as an example. The power gain reaches 10dB with more than 20GHz bandwidth. The synthesis costs only 25 hours, having a comparable result and a 9 times speed enhancement compared with directly using the EM simulator in combination with a global optimization algorithm.

The remainder of this chapter is organized as follows. Section 2.2 reviews the existing works for RF IC synthesis, and motivates the construction of the EMLDE algorithm. Section 2.3 introduces the basic mathematical and computational intelligence techniques used in this chapter. Section 2.4 briefly introduces the MMLDE method as a first step for EMLDE. Section 2.5 elaborates the EMLDE method. The experimental verifications are in Section 2.6. Section 2.7 concludes the chapter.

2.2 Review of Related Works and Challenges

2.2.1 *RF Integrated Circuit Synthesis*

Existing RF IC design automation methods focus on low-GHz synthesis [2-4,8-14] by employing lumped equivalent circuit models for passive components (e.g. transformer, inductor). The framework of most of these methods is shown in Figure 2.1. Compared with the low-frequency analog circuit sizing flow, a key part is the generation of the parasitic-aware model of the passive components. In RF IC designs at low-GHz frequencies, a simple lumped model is often extracted to mimic the behavior of the key passive components (transformer, inductor). Regression methods are then used to fit the (calibrated) EM simulation results (S-parameters) to the parasitic-included equivalent circuit models. The generated passive component models are accurate at low-GHz frequencies and computationally efficient.

To make the parasitic-aware model reliable in providing the correct performances for different design parameters, a strictly enforced layout template is often necessary. [10,11] use the parasitic corner, rather than a strict layout template, to improve the flexibility of the generated layout for circuits below 10GHz, yielding good results. In the development of the optimization kernel, evolutionary algorithms (EAs) are introduced in RF IC synthesis to achieve global search, getting very good results. [14] uses Particle Swarm Optimization (PSO) and [13] introduces the non-dominated genetic algorithm (NSGA) to RF IC synthesis in order to achieve multi-objective optimization.

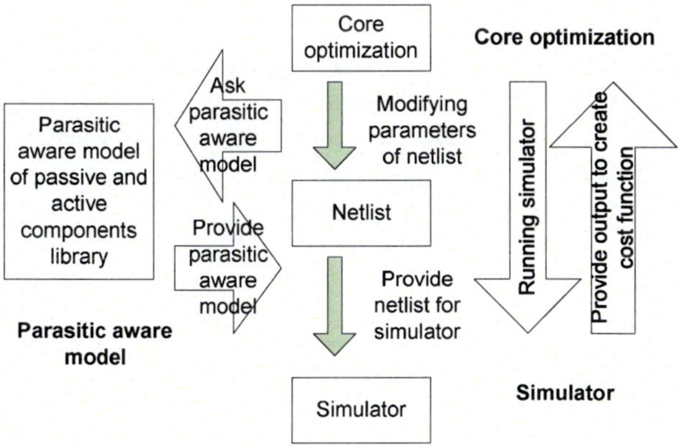

Fig. 2.1 Framework of parasitic-aware optimization for RF ICs (from [4])

Parasitic-aware lumped equivalent circuit models for passive components that accurately match the EM simulation results are difficult to find at frequencies between say 40GHz and above 100GHz due to the distributed effects at these mm-wave frequencies [6]. Hence, when employing lumped equivalent circuit models, available RF integrated circuit design automation methods are limited to low-GHz instances. Because the speed enhancement method for RF IC synthesis (using lumped models) cannot be extended to mm-wave integrated circuit synthesis, and because directly including the EM simulations in each performance evaluation is too CPU time intensive, no good efficient method for mm-wave integrated circuit synthesis exists today. The only way left to mm-wave circuit designers is the "experience and simulation verification" method, which is at odds with today's high-performance and tightening time-to-market requirements.

To summarize, the goal of this chapter is to fill the blank of efficient automated design of mm-wave-frequency integrated passive components and integrated circuits (linear amplifiers as an instance), achieving good accuracy while knowing an acceptable CPU time.

2.3 Basic Computational Intelligence Techniques

The methods presented in this chapter are based on computational intelligence techniques, i.e. evolutionary computation and machine learning techniques in specific. In the following, we will introduce three basic techniques: the Differential Evolution (DE) algorithm, the Gaussian Process (GP) machine learning and the Naive Bayes Classifier (NBC), which are the fundamentals for the presented algorithms MMLDE and EMLDE.

2.3.1 Differential Evolution

The DE algorithm [15] is selected as the search engine in MMLDE and EMLDE. The DE algorithm outperforms many other evolutionary computation (EC) algorithms in terms of solution quality and convergence speed. DE uses a simple differential operator to create new candidate solutions and a one-to-one competition scheme to greedily select new candidates.

The i-th candidate solution in the d-dimensional search space at generation t can be represented as

$$x_i(t) = [x_{i,1}, x_{i,2}, \cdots, x_{i,d}] \tag{2.1}$$

At each generation t, the *mutation* and *crossover* operators are applied to the candidate solutions, and a new population arises. Then, *selection* takes place, and the corresponding candidate solutions from both populations compete to comprise the next generation. The operators are now explained in detail.

For each target candidate solution, according to the *mutation* operator, a *mutant vector* is built:

$$V_i(t+1) = [v_{i,1}(t+1), \ldots, v_{i,d}(t+1)] \tag{2.2}$$

It is generated by adding the weighted difference between a given number of candidate solutions randomly selected from the previous population to another candidate solution. The mutation operation is therefore described by the following equation (DE/best/1/bin [23]):

$$V_i(t+1) = x_{best}(t) + F(x_{r1}(t) - x_{r2}(t)) \tag{2.3}$$

where indices r_1 and r_2 ($r_1, r_2 \in \{1, 2, \ldots, NP\}$, NP is the size of the population) are randomly chosen and mutually different, and also different from the current index i. Parameter $F \in (0, 2]$ is a constant called the scaling factor, which controls the amplification of the differential variation $x_{r1}(t) - x_{r2}(t)$. The base vector to be perturbed $x_{best}(t)$ is the best member of the current population, so that the best information can be shared among the population. To avoid stagnation and to improve the balance between exploration and exploitation, we use the random-scale search DE mutation operator. In this mutation, for the scaling factor we use a vector $\hat{F}$ composed of Gaussian-distributed random variables with mean value μ and variance σ: $\hat{F}_{i,j} = norm(\mu, \sigma), \quad i = 1, 2, \ldots NP, \quad j = 1, 2, \ldots d$. Equation (2.3) is therefore changed to equation (2.4):

$$V_i(t+1) = x_{best}(t) + \hat{F}_i(x_{r1}(t) - x_{r2}(t)) \tag{2.4}$$

After the *mutation* phase, the *crossover* operator is applied to increase the diversity of the population. Thus, for each target candidate solution, a *trial vector* is generated as follows:

$$U_i(t+1) = [u_{i,1}(t+1), \ldots, u_{i,d}(t+1)] \tag{2.5}$$

$$U_{i,j}(t+1) = \begin{cases} v_{i,j}(t+1), if\,(rand(i,j) \le CR)\ or\ j = randn(i), \\ x_{i,j}(t), otherwise, \end{cases} \tag{2.6}$$

where $rand(i, j)$ is an independent random number uniformly distributed in the range [0,1]. Parameter *randn(i)* is a randomly chosen index from the set $\{1, 2, \ldots, d\}$. Parameter $CR \in [0,1]$ is a constant called the crossover parameter, which controls the diversity of the population.

Following the *crossover* operation, the *selection* operation decides on the population of the next generation (t+1). In standard DE, $U_i(t+1)$ is compared to the initial target candidate solution $x_i(t)$ by a one-to-one-based greedy selection criterion. However, in MMLDE and EMLDE, we do not use this selection operator, but a variant instead, because we need to minimize the number of EM simulations. The new selection scheme will be introduced in Section 2.4.3 and 2.5.3.

2.3.2 Gaussian Process Machine Learning

Gaussian Process machine learning is the main learning machine used to train surrogate models in our work, due to its solid mathematical foundation. Typically, classification and prediction are the two main applications areas of machine learning methods. In EM simulation-based design automation, prediction is emphasized. In this subsection, the basics of Gaussian Process (GP) modeling and prediction are introduced briefly.

GP predicts a function value $y(x)$ at some design point x by modeling $y(x)$ as a stochastic variable with mean μ and variance σ. If the function is continuous, the function values of two points x_i and x_j should be close if they are highly correlated. For two points x_i and x_j, their correlation is defined as:

$$Corr(x_i, x_j) = \exp(-\sum_{l=1}^{d} \theta_l \mid x_{il} - x_{jl} \mid^{p_l}) \\ \theta_l > 0, 1 \le p_l \le 2 \tag{2.7}$$

where d is the dimension of x and θ_l is the correlation parameter which determines how fast the correlation decreases when x_{il} moves in the l direction. Parameter p_l is related to the smoothness of the function with respect to x_{il}. The values of μ, σ and θ are determined by maximizing the likelihood function of the observed data. Suppose that there are n observed data $x = (x_1, x_2 \cdots, x_n)^T$, and their corresponding function values are $y = (y_1, y_2 \cdots, y_n)^T$, then the optimal values of μ and σ can be found by setting the derivatives of the likelihood function to 0:

$$h = \frac{1}{(2\pi)^{n/2}(\sigma^2)^{n/2} \mid R \mid^{1/2}} \exp(-\frac{1}{2\sigma^2}(y - I\mu)^T R^{-1}(y - I\mu)) \tag{2.8}$$

where I is a $n \times 1$ vector of ones, and R is the correlation matrix:

$$R_{i,j} = Corr(x_i, x_j),\ i, j = 1, 2, \cdots n \tag{2.9}$$

By solving the equations, the $\hat{\mu}$ and $\hat{\sigma}^2$ are as follows:

$$\hat{\mu} = (I^T R^{-1} I)^{-1} I^T R^{-1} y \tag{2.10}$$

$$\hat{\sigma}^2 = \frac{(y - I\hat{\mu})^T R^{-1}(y - I\hat{\mu})}{n} \tag{2.11}$$

Using the GP model, the function value $y(x^*)$ at a new point x^* can be predicted as (x^* should be added in *R*, *r*):

$$\hat{y}(x^*) = \hat{\mu} + r^T R^{-1}(y - I\hat{\mu}) \tag{2.12}$$

where

$$r = [Corr(x^*, x_1), Corr(x^*, x_2), \cdots, Corr(x^*, x_n)]^T \tag{2.13}$$

The measurement of the uncertainty of the prediction, i.e. the mean square error (MSE) or $\hat{s}^2$, which is used to assess the model accuracy, can be described as:

$$MSE(x^*) = \hat{\sigma}^2[I - r^T R^{-1} r + (I - r^T R^{-1} r)^2 (I^T R^{-1} I)^{-1}] \tag{2.14}$$

2.3.3 *Naive Bayes Classifier*

The naive Bayes classification [16] is a supervised learning method. It is very efficient and outperforms many existing classification methods, even some newly developed methods [17]. A classifier is a machine that maps the input feature space *F* to the output class label space *C*. Naive Bayes classification learns from a training data set of input vectors (input features) and their corresponding classes. In the following, we introduce how the naive Bayes classifier works.

Assume that the input vector is *d*-dimensional, so we have feature variables from F_1 to F_d. Each input vector is classified to a class C_i ($i = 1, ..., n$). For a new input vector *x*, the class it belongs to is decided by the maximum probability of the hypothesis that *x* belongs to C_i, that is:

$$class(x) = \arg\max_{C_i} p(C = C_i \mid F_1 = x_1, \ldots, F_d = x_d) \tag{2.15}$$

The naive Bayes classifier assumes that each feature F_k ($k = 1, \ldots, d$) is conditionally independent of every other feature. Hence, the conditional probability $p(C \mid F_1, \ldots, F_d)$ can be simplified to:

$$p(C \mid F_1, \ldots, F_d) = \frac{p(C, F_1, \ldots, F_d)}{p(F_1, \ldots, F_d)} = \frac{p(C)\prod_{k=1}^{d} p(F_k \mid C)}{p(F_1, \ldots, F_d)} \tag{2.16}$$

where $p(F_1, \ldots, F_d)$ is common to all and does not affect the ranking of equation (2.15). In this work, we assume that the input vector values associated with each class are Gaussian distributed. According to the training data, the mean and variance of the data associated with each class can be calculated. Using the probability density function of the Gaussian distribution and plugging equation (2.16) into equation (2.15), the corresponding class for a new vector x can be calculated.

Although the basic assumption of independence of all the features is often not accurate enough, the naive Bayes classifier uses the maximum of posteriori rule [16]. Therefore, the classification is decided by the ranking, rather than by the accurate estimation of $p(C = C_i \mid F_1 = x_1, \ldots, F_d = x_d)$. This is the main reason why the naive Bayes classifier can still be very effective, even when using such a simplified assumption.

2.4 MMLDE: Efficient Synthesis of Integrated Passive Components at High Frequencies

The proposed MMLDE algorithm [6] aims at efficiently synthesizing integrated passive components at high frequencies. It is clear that to achieve the targets, MMLDE must use EM simulations (to be general for passive components at high frequencies), a global optimization algorithm (to obtain highly optimized solutions) and most importantly, surrogate modeling techniques (to enhance the efficiency). The method for the integration of the machine learning techniques into the global optimization and the EM simulation-based algorithm is the key of MMLDE.

2.4.1 Key Ideas of MMLDE

An initial Gaussian Process (GP)-based surrogate model is constructed first by using a small number of Latin Hypercube (LHS) [18] samples that uniformly cover the design space. This model can provide a very rough estimation of the performances of the passive component. In optimization, the constructed surrogate model and the prescreening method evaluate the potential of the candidate designs. The candidate designs are ranked and the one with the best potential is selected to perform the EM simulation. The new exact point will be used to update the surrogate model. We iteratively repeat this process until the termination condition is met. An important advantage of MMLDE is the use of the on-line surrogate model, which is constructed based on the available data in the optimization process. The promising solutions are selected meanwhile and guide further candidate solution generation. The advantages on efficiency and reliability of on-line surrogate model-based optimization compared with off-line surrogate model-based optimization [19] are described and compared in [6].

On the other hand, the challenge of on-line surrogate-model-based optimization is that the quality of the surrogate model is not always good, as it is improving gradually throughout the optimization process. If the training data is little in some area of the design space, especially in the beginning stage, not enough information can be provided to the learning machine, so the surrogate model might not be good enough. When direct using the performance values predicted by the surrogate model to judge the potentials of the candidate designs, the search may go to wrong directions and can finally be trapped in a local optimal point [6,20]. To solve this problem, the expected improvement (EI) prescreening with a GP-based surrogate model focusing on global search and an artificial neural network (ANN)-based surrogate model [21] focusing on local search is proposed. This combined method achieves a good ranking and a high probability of correct selection for promising candidates, even when a good enough surrogate model is not available. More details are in [6]. The expected improvement (EI) prescreening method plays a key role in MMLDE and EMLDE, and is introduced in the next subsection.

2.4.2 Expected Improvement Prescreening

This subsection introduces the expected improvement (EI) prescreening associated with Gaussian Process surrogate modeling. Before introducing EI, a first note is that the prescreening methods (besides directly using the predicted values) are only suitable for GP-based surrogate modeling, since the prescreening methods are also based on the Gaussian stochastic process.

The EI prescreening is calculated as follows:

$$E[I(x)] = (f_{\min} - y(x))\Phi\left(\frac{f_{\min} - y(x)}{\sqrt{MSE(x)}}\right) + \sqrt{MSE(x)}\,\phi\left(\frac{f_{\min} - y(x)}{\sqrt{MSE(x)}}\right) \tag{2.17}$$

where $f_{\min}$ is the current best function value in the population (the population with EM simulation results, not the generated population after evolutionary operators). $\phi(\cdot)$ is the standard normal density function, and $\Phi(\cdot)$ is the standard normal distribution function. $I(x)$ is the improvement of f. EI is essentially the part of the curve of the standard error in the model that lies below the best function value sampled so far. It can be seen that the EI prescreening considers both the predicted value and the possible prediction error. Therefore, the quality of a new candidate is evaluated in a global picture. More details are in [6].

2.4.3 The General Framework of MMLDE

Based on the above components, the overall MMLDE algorithm for the efficient synthesis of high-frequency RF passive components can now be constructed. The description is as follows. The parameter setting rules of MMLDE can be found in [6].

Step 0: Initialize the parameters, e.g. the generation threshold of repeatedly using GP/ANN when no improvement is shown, the DE algorithm parameters (e.g. CR), the GP parameters (e.g. the correlation function), the ANN parameters (e.g. the number of neurons, the training algorithm).

Step 1: Initialize the population by LHS sampling of the design space. The EM simulations are performed for the sampled design points.

Step 2: Check if the stopping criterion (e.g. a convergence criterion or a maximum number of iterations) is met. If yes, output the result; otherwise go to step 3.

Step 3: Judge to use the GP or ANN machine learning techniques (see [6]).

Step 4: Train the selected surrogate model according to the available samples (population).

Step 5: Use the available samples as the current population, and perform the mutation operation to obtain each candidate solution's mutant counterpart (equation (2.4)).

Step 6: Perform the crossover operation between each candidate solution and its corresponding mutant counterpart to obtain each individual's trial individual (equation (2.5) and (2.6)).

Step 7: According to the model selected in Step 3, use the EI or the predicted value to select the individual with the possible best potential and perform the EM simulation to it.

Step 8: Update the population by adding the point from step 7 and its performance. Update the best solution obtained so far. Update other parameters. Go back to Step 2.

2.4.4 Experimental Results of MMLDE

In this section, one example is shown to verify MMLDE. For more examples, please see [6].

The example integrated passive component is a 60GHz overlay transformer with octagonal shape in a 90nm CMOS process. The design variables are the inner diameter of the primary inductor (*dinp*), the inner diameter of the secondary inductor (*dins*), the width of the primary inductor (*wp*) and the width of the secondary inductor (*ws*). The ranges of the design variables are $dinp, dins \in [20,150]$, $wp, ws \in [5,10]$ (all in μm). The design specifications are the coupling coefficient $k > 0.85$, the quality factor of the primary inductor $Q_1 > 10$, the quality factor of the secondary inductor $Q_2 > 10$. The output load impedance is $25\,\Omega$, which is the input resistance of the following stage. The specifications of the input impedance (at 60GHz) are $\mathrm{Re}(Z_{in}) \in [10,20]$ and $\mathrm{Im}(Z_{in}) \in [10,25]$ (Ω), which is the required optimal load impedance of the driver stage. The optimization goal is to maximize the power transfer efficiency (*PTE*). The transformer synthesis results are shown in Table 2.1. The average *PTE* for 10 runs are provided for each method. The EM simulation tool is ADS-Momentum.

From Table 2.1, the results can be analyzed. MMLDE costs 2.3 hours on a single CPU node, which is very reasonable for practical use. Moreover, the result quality of MMLDE is comparable with the benchmark (method of using full EM simulations in combination with DE), but MMLDE is more than 10 times faster.

The standard deviation of MMLDE on the optimization goal, *PTE*, is 0.25%. The mesh density is set to 30 cells/wavelength and the Arc resolution is set to 45 degrees. A typical 60GHz transformer result of MMLDE is shown in Figure 2.2.

Table 2.1 Results of different methods for integrated transformer synthesis

	Reference method	MMLDE
PTE	89.0%	88.8%
N	965	87
T	24.7 hours	2.3 hours

From the experiments, it can be seen that MMLDE is an efficient method and is the first practical method for high-frequency (especially mm-wave) integrated passive component synthesis, achieving highly optimized results in a very efficient time.

2.5 EMLDE: Efficient Synthesis of mm-Wave Linear Amplifiers

The EMLDE method aims at synthesizing mm-wave circuits working at very-high frequencies (e.g. 100GHz). Taking amplifiers as an example, at these frequencies, power gain is often the main considered performance. Due to the rather limited f_T of the CMOS technologies, other performances, such as the efficiency, are often very low and do not make sense to be optimized. The EMLDE method is presented here for the synthesis of such amplifiers.

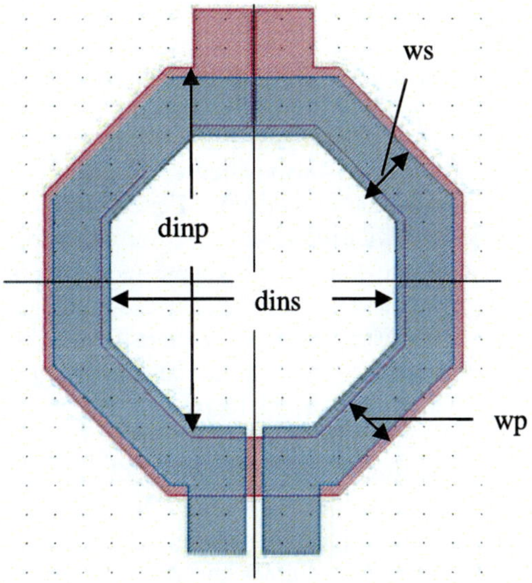

Fig. 2.2 A typical 60GHz transformer result generated by MMLDE

2.5.1 Overview of EMLDE

2.5.1.1 EMLDE: Goals, Challenges, Main Structure

EMLDE is based on combining electromagnetic (EM) simulation with evolutionary algorithms. Combining these two techniques, high-quality solutions can be obtained. However, like the synthesis of high-frequency integrated passive components, the challenge is that the EM simulations included in each performance evaluation are very CPU time expensive and that the standard EA needs more function evaluations compared with non-population-based optimization algorithms, although the optimization capability is much higher [22]. Hence, computational efficiency becomes the main problem. To address this problem, a natural idea is to use the on-line surrogate model construction and prescreening from MMLDE. However, the challenge with circuits instead of only passive components is the higher dimensionality. Therefore, a hierarchical machine learning-based evolutionary optimization mechanism is designed, which is the main framework of EMLDE. EMLDE aims to:

- develop the first synthesis method (layout included) for linear mm-wave RF amplifiers beyond 60GHz starting from a given circuit topology, specifications and some hints on layout (e.g. the metal layer to be used, the transistor layout template with different number of fingers);
- provide highly optimized results comparable to directly using an EA with EM simulations in the optimization loop, which is the best known method with respect to the solution quality aspect;
- use much less computational effort compared to using the standard EA, and as such make the computational time of the synthesis practical;
- be general enough for any technology and any frequency in the mm-wave frequency range.

Although MMLDE can very well solve low-dimensional computationally expensive optimization problems (e.g. transformer synthesis), the synthesis of mm-wave RF linear amplifiers brings new challenges, which make the MMLDE algorithm not workable for this problem. This is also called the "curse of dimensionality" in surrogate model assisted evolutionary algorithms. For the problem of RF amplifier synthesis, one stage of the amplifier often has 10-20 design variables (Section 2.6 provides an example). However, most GP-surrogate model assisted evolutionary algorithms normally can handle small-scale expensive optimization problems (e.g. 5 dimensions) very efficiently. Many works of expensive black-box optimization in the computational intelligence field focus on small-scale problems (e.g. [20],[23]). When the number of dimensions increases, two challenges appear. (1) Solution quality: an initial surrogate model that can roughly approximate the performance of the circuit is often difficult to construct with a reasonably small number of initial samples. Because the initial information is very limited, promising areas in the solution space are hard to be selected correctly, even with good prescreening methods. (2) Efficiency: a linear increase of the number of design variables causes an exponential increase of the search space, which requires more training data in the on-line optimization process. This also lowers the speed of the synthesis considerably, because more samples and iterations are needed and each of them is expensive. In addition, the computational effort to construct the GP model itself increases drastically with the number of design variables and the number of training data [24].

To address this problem, the new contributions in EMLDE focus on dimension reduction to transform the original high-dimensional problem to a lower-dimensional problem and on the efficient solution of this reformulated low-dimensional but more complex problem. The dimension reduction method is introduced in Section 2.5.1.4, and the proposed new algorithm to solve "hard to predict" low-dimensional expensive optimization problems is introduced in Section 2.5.2.

The general framework of the EMLDE method for mm-wave RF linear amplifier synthesis is shown in Figure 2.3. The key ideas and main blocks in the flow will be described in the next subsection. The core algorithms will be illustrated in a separate subsection.

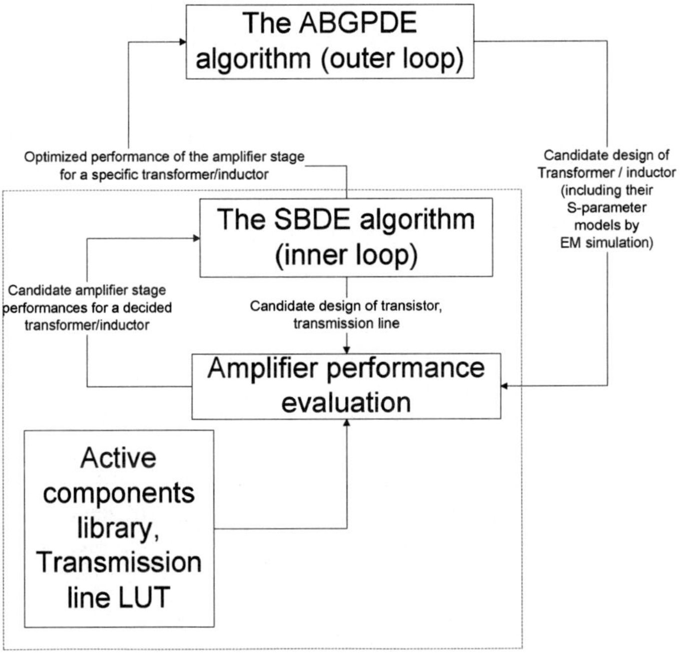

Fig. 2.3 The framework of the EMLDE method for the synthesis of high-frequency active circuits

2.5.1.2 The Active Components Library and the Look-Up Table for Transmission Lines

For the active components, EMLDE uses the parasitic-aware active device model library. For transmission lines, we use a look-up table (LUT) to get the values of the S-parameters, since the number of parameters of a transmission line is often not large and their S-parameters are highly linear. In the case of a linear RF amplifier optimizing performances according to S-parameters (e.g. power gain), the computationally expensive parts are: the parasitic extraction of active devices (transistors), the EM simulation of the long transmission lines and the EM simulation of transformers and inductors. When the performance optimization is based on linear analysis, usually the most critical problem is the impedance matching, and the transistors often have clear design rules. For example, a typical method is to use the minimum transistor length and a fixed width, while only the number of fingers is changed. In addition, the transistor layout is decided before any other components in many high-frequency amplifier designs. Hence, we first extract the parasitics of the transistors with different number of fingers but with fixed width and length beforehand and then directly use the extracted models in full-fletched optimization. Although optimizing the transistors (by changing the number of fingers) with already extracted models is not needed in the experiments of this chapter, this method is recommended when transistor sizing is necessary in synthesis. The extraction consumes some computational effort, but it is a one-time investment for each technology. Note that changing the number of fingers brings a discrete design variable in the optimization. A quantization technique [15] can be used to make the floating-point-based DE method also workable for mixed continuous and discrete optimization problems.

For the transmission lines, their S-parameters are highly linear. We first sweep the transmission lines with different line widths, lengths and distances between two lines when using differential transmission lines, and then build a look-up table (LUT). Through experiments, we found that the S-parameters generated by interpolation from the LUT differ little compared to the EM simulation results. The very small difference added less than 1% error on the circuit performances in our experiments. Like the active components, the data generation for the transmission lines is also a one-time investment. This solves the efficiency problem of performing on-line EM simulations for newly generated long transmission lines. Therefore, the most expensive part remaining in EMLDE is the EM simulation of transformers and inductors, which cannot be solved by the existing methods when going to mm-wave frequencies. The LUT method and the artificial neural network have been tested, but some of the S-parameters are very difficult to be trained and predicted with an acceptable accuracy. The solution will be presented in the next subsection.

2.5.1.3 Handling Cascaded Amplifiers

At high frequencies, to obtain a higher gain, the amplifier often includes multiple stages, which are cascaded together. For example, [25] designs the first fully differential 100GHz CMOS amplifier, which uses 6 cascaded stages to obtain about 10dB power gain. In manual design, the designer often copies the design of one stage to the other stages to construct the cascaded amplifier. This result is less optimal because the required impedance matching of each stage is different. In contrast, the synthesis method proposed here optimizes the cascaded amplifier stage by stage according to each stage's own impedance matching. In EMLDE, instead of dividing the circuit by active components like in manual design, transformers and inductors are the main objects for circuit division. Even for single-stage amplifiers but with a complex structure, dividing the circuit by key passive components is also workable in most cases. The division rules are: (1) one stage includes one and only one computationally expensive passive component; (2) the components in each stage must be connected together; (3) there should not exist components that do not belong to any stage, such as the input / output pads. More details of the division method will be shown by the example in Section 2.6.

2.5.1.4 The Two Optimization Loops

We now introduce our method to reduce the number of dimensions in one stage of the RF amplifier. Usually, the design parameters of a stage of the RF amplifier include the parameters of the transformers or inductors, the parameters of the transistors and the parameters of the connecting transmission lines. The overall circuit performance is decided by all of them. But with the help of the active components library and the look-up table for transmission lines discussed above, only the parameters of the transformers or inductors need expensive EM simulation. In addition, the number of parameters of a transformer or inductor is not large (often 4-5). Hence, a natural idea is to separate these design variables. Our method to reformulate the overall synthesis problem is as follows and is also shown in Figure 2.3.

The parameters of the transformers or inductors are set as the design variables (input), and the performances of the amplifier with the *decided* transformer or inductor and the *corresponding optimized* transistors and transmission lines are the output variables. In this way, the GP-based machine learning can be used for the outer optimization loop to decrease the number of expensive EM simulations. In other words, the original plain optimization problem is reformulated as a hierarchical optimization problem. The outer loop is the optimization of the transformer or inductor parameters, whose function values are the optimized performances of the amplifier stage, which is obtained by the inner optimization loop. The inner loop is the optimization of the transistors and transmission lines for the decided transformer or inductor provided by the outer loop. Although the inner loop needs more computational effort (an optimization is needed, rather than a single simulation), thanks to the efficient models for transistors and transmission lines, and the fast S-parameter circuit simulation, the evaluation of the inner function is very cheap. In addition, because of the independence of the candidates in the population of DE, parallel computation is used to further decrease the computational time. An 8-core CPU is used in this work. We use the selection-based differential evolution (SBDE) algorithm [26] for the inner optimization. Details of its settings are described in Section 2.5.2.

The price to pay for lowering the number of dimensions of the problem by decomposition is that the GP prediction and the expected improvement (EI) prescreening of the potential of a candidate design become more difficult. The reason is that the original performance is explicitly correlated to 10-20 or more variables, while in the new problem formulation it is predicted by 4-5 variables only and more than 10 variables are hidden. Hence, the problem to be predicted is more complex. Through experiments, we have found that only using the standard GP method and EI prescreening is not good enough to make the selection of the promising solutions effectively. This means that more iterations are necessary, which naturally leads to more EM simulations and more inner optimizations. Hence, we propose a new GP surrogate model assisted evolutionary algorithm: combining adaptive population generation, naive Bayes classification and Gaussian Process-based differential evolution (called ABGPDE). Using the ABGPDE algorithm, highly optimized designs can be obtained efficiently, with a solution quality comparable to directly using a global optimization algorithm with EM simulations in the function evaluations. In the test example of a 100GHz differential amplifier, the total synthesis time only costs 25 hours, and the speed is 9 times faster compared to directly using evolutionary algorithms with embedded EM simulations (needs 9 days), which makes the computational time very practical. ABGPDE will be described in Section 2.5.2.

2.5.2 Key Algorithms in the EMLDE Method

2.5.2.1 The ABGPDE Algorithm

The proposed ABGPDE algorithm is a surrogate model assisted evolutionary algorithm for low-dimensional expensive optimization problems, especially suitable for problems with difficult to predict data sets. In EMLDE, the ABGPDE algorithm solves the outer loop optimization. The inputs are the design parameters of the transformer or inductor (4-5 dimensions), and the output is the performance of a stage of the linear amplifier with its corresponding optimized transistors and transmission lines. The function evaluation includes the EM simulation of the transformer or inductor and the entire inner optimization loop.

2.5.2.1.1 The Structure of ABGPDE

Due to the dimension reduction, the basic structure of MMLDE can therefore be used in ABGPDE. The constraint handling method is the static penalty function method [27]. The differences with MMLDE are: (1) The inner optimization is included. Both in initialization and optimization, an inner optimization is performed for each passive component design to obtain the corresponding optimal performance of the amplifier stage. (2) Although the EI prescreening (see equation (2.17)) is still used, the evaluation of the potential of the candidates is different from MMLDE. (3) The population setting is different. In ABGPDE, there are two populations: one is the population containing all the simulated candidates, which is the same as MMLDE; the other is adaptively constructed in each iteration as described below.

It can also be mentioned that the EM simulations for the initial points can be done beforehand for a given technology, which is a one-time investment and can be used in the synthesis of all the stages. Only the inner optimization loop for the initial points needs to be done for each stage, because the best matching can be different from stage to stage.

2.5.2.1.2 Handling Difficult to Predict Data Sets

After lowering the original plain optimization problem with 10-20 design variables to a hierarchical optimization problem with 4-5 variables, the GP model predictions are more difficult. Through experiments, we found that using MMLDE, a satisfactory solution often needs many iterations, and each iteration is computationally expensive. The reason is that because of the complexity of the prediction problem, the number of wrong selections using the EI prescreening increases a lot.

We address this problem in two ways: (1) improving the potential evaluation of the candidates; (2) revising the EA.

For the EI prescreening, at the same time while achieving global search, it also bears the unavoidable risk of not selecting good candidates. For a candidate whose tail of the probability density function is smaller than f_{min} (the smallest function value found so far), the possibilities exist both of the candidate being truly a promising one or being not promising but having a large estimation variance. But EI cannot classify them. Hence, rather than improving the EI, we add the naive Bayes classification to help EI for this classification. If the function is continuous, the function values of two points x_i and x_j should be close if they are highly correlated. We use the naive Bayes classifier which only considers the input space x to make classifications. This is a good supplement to GP machine learning which considers both the input space and the output space. If a candidate has a high EI value but is classified into the unpromising points class, there is a high probability that the point has an unpromising function value but with a large estimation variance.

In ABGPDE, we use the mean performance of the current circuit being synthesized (e.g. the power gain of two stages of the circuit) for all the candidate designs of the current population as the threshold. The candidates with function value better than the threshold are classified as promising points; otherwise, they are classified as unpromising points. Both of them construct the training data. For a new population, we select the candidate solution with the highest EI value in the promising point class as the most promising one and evaluate it.

Although the naive Bayes classifier helps EI to evaluate the potential of the candidate solutions, it only contributes to the identification of promising solutions. On the other hand, the problem of how to make promising solutions being generated more efficiently is still not answered. The evolutionary algorithm (EA) we use for expensive optimization is different from the standard EAs. In standard EAs, the solution quality of the population is improving in the evolution process; so beneficial information to generate promising candidates keeps increasing in the consecutive populations. After some iterations, a high percentage of the information in the current population is beneficial to generate a candidate with good quality. In contrast, for surrogate model-assisted optimization, besides the initial samples, only one or few good new individuals are evaluated and added to the population in each iteration to increase the efficiency. Hence, in many occasions, the majority of the population are the initial samples. The goal of the initial samples is to cover the design space but many of them may not be good solutions.

Consequently, the percentage of beneficial information in the consecutive populations increases slowly. In evolutionary optimization, the new population is generated according to the information of the previous population by evolution operators. If the amount of beneficial information in the previous population is less, generating promising candidates is more difficult.

Our idea to solve this problem is to artificially increase the amount of beneficial information by constructing a new population for evolution. In each iteration, we rank all the candidates in the original population and select the top 75% candidates to enter the new population for evolution. The remaining 25% of the new population is filled by randomly selecting the candidates from the top 75% candidates of the original population. This operation may sacrifice the diversity a little bit but causes fast convergence to a satisfactory result, which fits the needs for expensive optimization problems. After all, in EMLDE, because all the simulated candidates are included in the original population, the diversity is quite high. Hence, if we replace some low-quality candidates, the diversity does not decrease too much.

Based on the above ideas, the ABGPDE algorithm is constructed. The flow diagram and the entire EMLDE algorithm description will be shown in Section 2.5.3.

2.5.3 The Embedded SBDE Algorithm

The SBDE algorithm is used for the inner optimization loop of the synthesis system (see Figure 2.3). The inputs are the transistor parameters, transmission line parameters and DC voltages. The used transformer or inductor and its EM simulation result are provided from the outer loop. The output is the performance of one stage of the RF amplifier with optimized transistors, transmission lines and biasing voltages.

The construction of SBDE is quite simple. The optimization core is the standard DE algorithm (see Section 2.3.1) with a change of the selection operator. Constraints (e.g. bandwidth ≥ 20GHz) are handled by using the selection rules in [28]. The selection rules are: (1) Given two feasible solutions, select the one with the better objective function value; (2) Given two infeasible solutions, select the solution with the smaller constraint violation; (3) If one solution is feasible and the other is not, select the feasible solution. More details are in [26].

Although the S-parameter simulation for the circuit is fast, the inner optimization needs to evaluate a full population in each iteration and the optimization needs several iterations. Because the evaluation of different candidate designs in a population is independent from each other in SBDE, we use parallel computation. In our implementation, an 8-core CPU is used. Experiments for the test circuit below show that the inner optimization of one candidate transformer design can be finished in 5-6 minutes. This time consumption is quite general for different circuits, because the time cost of the inner loop is mainly dominated by the S-parameter circuit simulation, which is correlated to the size of the simulated circuit stage. Using the same hardware and the circuit division method (see Section 2.5.1.4) based on passive components, the size of each stage often does not vary much, even for different amplifiers.

2.5.4 The EMLDE Method

Based on the above techniques, the entire EMLDE method is now described.

2.5.4.1 The Flow Diagram of EMLDE

The general flow of the high-frequency linear RF amplifier synthesis system is shown in Figure 2.3. The flow diagram of the EMLDE algorithm is shown in Figure 2.4. ABGPDE and SBDE are included in EMLDE. The parameter setting rules for EMLDE can be found in [7].

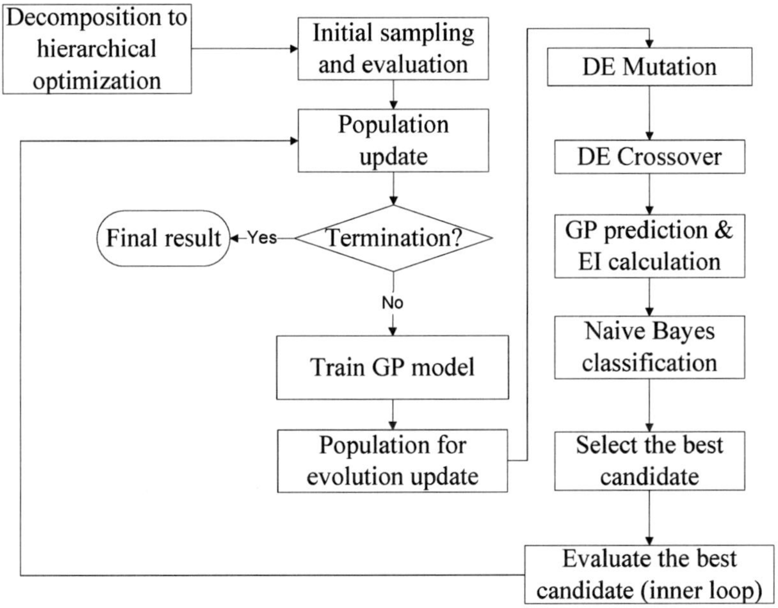

Fig. 2.4 Flow diagram of the EMLDE method for the synthesis of high-frequency active circuits

The EMLDE algorithm works as follows.

Step 0: Decompose the problem of optimizing an amplifier stage into the hierarchical optimization with outer and inner loops according to the method described in Section 2.5.1.4.

Step 1: Initialize the parameters, e.g. the DE algorithm parameters.

Step 2: Initialize the population by LHS sampling of the design space and perform EM simulation to the samples (can be done beforehand for a given technology). Perform the inner optimization loop for the samples. SBDE, the active component library and the transmission line LUT are used in this step.

Step 3: Update the population A by adding newly generated samples and their performances. In the first iteration, the added samples are from step 2; afterwards, they are from step 11. Update the best solution obtained so far.

Step 4: Check if the stopping criterion (e.g. a convergence criterion or a maximum number of iterations) is met. If yes, output the result; otherwise go to step 5.

Step 5: Train the GP surrogate model according to population A.

Step 6: Construct the population for evolution (population B) as described in Section 5.2.1.2.

Step 7: Use population B and perform the DE mutation operation (see equation (2.4)) to obtain each candidate solution's mutant counterpart.

Step 8: Perform the crossover operation between each candidate solution and its corresponding mutant counterpart (see equation (2.5) and (2.6)) to obtain each individual's trial individual.

Step 9: Calculate the EI value of all the trial individuals from step 8.

Step 10: Use the population A as the training data, perform naive Bayes classification as described in Section 2.3.3 to all the trial individuals from step 8.

Step 11: Select the individual with the best potential according to the selection rule of SBDE and evaluate it using the same way as in step 2. Go back to Step 3.

2.6 Experimental Verification of the EMLDE Method

2.6.1 Example and Settings

The EMLDE method is now demonstrated for the synthesis of a 100GHz three-stage transformer-coupled fully differential amplifier [25] in a 90nm CMOS technology. One stage of the circuit configuration is shown in Figure 2.5. Using the same configuration but different sizing for each stage, the different stages are cascaded together. The optimization goal is the power gain (S21 at 100GHz); the constraints are bandwidth ≥ 20GHz and the Rollet stability factors (K factors) [29] >1. Note that at 100GHz, design for high gain is difficult due to the limitation of the f_T of the 90nm technology. Achieving similar performances, [25] uses six stages (measurement result). Although simulation results and measurement results cannot be compared directly, the manual design result is a good reference to verify the high solution quality of EMLDE.

The transmission line used is a high-Q slow-wave coplanar transmission line [30]. The differential lines (CPW line) are on the top metal layer and the floating metal strips are on the lower metal layer. For the transformer, the top metal layer is used. All the transistors have the same size to make sure that each stage can drive the next stage. All the transistors have 1 μm width, 90nm length and 15 fingers as in [25]. In a cascaded multi-stage RF amplifier with the same transistor size, the optimal design parameters of the transformers often do not differ much from one stage to the other. Indeed, the typical manual design method of copying the design of one stage to construct the whole amplifier achieves less optimal results. Because of this, all the passive components need to be re-synthesized. But after performing the inner optimization on the initial samples in the synthesis of the previous stage, we can delete a few samples which have very bad performances when synthesizing the next stage. The design variables are as follows. For transformers, the design variables are the inner diameter of the primary inductor (*dinp*), the inner diameter of the secondary inductor (*dins*), the width of the primary inductor (*wp*) and the width of the secondary inductor (*ws*) and the spacing between the two ports (*sp*). For transmission lines, the design variables are the metal width (*lw*), the metal length (*ll*), and the spacing between the differential lines (*ls*). Two DC voltages (V_D and V_{Bias}) are included in each stage. The overall input / output load impedance is $50\,\Omega$. The ranges for the design variables are in Table 2.2, which are provided by the designer. There are in total 51 design variables for the three-stage amplifier.

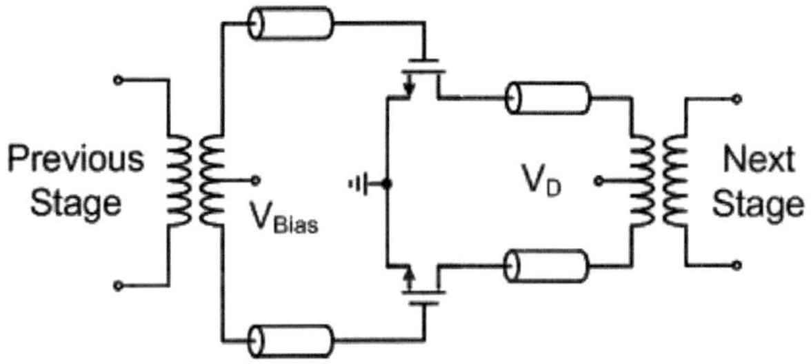

Fig. 2.5 One stage of the 100GHz amplifier (the full amplifier needs three such stages)

The algorithm parameter settings are as follows. In ABGPDE, the parameters are the same as in the MMLDE settings. For the GP model used in ABGPDE, the DACE toolbox [31] is used. For SBDE, we set the population size to 20 and the number of iterations to 20. The purpose of this setting is to call for a good balance between the solution quality and the efficiency of the inner optimization. EMLDE stops when the performance cannot be improved for 20 consecutive generations or when the number of outer iterations reaches 90 (including initial points). The examples are run on a PC with Intel 2.66GHz dual Xeon 2×6 core-CPU (only 8 cores are used) under the Linux operating system. All the time consumptions mentioned in the experiments are clock time. ADS Momentum is used as the EM simulator. Note that in all the methods investigated, each EM simulation is parallelized automatically by Momentum using the 8 cores. Synopsys HSPICE is used as the circuit simulator with S-parameter models.

Table 2.2 Design parameters and their ranges

Parameters	Lower Bound	Upper Bound
dinp,dins (μm)	30	110
wp, ws (μm)	2	10
sp (μm)	8	23
lw (μm)	1	10
ll (μm)	2	80
ls (μm)	7	23

We synthesize the amplifier stage by stage, starting from the output stage forward to the input stage. For the current stage being optimized, the S-parameter models of the passive components (transformer, transmission lines) are separated for HSPICE simulation, while the already synthesized stages are described by a single S-parameter model integrating all the passive components. For example, when optimizing stage 2, the transformer and the transmission lines have their own S-parameter models to enter the HSPICE simulation. For stage 3, which is already synthesized at that moment, the pad, transformer and transmission lines are connected together to perform an EM simulation, whose S-parameters result will be used when synthesizing stage 2. Because the overall amplifier includes four parts (see Figure 2.6) to synthesize, and the matching impedances of each part are different, four test problems are included in this example. The performance of the whole amplifier is affected by all of the four test problems, which shows the robustness of EMLDE.

2.6.2 Example: Three-Stage Linear Amplifier Synthesis

We first use the method of the DE algorithm with EM simulation for the transformers as the performance evaluation. No machine learning method is used. This method can provide the best result, which serves as the reference result, but this method is of course very CPU time expensive. Parallel computation of the HSPICE simulations is not included. The synthesized circuit is shown in Figure 2.6 and the simulation results are shown in Figure 2.7. The power gain is 10.53dB and the time consumption is about 9 days.

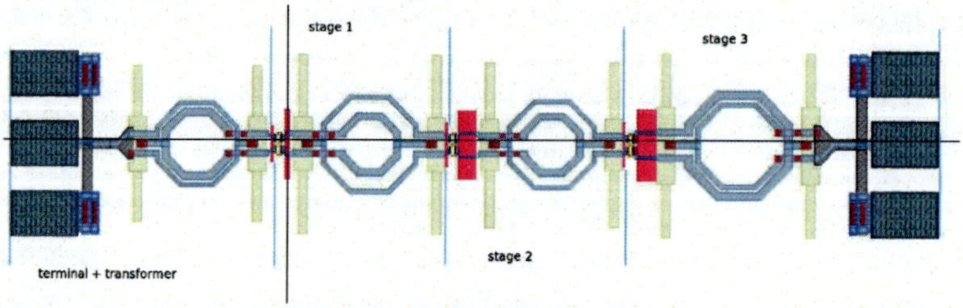

Fig. 2.6 The amplifier synthesized without machine learning

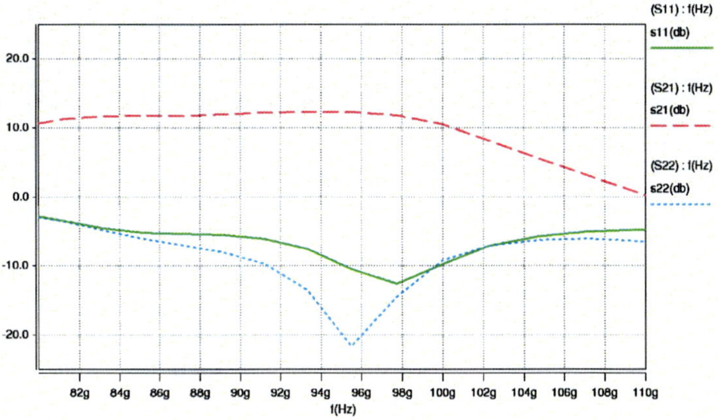

Fig. 2.7 S-parameters curve of the amplifier from Figure 2.6 from 80GHz to 110GHz (reference method – 9 days)

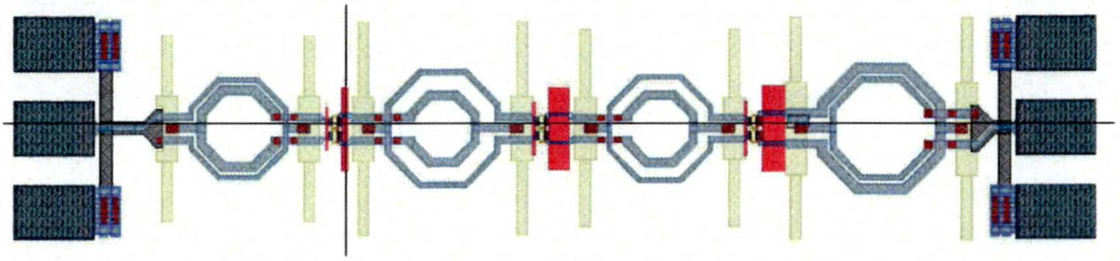

Fig. 2.8 The amplifier synthesized by EMLDE

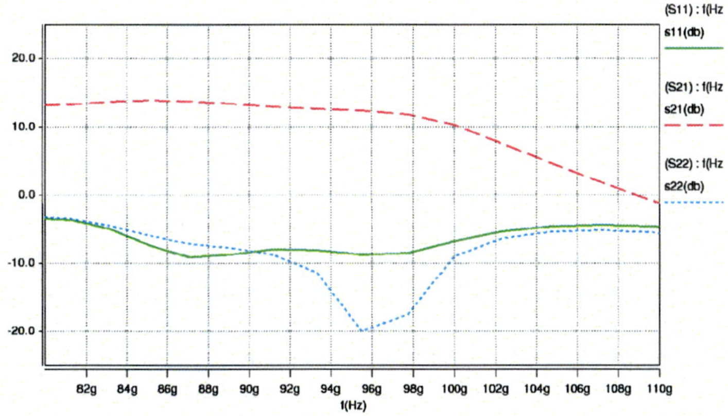

Fig. 2.9 S-parameter curve of the amplifier synthesized by EMLDE (Figure 2.8) from 80GHz to 110GHz (EMLDE – 25 hours)

Then, we use the EMLDE-based synthesis system. The synthesized circuit is shown in Figure 2.8 and the simulation results are shown in Figure 2.9. The power gain is 10.41dB (i.e. slightly less than 10.53dB) but the CPU time cost is only 25 hours. The constraints are satisfied for both methods.

It can be seen that the presented high-frequency RF linear amplifier synthesis system can achieve a result comparable to directly using the DE algorithm and EM simulation, which is the reference benchmark. We can also see the high solution quality from the mm-wave frequency RF IC design aspect. It is well known by designers that achieving 3 dB power gain per stage requires a very good matching beyond 60GHz, and that the higher the working frequency, the more difficult it is to achieve high gain. The result shows that the average power gain of each stage reaches nearly 3.5 dB in the synthesized amplifier at 100GHz which is impressive considering the loss of the passive components and the influence of the pads. In terms of computational efficiency, about 9 times speed enhancement is achieved by EMLDE. The time cost of 25 hours is very reasonable for practical use.

In this synthesis, a total of 48 EM simulations are used (excluding the 49 initial sampling points which can be done beforehand in a given technology). The time spent on the EM simulations is 2.5 hours. The inner loop optimization costs 22.7 hours. When directly using the DE algorithm and EM simulation, nearly 4000 EM simulations are needed. It can be seen that EMLDE decreases the number of expensive EM simulations carried out by about 80 times. Although there are much more circuit simulations in EMLDE, the linear circuit simulation is cheap. We can also conclude that the more complex the key passive components, which need more EM simulation time, the higher the advantage of EMLDE.

2.7 Conclusion

This chapter has reported the efficient high-frequency synthesis methods for the synthesis of integrated passive components (with a method called memetic machine learning-based differential evolution (MMLDE)) as well as for the synthesis of mm-wave-frequency linear amplifiers (with a method called Efficient Machine Learning-based Differential Evolution (EMLDE)). The common ideas are to introduce machine learning into evolutionary computation, in order to construct a surrogate model assisted evolutionary algorithm. In MMLDE, by using memetic computation integrating Gaussian Process machine learning with expected improvement prescreening and an artificial neural network with predicted values, the search quality and efficiency have considerably been enhanced. In EMLDE, by using the decomposition method, which reformulates the problem into a hierarchical structure, and using the ABGPDE algorithm to solve the low-dimensional but more complex expensive optimization problems, the "curse of dimensionality" of mm-wave circuit synthesis has been addressed. MMLDE and EMLDE have been shown to provide results that are comparable to directly using a global optimization algorithm in combination with EM simulations as performance evaluation, which is the best framework in terms of solution quality, but at far lower (nearly an order of magnitude) computational cost. The objectives of achieving high optimization ability, high efficiency and high generality in the automated synthesis of high-frequency circuits are therefore met.

Acknowledgements. This research has been supported by a bilateral agreement scholarship of Katholieke Universiteit Leuven, Belgium and Tsinghua University, P. R. China. The authors sincerely thank Dixian Zhao, Noel Deferm, and their supervisor, Patrick Reynaert from ESAT-MICAS for providing the examples and verifications. The authors also thank Brecht Machiels, Ying He, Borong Su, Wan-Ting Lo and Bohan Yang, previous students of ESAT-MICAS, for supporting works.

References

[1] Niknejad, A.M., Hashemi, H.: mm-Wave Silicon Technology: 60GHz and Beyond. Springer (2008)
[2] Nieuwoudt, A., Massoud, Y.: Variability-Aware Multilevel Integrated Spiral Inductor Synthesis. IEEE Transactions on Computer-Aided Design of Integrated Circuits and Systems 25, 2613–2625 (2006)
[3] Allstot, D.J., Choi, K., Park, J.: Parasitic-Aware Optimization of CMOS RF Circuits. Kluwer Academic Publishers (2003)
[4] Choi, K., Allstot, D.J.: Parasitic-aware design and optimization of a CMOS RF power amplifier. IEEE Transactions on Circuits and Systems - I, 16–25 (2006)
[5] Fogel, D.B.: Evolutionary computation: toward a new philosophy of machine intelligence. Wiley-IEEE (2006)
[6] Liu, B., Zhao, D., Reynaert, P., Gielen, G.: Synthesis of Integrated Passive Components for High-Frequency RF ICs Based on Evolutionary Computation and Machine Learning Techniques. IEEE Transactions on Computer-Aided Design of Integrated Circuits and Systems 30, 1458–1468 (2011)
[7] Liu, B., Deferm, N., Zhao, D., Reynaert, P., Gielen, G.: An Efficient High-Frequency Linear RF Amplifier Synthesis Method Based on Evolutionary Computation and Machine Learning Techniques. IEEE Transactions on Computer-Aided Design of Integrated Circuits and Systems 31, 981–993 (2012)
[8] Tulunay, G., Balkir, S.: A synthesis tool for CMOS RF low-noise amplifier. IEEE Transactions on Computer-Aided Design of Integrated Circuits and Systems 27, 977–982 (2008)
[9] Ramos, J., Francken, K., Gielen, G., Steyaert, M.: An efficient, fully parasitic-aware power amplifier design automation tool. IEEE Transactions on Circuits and Systems - I 52, 1526–1534 (2005)
[10] Agarwal, A., Vemuri, R.: Layout-aware RF circuit synthesis driven by worst case parasitic corners. In: Proc. of ICCD, pp. 444–449 (2005)
[11] Zhang, G., Dengi, A., Rohrer, R.A., Rutenbar, R.A., Carley, L.R.: A synthesis flow toward fast parasitic closure for radio-frequency integrated circuits. In: Proc. of DAC, pp. 155–158 (2004)
[12] De Ranter, C.R.C., Van der Plas, G., Steyaert, M., Gielen, G., Sansen, W.: CYCLONE: automated design and layout of RF LC-Oscillators. IEEE Transactions on Computer-Aided Design of Integrated Circuits and Systems 21, 1161–1170 (2002)
[13] Chu, M., Allstot, D.J.: Elitist nondominated sorting genetic algorithm based RF IC optimizer. IEEE Transactions on Circuits and Systems - I 52, 169–174 (2005)
[14] Park, J., Choi, K., Allstot, D.J.: Parasitic-aware design and optimization of a fully integrated CMOS wideband amplifier. In: Proc. of ASP-DAC, pp. 904–907 (2003)
[15] Price, K.V., Storn, R.M., Lampinen, J.A.: Differential Evolution. A Practical Approach to Global Optimization. Springer, Heidelberg (2005)
[16] Larose, D.: Data Mining Methods and Models. John Wiley & Sons (2006)

[17] Caruana, R., Niculescu-Mizil, A.: An empirical comparison of supervised learning algorithms. In: Proc. of the International Conference on Machine Learning, pp. 161–168 (2006)
[18] Stein, M.: Large Sample Properties of Simulations Using Latin Hypercube Sampling. Technometrics, 143–151 (1987)
[19] Mandal, S.K., Sural, S., Patra, A.: ANN- and PSO-Based Synthesis of On-Chip Spiral Inductors for RF ICs. IEEE Transactions on Computer-Aided Design of Integrated Circuits and Systems 27, 188–192 (2008)
[20] Jones, D.R., Schonlau, M., Welch, W.J.: Efficient Global Optimization of Expensive Black-Box Functions. Journal of Global Optimization 13, 455–492 (1998)
[21] Wasserman, P.D.: Neural computing: theory and practice. Van Nostrand Reinhold, New York (1988)
[22] Liu, B., Fernandez, F.V., Gielen, G., Castro-Lopez, R., Roca, E.: A Memetic Approach to the Automatic Design of High Performance Analog Integrated Circuits. ACM Trans. on Design Automation of Electronics Systems 14, Article 42 (2009)
[23] Jones, D.R.: A Taxonomy of Global Optimization Methods Based on Response Surfaces. Journal of Global Optimization 21, 345–383 (2001)
[24] Emmerich, M.T.M., Giannakoglou, K.C., Naujoks, B.: Single- and multiobjective evolutionary optimization assisted by Gaussian random field metamodels. IEEE Transactions on Evolutionary Computation 10, 421–439 (2006)
[25] Deferm, N., Reynaert, P.: A 100GHz transformer-coupled fully differential amplifier in 90nm CMOS. In: Proc. of RF IC, pp. 359–362 (2010)
[26] Zielinski, K., Laur, R.: Constrained single-objective optimization using differential evolution. In: Proc. IEEE Congress on Evolutionary Computation, pp. 223–230 (2006)
[27] Michalewicz, Z.: A Survey of Constraint Handling Techniques in Evolutionary Computation Methods. In: Proc. of the 4th Annual Conference on Evolutionary Programming, pp. 135–155 (1995)
[28] Deb, K.: An Efficient Constraint Handling Method for Genetic Algorithm. Computer Methods in Applied Mechanics and Engineering 2, 311–338 (2000)
[29] Rollett, J.: Stability and power-gain invariants of linear two ports. IRE Transactions on Circuit Theory 9, 29–32 (1962)
[30] Kaddour, D., Issa, H., Franc, A.L., Corrao, N., Pistono, E., Podevin, F., Fournier, J.M., Duchamp, J.M., Ferrari, P.: High-Q slow-wave coplanar transmission lines on 0.35um CMOS process. IEEE Microwave and Wireless Components Letters 19, 542–545 (2009)
[31] Lophaven, S.N., Nielsen, H.B., Sondergaard, J.: DACE: A MATLAB Kriging Toolbox. Technical University of Denmark, http://www2.imm.dtu.dk/~hbn/dace/ (accessed December 1, 2011)

Chapter 3
Self-Healing Circuits Using Statistical Element Selection

Vanessa H.-C. Chen, Gokce Keskin, and Lawrence T. Pileggi

Abstract. Due to the ongoing aggressive scaling of integrated circuit technologies, designers are challenged by creating robust analog and mixed-signal circuit designs. The increasing random intra-die variations of small feature sizes in advanced CMOS nodes severely limit the benefits of scaling for analog/mixed-signal circuits with the diminishing voltage headroom. This chapter describes the details of the statistical element selection (SES) methodology that relies on the combinatorial growth in number of subsets. With selectable circuit elements, the randomness can be used to provide post-manufacturing configuration to achieve specifications. The calibration methodology is demonstrated with two silicon results in 65nm CMOS technology. One test chip consists of an array of digitally calibrated comparators with built-in combinatorial redundancy. Over 99.5% of the comparators reach the given offset requirement compared to 15% for Pelgrom-type sizing. The other test chip is an 8-bit, 1.5GS/s flash ADC. The prototype achieves 37dB of SNDR with 1.3GHz ERBW for 35mW power consumption and 0.42pJ/conv-step of figure of merit.

3.1 Introduction

Aggressive scaling of integrated circuit technologies has created significant challenges for robust design of analog and mixed-signal circuits. The digitalization of mixed-signal blocks benefits most from the technology scaling; however, large-size transistors are needed in analog/mixed-signal circuits to address matching issues for nano-scale devices using traditional sizing methods [1]. In the presence of large-scale random variations, mismatch becomes a significant design challenge

Vanessa H.-C. Chen · Lawrence T. Pileggi
Department of Electrical and Computer Engineering
Carnegie Mellon University, Pittsburgh, PA, USA
e-mail: {vanessachen,pileggi}@cmu.edu

Gokce Keskin
Formerly Carnegie Mellon, now with Intel Corporation, Hillsboro, OR, USA
e-mail: gokce.keskin@intel.com

M. Fakhfakh et al. (Eds.): *Analog/RF & Mixed-Signal Circuit Sys. Design*, LNEE 233, pp. 53–75.
DOI: 10.1007/978-3-642-36329-0_3

and alternative device sizing approaches must be considered. Manufacturing variations can generally be classified into two categories, systematic and random. Many of the dominant systematic variations can be predicted and addressed by using careful circuit design and layout techniques. However, random variations of the identically designed devices are unpredictable and can cause significant mismatch among devices.

The traditional approach for reducing the local random mismatch is to increase the device size following Pelgrom's seminal paper [1]. The standard deviation of random mismatch characterized by the Pelgrom model is inversely proportional to the square root of the device area, which corresponds to increased area and current for improved matching. With increased device sizing, the power consumption increases due to the larger currents, but can increase further due to the additional power required to achieve the bandwidth targets with the increase in device loading capacitance. For some designs, a better tradeoff between power consumption and device mismatch is achieved using digital calibration techniques, such as redundancy [2]. Redundancy was demonstrated for flash ADCs in [2], for example, as an implementation with extra comparators followed by selection of the best 2^N comparators for an N-bit design.

With the additional variability, particularly that due to random threshold mismatch, that is incurred with scaling to and below the 65nm CMOS node, even digital calibration and redundancy methods can require significant overhead costs. Statistical Element Selection (SES) was recently proposed to alleviate the impact of extreme random variability via a combinatorial implementation of redundancy [3]. The basic concept is to choose one subset from among $2^N - 1$ available subsets. Each analog component that is to be matched is formed from N identically-designed subcomponents (e.g., branches of input transistors in a comparator). But rather than grouping the branches into predetermined blocks for redundancy selection, each pair is allowed to be individually selected. The combinatorial redundancy (CR) can be implemented efficiently when the subcomponents are enabled by digital selection. The search space of combinatorial redundancy is much larger than that of redundancy, and the exponential increase in the number of combinations results in a significant improvement in matching at the increased cost of statistical selection.

In this chapter the SES methodology is analyzed in detail and described for the implementation of low-power self-healing circuits and systems. Two silicon results in 65nm CMOS bulk process demonstrate the potential for improved power efficiency with SES. The primary sources of variation in modern CMOS processes and their effects on analog circuits are first reviewed in the following section.

3.2 Process Variations

Manufacturing variations are a significant problem for both digital and analog circuits in advanced CMOS process nodes and are expected to grow in importance with each new generation. Variations in modern CMOS nodes can generally be classified into two categories: systematic and random. Many of the dominant systematic variations can be predicted and addressed by using careful circuit design and layout techniques. Random variations are unpredictable and can cause significant mismatch among devices on the same die. It is the randomness of mismatch that is actually exploited by the CR/SES approach to tune the circuits after manufacturing.

3.2.1 Systematic Variations

Systematic variations can be broadly classified into two sub-groups [4]:

- Across-field effects that are caused by lithography or etching: Location of the die on the wafer can lead to a systematic shift in device parameters. All devices in the same vicinity are affected the same way due to these effects.
- Layout dependent effects that result in different characteristics of identical devices in the same vicinity in the wafer: An example is variation due to the well proximity effect where threshold voltage of a MOSFET close to an n-well can be different from an identical MOSFET far away from n-wells. Other major sources include those due to the polysilicon surrounding of the gates and STI stress [5]–[7].

Restricted design rules with fixed gate lengths, high regularity in diffusion, poly and metal layers, single poly orientation and lithography solutions such as double patterning and optical proximity correction are already proposed techniques to alleviate the systematic effects in leading-edge CMOS processes [8][9]. Although these methods are mainly discussed for logic gates and memories, analog circuits will ultimately require similar patterning rules for controlling variations.

3.2.2 Random Variations

Random variations are due to unpredictable and unrepeatable sources of variation in manufacturing. Random dopant fluctuation (RDF) in the transistor channel is an example [10]. Several new technologies such as undoped channels, high-k metal gates, thin SOI, and Fin-FETs are being evaluated, but tens of millivolts of variation in threshold voltage is still expected [11]–[15].

Another source of random variation is line edge roughness (LER). Microscopic deviations in the poly line forming the gate can lead to uneven channel length across the width of the device. These variations can lead to an effective difference in the conductance constant and adversely affect matching.

3.2.3 Mismatch Correction Methods

Random sources of variation cannot be alleviated by following restricted design rules, and hence become the primary challenge in matching-limited designs. Minimization of mismatch is desirable in order to scale more aggressively and at the same time to improve the power efficiency. Increasing the device size is the conventional way to reduce the local random mismatch. The standard deviation of input offset of a differential pair is characterized by the Pelgrom model as

$$\sigma(V_{os}) = \sqrt{\frac{A_{VT}^2+0.25(V_{GS}-V_{th})^2 A_\beta^2}{WL}} \tag{3.1}$$

where A_{VT} and A_β are process-related coefficients, W and L are width and length of the differential pair.

Assuming that the input offset of the differential pair is dominant, the input offset is inversely proportional to the square root of the device area. However, large area and hence more current are required for better matching property. Therefore, the power consumption increases rapidly as the number of analog circuit elements increases.

In order to achieve better tradeoff between power consumption and device mismatch for an input differential pair, digital calibration techniques with redundancy elements or digital-to-analog converters (DACs) are used. Redundancy was demonstrated for flash ADCs as the implementation of an excess of comparators followed by selection of the best comparator [2] as shown in Fig. 3.1. *K* identical comparators are implemented, each with a unique offset taken from the normal distribution. The comparator with the minimum offset found within the set of *K* comparators is used. Redundancy can be extended to using more than 1 comparator from the set of *K* comparators. In flash ADCs the redundant comparators must be spread out across the range of reference voltages. If the comparators have σ_{offset} on the order of 1 LSB, the probability distribution functions (PDFs) of the offsets of comparators with neighboring reference voltages will overlap considerably. The overlapping PDFs allow for comparator reassignment, where a comparator nominally designed for one reference voltage is chosen to act as the comparator for a second reference voltage. This method significantly increases the probability that at least one redundant element with small area can satisfy the given input offset specification.

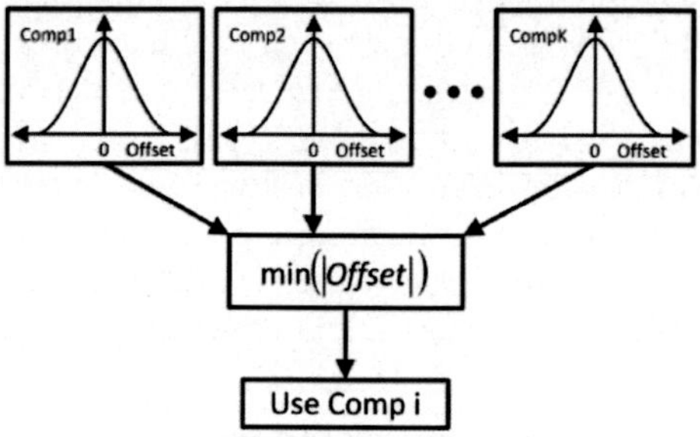

Fig. 3.1 Redundancy for comparators: choosing 1 of K comparators

DAC-based calibration can be implemented with low overhead by adding a thermometer-coded current DAC tapping into the A and B nodes as shown in Fig. 3.2 [16]. The injected current changes the offset voltage. Each unit cell could consist of only three near-minimum size transistors: two switches to control polarity and a current source. The DAC used in DAC-based calibration must be able to calibrate the comparator to a given accuracy. The calibration requirements result in a range and resolution specification for the DAC. To provide a high yield, the DAC must provide enough full scale range to correct for the expected range of random input offsets.

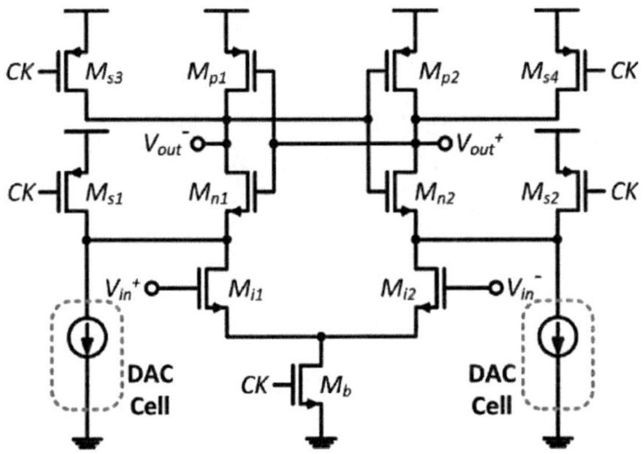

Fig. 3.2 Current DAC calibrated comparator

3.3 Statistical Element Selection

3.3.1 Basis

The basic concept of Statistical Element Selection (SES) [3] is to use N identically laid-out elements for a critical circuit block (e.g., branches of input transistors in a comparator) and choose one subset from among the $2^N - 1$ available subsets such that will satisfy the desired specification (e.g., input offset voltage). Consider the differential amplifier in Fig. 3.3, with N pairs of input NMOS transistors that are labeled as $M_{1a}\backslash M_{1b}$ through $M_{Na}\backslash M_{Nb}$. Each pair has its own tail NMOS transistor with gates tied to digital control signals: Sel_1 through Sel_N. Each pair can be turned on or off as desired by $Sel < 1{:}N >$. Each transistor has different characteristics due to the random variations, and the mismatches between the pair transistors result in non-ideal effects such as input offset voltage.

With sizing for random variations based on the approach in [1], all branches from 1 to N are selected. All signals $Sel < 1{:}N >$ are connected to the same line during the design phase. The larger area by the selection of all mismatched branches results in a lower amount of effective variation, and yields an improvement of $1/\sqrt{N}$ in matching standard deviation [1]. For the redundancy technique, branches are grouped into predetermined identical blocks during the design phase. Only one block can be selected at a time during post-manufacturing calibration. For example, assume that each pair in Fig. 3.3 forms one block as a result of total

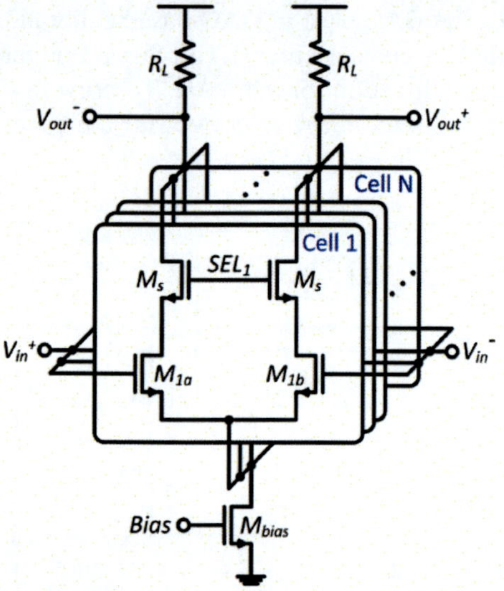

Fig. 3.3 SES-based differential amplifier

N blocks. Among the available *N* combinations, the one with the best offset specification is selected. If $N/2$ branches form one block, there are only 2 combinations to select from during calibration.

Rather than grouping the branches into predetermined blocks, each pair is allowed to be individually selected in CR/SES. This is essentially a finer grain redundancy that can be used for efficient digital selection of the "elements." If total of $N/2$ pairs are desired, the selection can be made among the $\binom{N}{N/2}$ subsets, which are formed using the control signals. The search space is much larger than that of the component redundancy described in [2]. If $N = 16$ and 8 pairs form one block, only two blocks are available for selection with redundancy elements. However 12,870 combinations are available as any subset of size 8 can be selected with SES. The subset size is not constrained to $N/2$, and any subset among $2^N - 1$ can be selected. This exponential increase in the number of combinations results in a significant improvement in finding a low offset combination.

Assume that the input offset of *i*th input pair is $V_{os,i}$ and the transconductances of all pairs are the same, then the input offset voltage of the differential amplifier is:

$$V_{os} = \frac{1}{N} \times \sum_{i=1}^{N} V_{os,i} \tag{3.2}$$

where $V_{os,i}$ is the input offset of the *i*th input pair. If we consider the case that only a subset of the *N* pairs is chosen (*N* times redundancy), the resulting input offset voltage is:

$$V_{os} = \frac{1}{\sum_{i=1}^{N} k_i} \times \sum_{i=1}^{N} k_i V_{os,i} \tag{3.3}$$

where $k_i = 1$ if the *i*th pair is chosen, and $k_i = 0$ otherwise.

V_{th} mismatch generally dominates the input offset of the differential pair, and we can write $V_{os,i} = \Delta V_{th,i}$ [17]. In addition, assume that the input transistors are the dominant source of mismatch, and that the $V_{os,i}$ distribution is centered at 0 and can be estimated as a Gaussian normal distribution with $\mathcal{N}(0, \sigma_{os,i}^2)$. Using (3.3), we can determine that the input offset voltage of the amplifier is $\mathcal{N}(0, \sigma_{os}^2)$, where $\sigma_{os} = \sigma_{os,i}/\sqrt{\sum_{i=1}^{N} k_i}$. This follows a close resemblance to the result found in [1], where matching of MOS devices in close proximity has been shown to improve by $1/\sqrt{Area}$.

As in all circuit designs, the main goal of SES is to achieve a target specification such as input offset voltage with arbitrarily high probability (e.g., 99.5%) with lowest possible power and area. The basic parameters to determine are:

- Total number of selectable elements (*N*)
- The number of elements selected (*k*)
- Size of each element
- Total number of sets among $\binom{N}{k}$ that will be tried, determining calibration time

Since different circuits and applications require different trade-offs among these parameters, a methodology to determine the values of the basic parameters is described in the following section.

3.3.2 Methodology

Figure 3.4 shows a latch type SES-based comparator where the dark sections are replicated *N* times [18]. Assume that each selectable element on Fig. 3.4 has an offset distribution that follows normal $\mathcal{N}(0, \sigma_{os,i}^2)$ and only one element among the *N* is selected. The probability that this element has an absolute offset smaller than a given specification *spec* is

$$p_{success} = erf\left(\frac{spec}{\sigma_{os,i} \times \sqrt{2}}\right) = 1 - p_{fail} \tag{3.4}$$

p_{fail} denotes the probability that this element will fall out of the given offset specification (*spec*). To ensure good linearity of the ADC, *spec* should be less than ±0.5LSB. Since the offset of each element is independent, one can calculate the probability that each and every one of the available *N* elements will fall out of the desired offset specification as:

$$p_{fail,total} = \left(p_{fail}\right)^N \tag{3.5}$$

This is a classical example of the *N*-time redundancy method. Let us now consider that all *N* elements are chosen. In this case, the offset distribution follows $\mathcal{N}(0, \sigma_{os,i}^2/N)$. The probability that the offset is within *spec* (denoted by $p_{fail,N}$) can be calculated simply by substituting $\sigma_{os,i}$ in (3.4) with $\sigma_{os} = \sigma_{os,i}/\sqrt{N}$. This is a classical example of Pelgrom-type sizing to reduce random variability and results in lower failure probability than using only a single element.

Redundancy and Pelgrom-type sizing are the two extremes for SES. Rather than selecting one at a time (redundancy), or all at once (sizing), k elements among N are selected at a time ($1 \leq k \leq N$). Fig. 3.5, which is generated using 1×10^6 Monte Carlo samples in MATLAB, shows the failure probability ($p_{fail,total}$) as k is varied as $N = 20$, $\sigma_{os,i} = 1$ and offset specification (*spec*) is 10^{-2}. In other words, an absolute offset less than 1/100 of the standard deviation of each element is tried to achieve. p_{fail} for each element can be calculated from (3.4) using $\sigma_{os,i} = 1$ and $spec = 10^{-2}$.

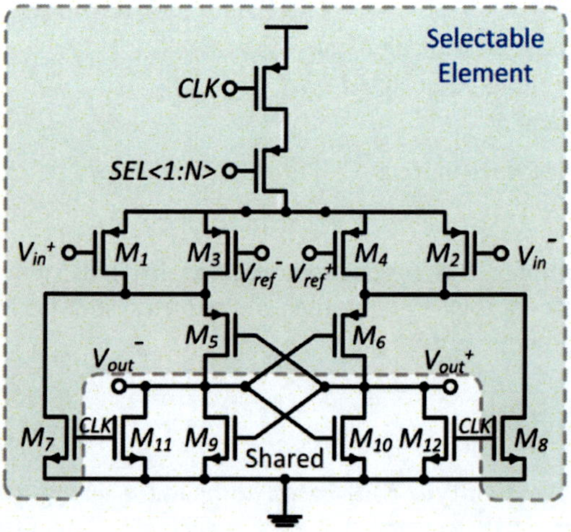

Fig. 3.4 SES-based latch type comparators

The leftmost point in the contour shows the case of redundancy, where we have 20 independent subsets of only one element each ($k = 1$). The failure probability at this point in the contour can be calculated simply by $p_{fail,total} = \left(p_{fail}\right)^{20}$. The rightmost point corresponds to the case where we have only one subset of 20 elements (select all elements, $k = N = 20$). Probability of failure for this subset, p_{fail}, can be calculated from (3.4) again, with $spec = 10^{-2}$ and $\sigma_{os,i} = 1/\sqrt{20}$; because we know that standard deviation decreases by $1/\sqrt{Area}$. The failure probability at the right end of the contour is simply $p_{fail,total} = p_{fail}$, since there is only one subset of size 20. This point corresponds to Pelgrom-type sizing ($k = N = 20$). Clearly, orders of magnitude of improvement in failure probability is achievable compared to both redundancy and Pelgrom-type sizing if we allow k to be anywhere between these two extremes; i.e. $1 < k < N$.

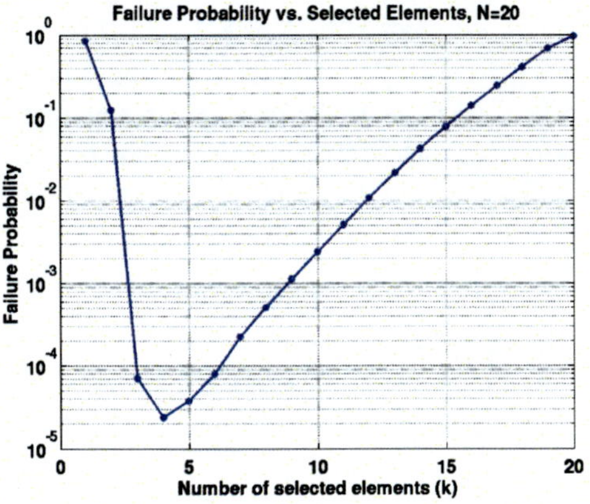

Fig. 3.5 Failure probability for $N = 20,\ \sigma_{os,i} = 1,\ spec = 10^{-2}$

Minimum failure probability is observed as $k = 4$; however, this may not be the optimum point because 16 unused elements are contributing to the parasitics. In the comparator example, that would slow down the circuit. In most cases it is desirable to minimize the number of unused elements, or simply maximize the k/N ratio while achieving the required offset specs.

Fig. 3.6 shows the plots when both N and k are varied. Each blue contour corresponds to a different N value ($1 < N < 20$), and the x-axis shows how many elements (k) are selected among N ($k \leq N$). As the previous case, each selectable element follows $\mathcal{N}(0,1)$ and $spec = 10^{-2}$. A good way to visualize the improvement in failure probability is to look at a vertical line at a given k (shown for $k = 10$), and determine the intersection points between this line and each contour. We increase N until we reach the failure probability target $p_{fail,total}$ (shown for $p_{fail,total} = 10^{-2}$). In this example, target is reached when $k = 18$. Any N above 18 can be chosen, but it results in the expense of increasing the number of unused elements.

A MATLAB script can search through the data, find the appropriate (N, k) pairs, and produce the highest k/N ratio for each k. In Fig. 3.6, these points have been marked with circles for each k where the $p_{fail,total}$ specification can be met. Although not fully monotonic due to the discrete nature of the problem, we observe higher k/N ratios as k increases. In other words, red circles to the right have, in general, better utilization of elements compared to the ones on the left.

Fig. 3.7 shows a comparison of these methods as N (normalized area unit) is varied. Each selectable element offset is assumed to follow a normal distribution $\mathcal{N}(0,90mV)$ with $spec = 2mV$. Only half of all elements are allowed to be selected for CR/SES ($k = N/2$). For the redundancy method, each element forms one block (N redundant blocks). The yield for CR/SES can be estimated by MATLAB Monte Carlo simulation (10,000 samples). The yield for DAC-based calibration is

$$p_{success} = erf\left(\frac{spec + LSB_{DAC} \times (2^{n-1} - 1)}{\sigma_{os,i} \times \sqrt{2}}\right) \tag{3.6}$$

where is the target offset, LSB_{DAC} is the calibration step size, and n is the DAC resolution. $LSB_{DAC} = LSB_{ADC}$ as discussed in [19]. The others are calculated from their yield equations. Dramatic improvement in success probability can be seen with CR/SES compared to the other methods.

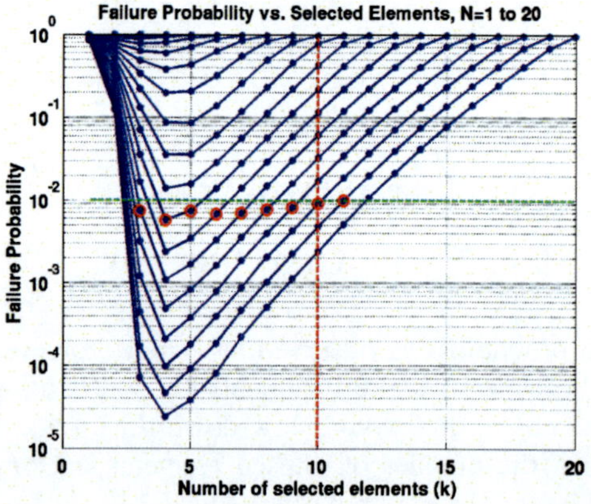

Fig. 3.6 Failure probability for $N = 1\ to\ 20,\ \sigma_{os,i} = 1,\ spec = 10^{-2}$

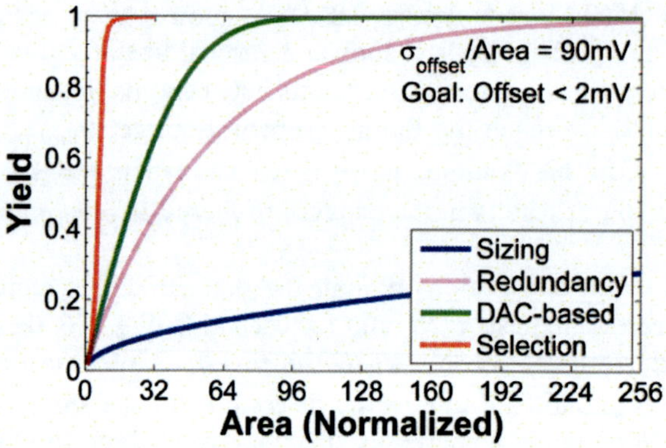

Fig. 3.7 Comparison of SES, redundancy, scaling and DAC-based calibration

Although the previous scenario is informative, it might not be completely realistic. In most cases, designers are not restricted to choose a fixed element size; they can choose among fewer but larger elements (e.g., $\mu_{element} = 0$ and $\sigma_{element} = \sigma_{os,i} < 1$). For the comparator example, assume that all transistors in the replicated section have a minimum length (L). Consider the following two cases:

- Case 1: N_1 total elements; in each element, all the transistors have width W_1, giving a standard deviation of σ_1. We are selecting k_1 elements among N_1.
- Case 2: N_2 total elements; in each element, all the transistors have width W_2, giving a standard deviation of σ_2. We are selecting k_2 elements among N_2.

For a fair comparison, assume that the total area in two cases is the same; i.e. $N_1 \times W_1 = N_2 \times W_2$, ignoring routing area and the storage for configuration bits. We want to determine which case has better resource utilization (has higher k/N ratio). In order to achieve this goal, we first regenerate the plot in Fig. 3.6 for different $\sigma_{element}/spec$ ratios to normalize it to *spec*. Fig. 3.8 shows these individual plots forming the slices of a "decision cube." Using the decision cube, the designer can evaluate tradeoffs between differing element sizes for a given *spec*. Each slice of the cube corresponds to a different element size.

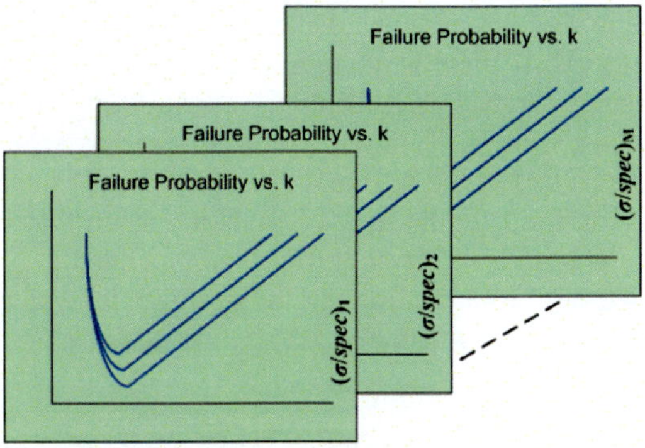

Fig. 3.8 The decision cube.

The decision cube is built only once for a predetermined range of (normalized) $\sigma_{element}/spec$ ratios (where *spec* is the offset specification). Each $\sigma_{element}/spec$ plot forms one slice of the cube. Since the cube is built on the normalized values (the $\sigma_{element}/spec$ ratio), it only needs to be built once. The same cube can be used for different designs with different resolutions or process technologies. In most practical applications, desired $\sigma_{element}/spec$ ratios would be from 10^1 to 10^3. An arbitrary number of slices can be formed between these points, but 100 slices are generally enough to converge on a decision of (N, k, σ) triplets that will satisfy the failure probability ($p_{fail,total}$) target. A simple design recipe is:

1. Specify the offset specification *spec*.
2. Specify the failure probability target $p_{fail,total}$ for each comparator. For example, if we would like to find a configuration that will satisfy the spec 99.5% of the time, $p_{fail,total} = 5 \times 10^{-3}$.
3. Specify the standard deviation of the offset for each type of selectable element ($\sigma_{element,i}$). For example, assume that the basic selectable element is a single transistor. The first selectable element could be a transistor with width W_1 and standard deviation $\sigma_{element,1}$, and the second selectable one could

be with width W_2 and standard deviation $\sigma_{element,2}$. These values can be determined by running circuit simulations for the design in the given process technology.

4. Calculate the ratio $\sigma_{element,i}$ for each selectable element type.
5. Input the results in steps 2 and 4 to a MATLAB script. For each selectable element type ($\sigma_{element,i}$), the script will produce all the (N, k) pairs that will satisfy the requirements in steps 1 and 2 using the cube in Fig. 3.8. Since the decision cube is pre-built, this is an efficient process step.
6. Now choose between the $(N, k, \sigma_{element,i})$ triplets that satisfy the requirements in steps 1 and 2 for the specific application. We have observed that in many cases, selecting half the total available elements ($k = N/2$) results in a good trade-off between resource utilization and the number of configuration bits.

The decision cube in Fig. 3.8 assumes that all available subsets are searched for a given set of N elements. If there is enough on-chip processing power available to perform an intelligent search, it is possible to search through all $2^N - 1$ available combinations. An easier but less optimal option is a greedy search, where random combinations are uploaded to the differential amplifier until a successful combination is found. We can limit the number of trials to ensure that calibration time is not very long. The maximum allowable trials can be added as a fourth dimension to the decision cube on Fig. 3.8, allowing the designer to evaluate the calibration time trade-off in addition to the $(N, k, \sigma_{element,i})$ triplets.

3.4 Comparator Array in 65nm Bulk CMOS Technology

3.4.1 Design Architecture

A test chip consisting of comparators in 65nm bulk CMOS was designed and fabricated to verify the modeling results. The comparator in Fig. 3.3 with 32 selectable elements has been used as the basic building block, out of which 16 are chosen. Each die includes 255 comparators, intended to be used for an 8-bit ADC. The architecture of the test chip and the timing diagram for calibration is given in Fig. 3.9.

The number of available selectable elements, the subset size and the size of each element are determined by using the methodology in the previous section. Maximum allowed calibration steps per comparator is chosen as 10,000. The full scale range (FSR) of the intended 8-bit ADC is 1V, giving a least significant bit (LSB) of 3.9mV. A comparator is defined as “within the specification” if at least one combination among the 10,000 steps results in an input offset voltage amplitude smaller than 0.5LSB. The design point is chosen so that all 255 comparators will be within the specification with 99.5% probability. During the design of the comparator, transistors in the shared block are sized such that their effect on the overall offset is much smaller than the replicated transistors. It results in efficiently controlling the offset of the comparator with the selection of the input differential pair.

The configuration bits for the 255 comparators are stored in flip flops. The differential output (2 bits) of each comparator is stored in 2 scan flip-flops. The timing diagram of the calibration process is also shown in Fig. 3.9. In region 1, configuration bits are scanned into the select flip-flops by using Scan In input and running Select Clk. Scan Enable is held low during this period. After all the selection bits are scanned in, Core Clk for the latch type comparator runs a few times to allow the outputs of the comparators to settle and clear any metastability in the latches (region 2). Comparator outputs are then loaded to the output scan flip-flops, which are subsequently put into scan-out mode by raising the Scan Enable signal (region 3). The differential output for each comparator is then read from Scan Out by toggling Scan Clk. The inputs ($V_{in\pm}$) are swept in small steps and the outputs of the comparators are read approximately 50 times through the output scan chain. At each input step, the number of times that each comparator outputs a value of 1 is collected. The input voltage vs. number of 1's curve is then fitted to a Gaussian cumulative distribution function. The mean of this distribution is used as the input offset voltage of the comparator for the given configuration.

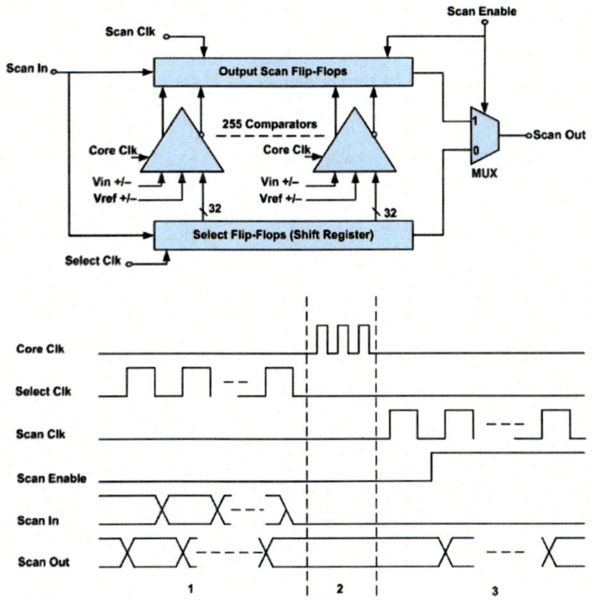

Fig. 3.9 Comparator array test chip architecture

3.4.2 *Testing Setup*

The test setup for the measurements is shown on Fig. 3.10. The setup is automated using built-in MATLAB toolboxes on the PC. Keithley 2400 sourcemeters with high precision are used for input voltages, and Agilent E3648A DC sources are used to supply the power to the core, I/Os, and the voltage references for the resistor ladder on the die. Core power supply is set at 0.8V and both ends of the resistor ladder are set at 0.4V. The chip is bonded in a QFN package and connected to a PCB using a compatible socket. Using the test socket, packaged die can be changed easily for statistical data collection.

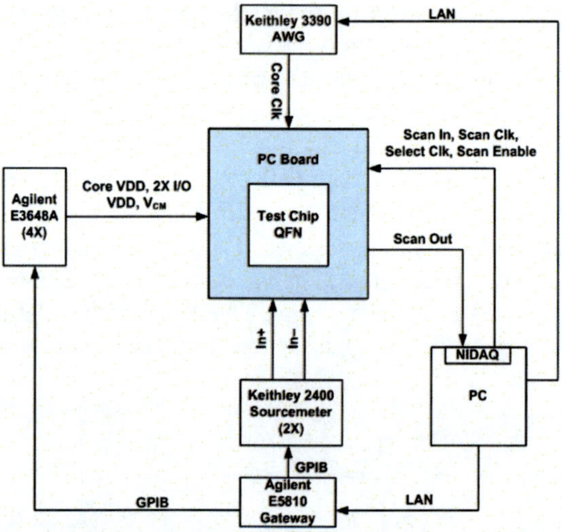

Fig. 10 Testing setup for the comparator array

Only a maximum of 10,000 calibration steps per comparator is allowed among more than 600×10^6 available combinations for each comparator. Since it is not possible to go through each of the 10,000 combinations per comparator due to measurement time constraints, the following method is applied to find the best sets:

1. Randomly determine 10,000 subsets of size $k = 16$ that each comparator can be configured to. These are the same for all comparators.
2. Determine a number of these subsets (X among 10,000) to be loaded to each comparator. Store these subsets in a selection matrix:

   - $$Sel_{X \times N} = \begin{pmatrix} S_{1,1} & S_{1,2} & \cdots & S_{1,N} \\ S_{2,1} & S_{2,2} & \cdots & S_{2,N} \\ \vdots & \vdots & \ddots & \vdots \\ S_{X,1} & S_{X,2} & \cdots & S_{X,N} \end{pmatrix} \quad (3.7)$$

   Each row of the matrix contains the configuration bits for each of the $N = 32$ elements. If element e of subset s is selected, $S_{s,e} = 1$, and 0 otherwise. The sum of each row is $k = 16$.
3. Measure the offset for each subset in the selection matrix and store it in a measured offset vector:

$$MO_{X\times 1} = \begin{pmatrix} mo_1 \\ mo_2 \\ \vdots \\ mo_X \end{pmatrix} \quad (3.8)$$

4. Find the estimated offset of each element in each comparator using the least squares solution in MATLAB:

$$IO_{N\times 1} = \frac{1}{k} \cdot (Sel_{X\times N})^{-1} \cdot MO_{X\times 1} \quad (3.9)$$

5. Use the estimated offsets of each element to find the *T* best subsets among the 10,000 subsets that are predicted to have less than 0.5LSB offset.
6. Upload the *T* subsets for each comparator to the test chip, and record the measured offset for each trial. For each comparator, select the subset that gives the lowest measured offset.

One can select the number of training sets (*X*) greater than the number of unknown variables (*N*) to build a linear model approximating the response. Although the linear model has error associated with it, *X* can be increased to build a more robust fit. Furthermore, using *T* best guess combinations for each comparator greatly increases the probability of finding a good combination.

3.4.3 Measurement Results

The chip photo of the comparator array fabricated in a 65nm CMOS technology is shown in Fig. 3.11. Comparator offsets from 13 different die (3315 comparators) were measured and calibrated using the methodology described above. $X = 100$ training sets and $T = 20$ best guess combinations for $N = 32$ elements are used. Measurement results show that over 85% of the tested best guesses satisfy the specification. Using this result, the probability that all 20 guesses will fail for one comparator is $(1 - 0.85)^{20} (< 10^{-12})$, verifying that the offset estimation procedure outlined in the previous section is practically feasible.

Fig. 3.12 shows the histograms before and after SES has been applied to the comparators. Fig. 3.12(a) shows the offset histogram when all 32 laid out elements are turned on (scaling). Fig. 3.12(b) shows the resulting histogram after SES has been applied to find the best subset among 10,000 allowable sets for each comparator and the subset size is $k = 16$. Close to two orders of magnitude of improvement in σ_{offset} is observed, from 11.21mV to 0.35mV. Figure 3.13 shows the histograms for 2X and 4X redundancy as applied to the comparators. For 2X redundancy, 32 elements in each comparator are divided into two blocks of 16 (the first 16, and the last 16). The block with the lower offset is selected. For 4X redundancy, 32 elements are divided into four blocks of 8 elements, and the lowest offset combination among the 4 is selected. Although improvement is observed compared to selection of all elements (32/32), component-level redundancy lags behind SES in terms of performance. Redundancy results are collected from 5 dies (1275 comparators).

Fig. 3.11 Die photo of the comparator array in 65nm CMOS

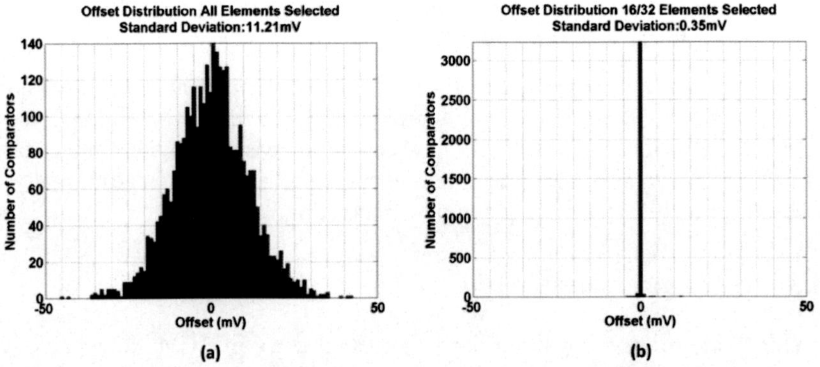

Fig. 3.12 Measured offset histograms (a) before and (b) after SES (3315 comparators)

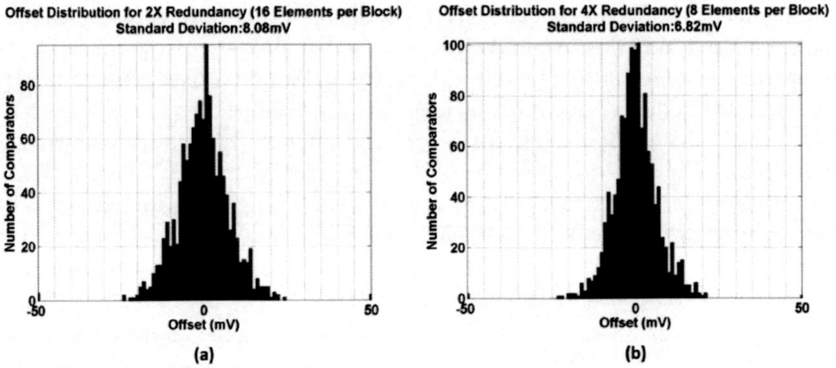

Fig. 3.13 Measured offset histograms for (a) 2X and (b) 4X redundancy (1275 comparators)

Success probability, defined as the number of comparators that have less than ±0.5LSB offset, is 15% for "select all" (32/32). Success probability for 2X and 4X redundancy is 25% and 28%, respectively. SES attains 99.5% success for select 16 over 32 as expected by modeling. Figure 3.14 shows the SES success

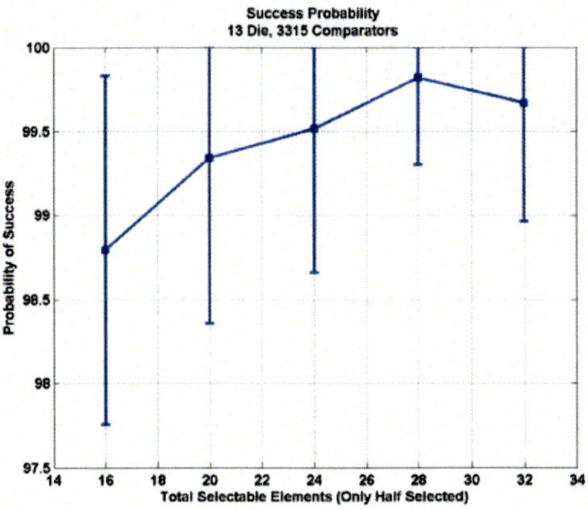

Fig. 3.14 Measured Success Probability for SES, *N*=16 to 32 (3315 comparators)

probability and 95% confidence intervals as *N* is varied from 16 to 32. $k = N/2$ for each *N*. Number of tested comparators is 3315. Success probability increases monotonically with increasing *N*, and above 98% success is observed for all cases.

3.5 An 8-bit 1.5-GHz Flash ADC in 65nm CMOS Process

3.5.1 Flash ADC Architecture

An 8-bit flash ADC was designed in the same process as the test chip in Section 4 but using a slightly different comparator topology [20]. Figure 3.15 shows the flash ADC architecture, depicted as single-ended for clarity. The front-end T/H consisting of a bootstrapped NMOS switch [21], a hold capacitor, and a well-tied NMOS source follower buffer as shown in Fig. 3.16 minimizes sampling time errors caused by clock and input distribution skew. The bootstrapped switch provides constant-V_{GS} while sampling, avoiding distortion caused by input-dependent charge injection. Two copies of the T&H are used pseudo-differentially. A well-tied NMOS source follower was chosen to meet the > 8-bit distortion requirement. The reference ladder is a single string of polysilicon resistors and folded once. The thermometer-to-binary encoder design is a bubble detector followed by an OR-gate one-hot-to-binary encoder and provides the 8-bit output. The bubble detector was chosen because it is lower cost than more complex encoders—such as the ones-counter—and accommodates individual high-offset comparators with a maximum error of 1 LSB. The bubble detector works well with the distribution of offsets after selection. 255 SES-based comparators convert the sampled analog signal to digital thermometer code. The CR/SES-based comparators are designed with 16 elements. Selection was performed using scan chains similar to Fig. 3.9.

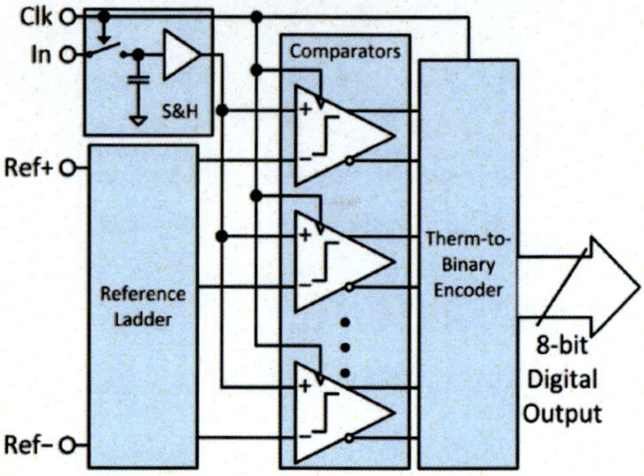

Fig. 15 8-bit flash ADC architecture

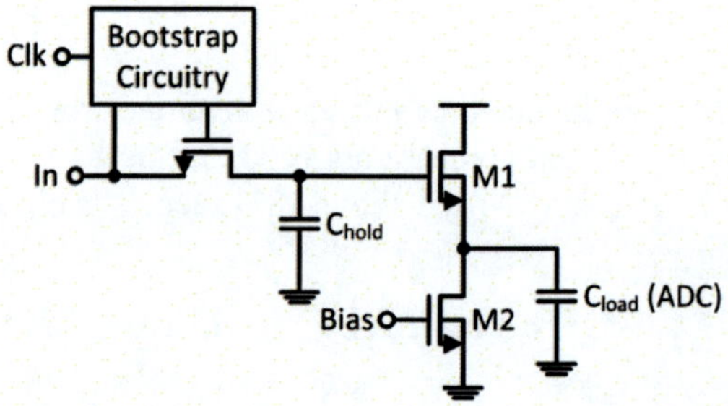

Fig. 3.16 Front-end track-and-hold

3.5.2 Comparator Design

The comparator design is a pseudo-differential NMOS-input latch-based dynamic comparator, as shown in Fig. 3.17. The design was chosen to reduce kickback from the comparator at the expense of common mode and power supply rejection. Identical copies of transistors M1 to M4, which are the dominant sources of offset, are connected in parallel, each branch forming a selectable element. M15 and M16 are added to select/deselect elements by the Select signals. M5 to M14 are shared, as they have minor offset contribution in simulation. All transistors in the selectable elements (M1 to M4, M15, and M16) are designed with minimum length. Widths are slightly larger than the minimum available by the process to optimize the manufacturability of the design. Other transistors (i.e., M5 to M14) are sized to balance speed and power while minimizing their contributions to offset. The unselected elements in this design have little effect on power or speed: in pre-layout simulation, removing the unused elements changes the delay and the power by less than 1%. The comparison time is primarily controlled by the pulldown current, which is set by the sizing of M1 to M4 in the standard comparator or by the number of selected elements in the SES-based comparator.

This comparator is designed with 16-choose-5 elements to achieve very low offset with minimal area and a reasonable balance of speed and power. As a consequence of the small size of input transistors (M1–M4), estimated individual element $\sigma_{offset} = 92\text{mV}$. Combinatorial redundancy meets tight offset specifications even with extremely small devices, and helps limit the amount of input parasitic capacitance due to the unused branches. A major advantage of this topology is the drastically reduced kickback into the ADC input and the reference ladder. Drain voltages of the unused pairs do not swing during normal operation, and the kickback due to the parasitic gate to drain capacitance of input transistors in unused pairs is not significant. After SES, the simulated yield is > 99.9% for offset < 2mV and $\sigma_{offset} = 0.3\text{mV}$.

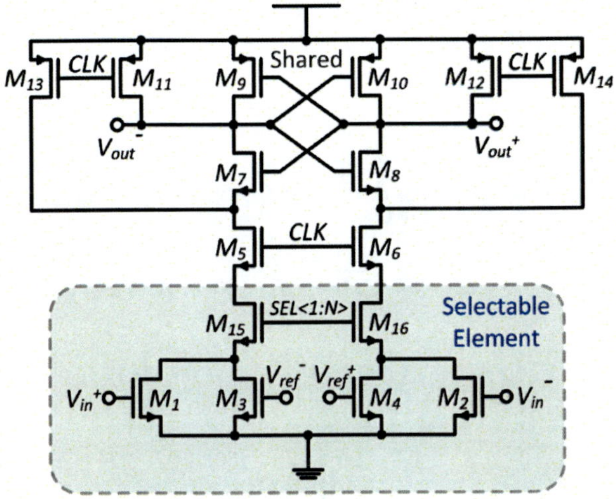

Fig. 3.17 NMOS-input comparator

3.5.3 Measurement Results

The 8-bit flash ADC is fabricated in 65 nm digital CMOS process and occupies an active area of 0.5 mm^2 as shown in Fig. 3.18. Probe testing was conducted at 1.5 GS/s and 1-V power supply on the test chip. The FSR is 1.06 V_{ppd}. The total power consumption (excluding I/O) is 35 mW, the T/H using 20 mW and the ADC core 15 mW.

After selection, the INL is 1.32 LSB and the DNL is 1.23 LSB as shown in Fig. 3.19. 232 of 255 comparators (91%) have offset of < 0.5LSB. The comparator yield is slightly lower than expected and may be due to error from comparator noise. For offset of < 1LSB, the yield is 250 of 255 comparators (98%). The ADC is monotonic with minimal-sized comparators. Fig. 3.20 shows the measured SNDR and SFDR with respect to variant input frequencies at 1.5 GS/s. The SNDR is 37 dB at low frequency, corresponding to 5.8 bits ENOB. Average comparator noise is 5 mVrms (1.3 LSB), significantly degrading SNDR. The ERBW is 1.3 GHz, and the SFDR is 43 dB at Nyquist rate.

Fig. 3.18 8-bit flash ADC die photomicrograph

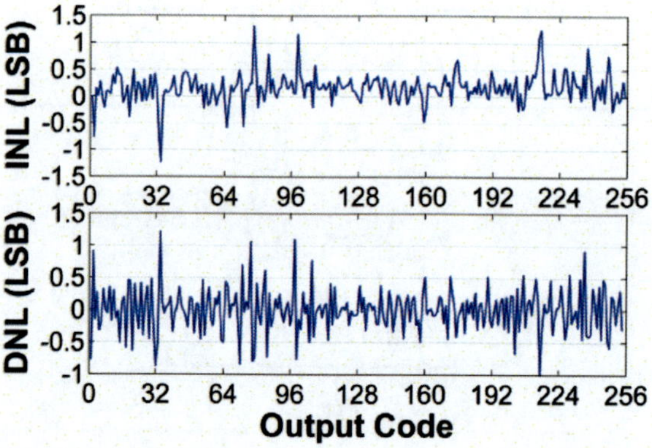

Fig. 3.19 INL and DNL after selection

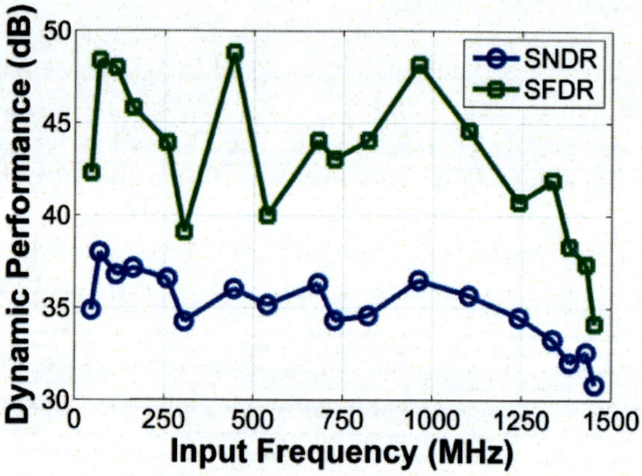

Fig. 3.20 SNDR and SFDR v.s. input frequency at 1.5 GS/s

A commonly used figure of merit (FoM) is

$$FoM = \frac{Power}{2^{ENOB} \times min(f_s, 2 \times ERBW)} \tag{3.9}$$

where ENOB is the effective number of bits, f_s is the sampling rate, and ERBW is the effective resolution bandwidth [22]. To avoid overstating the capabilities of the ADC, $min(f_s, 2 \times ERBW)$ is used. The FoM is 0.42 pJ/conv-step. Compared to other recently-published ≥ 1GHz flash ADCs, this work has comparable performance to state-of-the-art averaging, interpolating, and folding flash ADCs despite having a basic flash ADC architecture with the digital-like sized comparator design. The limiting factors on the performance of this ADC are the comparator noise and the T/H power consumption. From the model and the previous noise simulations, the comparator noise is responsible for a loss in ENOB of approximately 1.7 bits. The noise severely limits the ENOB and can be improved by redesigning the comparator. A noise-robust thermometer-to-binary encoder can improve the ENOB by approximately 0.8 bits without changing the comparator noise. The T/H is a simple, power-hungry design and more complex designs have been proposed with the potential of reduced power. As the T/H consumes 57% of the total power, a significant improvement in power is possible.

3.6 Conclusion

A digital calibration methodology based on statistical element selection offers an effective approach to coping with random variation impact on mismatch. While SES has been demonstrated in this chapter for comparator offset calibration in flash ADCs, it is a general approach that can be applied to matching of any two or more components on a chip. Monte Carlo simulation of the CR/SES approach for any matching design problem can offer similar orders of magnitude of improvement as compared to other calibration schemes.

The 65 nm digital bulk CMOS testchips validated the mathematical models, and the benefits from random process variations will only be more pronounced at 45 nm nodes and beyond. The efficacy of CR/SES will increase as more focus and effort is placed on the co-design of the circuits and the combinatorial redundancy sub-components.

Acknowledgment. The authors acknowledge the support of the C2S2 Focus Center, one of six research centers funded under the Focus Center Research Program (FCRP), a Semiconductor Research Corporation entity. The authors thank J. Proesel, J.-O. Plouchart, M. Soyuer, and D. Friedman of IBM Research for their support, and R. Carley and J. Paramesh of CMU for discussion and assistance. Fabrication was provided by IBM.

References

[1] Pelgrom, M.J.M., Duinmaijer, A.C.J., Welbers, A.P.G.: Matching Properties of MOS Transistors. IEEE J. Solid-State Circuits 24(5), 1433–1439 (1989)

[2] Flynn, M.P., Donovan, C., Sattler, L.: Digital Calibration Incorporating Redundancy of Flash ADCs. IEEE Trans. Circuits Syst. II, Analog Digit. Signal Process. 50(5), 205–213 (2003)

[3] Li, X., Taylor, B., Chien, Y., Pileggi, L.T.: Adaptive Post-Silicon Tuning for Analog Circuits: Concept, Analysis and Optimization. In: IEEE/ACM Int. Conf. on Computer-Aided Design Dig. Tech. Papers, pp. 450–457 (November 2007)

[4] Agarwal, K., Nassif, S.: Characterizing process variation in nanometer CMOS. In: Proc. ACM/IEEE Design Automation Conf., pp. 396–399 (June 2007)

[5] Scott, G., et al.: NMOS drive current reduction caused by transistor layout and trench isolation induced stress. In: IEEE Int. Electron Devices Meeting (IEDM) Dig. Tech. Papers, pp. 827–830 (December 1999)

[6] Fukutome, H., et al.: Carrier profile designing to suppress systematic variation related with device layout by controlling STI-enhanced dopant diffusions correlated with point defects. In: IEEE Int. Electron Devices Meeting (IEDM) Dig., pp. 53–56 (December 2009)

[7] Gupta, P., Heng, F.-L.: Toward a systematic-variation aware timing methodology. In: ACM/IEEE Design Automation Conf. Dig. Tech. Papers, pp. 321–326 (June 2004)

[8] Webb, C.: 45 nm design for manufacturing. Intel Technology J. 12, 121–130 (2008)

[9] Jhaveri, T., et al.: Co-optimization of circuits, layout and lithography for predictive technology scaling beyond gratings. IEEE Trans. Comput.-Aided Design Integr. Circuits Syst. 29(4), 509–527 (2010)

[10] Watt, J.T., Plummer, J.D.: Dispersion of MOS capacitance-voltage characteristics resulting from the random channel dopant ion distribution. IEEE Trans. Electron Devices 41(11), 2222–2232 (1994)

[11] Sugii, N., et al.: Local variability and scalability in silicon-on-thin-BOX (SOTB) CMOS with small random-dopant fluctuation. IEEE Trans. Electron Devices 57(4), 835–845 (2010)

[12] Matsukawa, T., et al.: Comprehensive analysis of variability sources of FinFET characteristics. In: IEEE Symp. VLSI Technology Dig., pp. 118–119 (June 2009)

[13] Chiang, M.-H., Lin, J.-N., Kim, K., Chuang, C.-T.: Random dopant fluctuation in limited-width FinFET technologies. IEEE Trans. Electron Devices 54(8), 2055–2060 (2007)

[14] Li, Y., Hwang, C.-H., Li, T.-Y., Han, M.-H.: Process-variation effect, metal-gate work-function fluctuation, and random-dopant fluctuation in emerging CMOS technologies. IEEE Trans. Electron Devices 57(2), 437–447 (2010)

[15] Cheng, K., et al.: Extremely thin SOI (ETSOI) CMOS with record low variability for low power system-on-chip applications. In: IEEE Int. Electron Devices Meeting (IEDM) Dig., pp. 49–52 (December 2009)

[16] Ellersick, W., Yang, C.-K., Horowitz, M., Dally, W.: GAD: A 12-GS/s CMOS 4-bit A/D converter for an equalized multi-level link. In: Symp. VLSI Circuits Dig. Tech. Papers, pp. 49–52 (1999)

[17] Kinget, P.R.: Device mismatch and tradeoffs in the design of analog circuits. IEEE J. Solid-State Circuits 40(6), 1212–1224 (2005)

[18] Keskin, G., Proesel, J., Pileggi, L.: Statistical modeling and post manufacturing configuration for scaled analog CMOS. In: Proc. IEEE Custom Integrated Circuits Conf., pp. 1–4 (September 2010)

[19] Chen, C.-Y., Le, M.Q., Kim, K.Y.: A low power 6-bit flash ADC with reference voltage and common-mode calibration. IEEE J. Solid- State Circuits 44(4), 1041–1046 (2009)
[20] Proesel, J., Keskin, G., Plouchart, J.-O., Pileggi, L.: An 8-bit 1.5 GS/s flash ADC using post-manufacturing statistical selection. In: Proc. IEEE Custom Integrated Circuits Conf., pp. 1–4 (September 2010)
[21] Abo, A., Gray, P.: A 1.5-V, 10-bit, 14.3-MS/s CMOS pipeline analog-to-digital converter. IEEE J. Solid-State Circuits 34(5), 599–606 (1999)
[22] Verbruggen, B., et al.: A 2.2 mW 1.75 GS/s 5 bit folding flash ADC in 90 nm digital CMOS. IEEE J. Solid-State Circuits 44(3), 874–882 (2009)

Chapter 4
Improving Design Feature Reuse in Analog Circuit Design through Topological-Symbolic Comparison and Design Concept Combination

Cristian Ferent and Alex Doboli

Abstract. This chapter presents a novel circuit synthesis flow based on concept comparison, combination, learning, and re-use. The discussion overviews a technique for systematically comparing two analog circuits. The comparison data presents the similar and distinguishing performance characteristics of two circuits with respect to DC-gain, bandwidth, common-mode rejection ratio (CMRR), noise, and sensitivity. The comparison data is important for getting insight about the common and unique benefits of a circuit, selecting fitting circuit topologies for system design, and circuit topology refinement and synthesis. The technique matches the topologies and nodal symbolic expressions of the compared circuits to find nodes with similar electric behavior. The impact on performance of the unmatched nodes is used to express the differentiating characteristics of the circuits. Experiments illustrate the comparison technique for a pair of analog circuits.

4.1 Introduction

Analog and mixed-signal (AMS) systems and circuits are important components in many modern systems integrated or interacting with the physical world, such as applications in telecommunication, environmental sensing, healthcare, and smart infrastructures. The time and cost effort to design and verify AMS systems is expected to grow due to insufficient CAD support for designing, optimizing, and validating the systems. Recent studies suggest that the current productivity gap is around 100x, and sometimes as high as 1000x, as compared to what is needed to design next-generation electronic

Cristian Ferent · Alex Doboli
State University of New York at Stony Brook, Department of Electrical and Computer Engineering
e-mail: {cferent,adoboli}@ece.sunysb.edu

M. Fakhfakh et al. (Eds.): *Analog/RF & Mixed-Signal Circuit Sys. Design*, LNEE 233, pp. 77–99.
DOI: 10.1007/978-3-642-36329-0_4

systems [14]. It has been also observed that validation and verification of new designs is more difficult because automated tools create solutions that are less similar in style to what human designers develop. Generating solutions that are more "readable" by human designers is not a main priority for many current synthesis and optimization tools. Some of the envisioned solutions to address the productivity gap challenges include developing (i) novel design methodologies and techniques based on higher levels of abstractions, (ii) design methods utilizing parameterized topologies and modules, (iii) new synthesis, evaluation/simulation, and layout design algorithms capable of better optimization results in shorter time, and (iv) novel techniques to enable superior design reuse. Design feature reuse in analog circuit design has been arguably less studied as compared to the other avenues.

CAD tools for design reuse must automatically identify the design intent, strategies, and improvements that are embedded in previously developed solutions, and then analyze how improvements can tackle a new application's performance requirements and/or constraints of the utilized fabrication processes. This insight can then be reused to produce novel design solutions. Design feature reuse is envisioned to be very effective as many industrial companies and research groups already possess large sets of previous designs, including IPs. Besides, many modern tools for transistor sizing, layout design, and circuit topology selection and synthesis use design constraints to increase the likelihood of creating feasible solutions and/or to improve the convergence of the algorithms. The insight extracted from the available designs can be also utilized as constraints for existing synthesis tools, or as templates to guide a manual design process. A central component for getting the needed insight to enhance design feature reuse is developing a procedure to systematically compare the topological and performance attributes of analog circuits to understand how similar and different circuit features introduce new design variables and trade-offs.

This chapter introduces a novel circuit synthesis methodology based on concept comparison, combination, learning, and re-use. A central component of the methodology is a technique to systematically compare two analog circuits. Using a dual topological and symbolic matching scheme, the comparison method identifies the similar and distinguishing design features of an analog circuit as compared to another circuit, and how the features impact the nodal behavior and performance attributes, like DC-gain, bandwidth, noise, CMRR, and sensitivity. A set of constraints relate the behavioral descriptions to performance attribute modification. The final step of the comparison method characterizes how topological and behavioral differences modify trade-offs in a design, availability of free (orthogonal) variables to set performance attributes, achievable performance values, and hardness to find the design parameters. The chapter also overviews some of the current AMS synthesis methods, and argues for the need of developing novel approaches that can understand and learn new design features from existing solutions and designs created during the synthesis process.

In addition to AMS synthesis, the symbolic comparison technique could also be useful in various other design activities, like incremental circuit topology design, circuit design retargeting for new fabrication processes, circuit parameter sizing, and topology selection.

The chapter has the following structure. Section 4.2 overviews related work. Section 4.3 presents the synthesis flow. Section 4.4 details the proposed method for circuit comparison. Section 4.5 offers experimental results. Conclusions end the chapter.

4.2 Related Work

An overview of analog synthesis tools based on their architecture selection approach is presented in [18]. Tools are analyzed with respect to four criteria: (i) the abstraction level at which design decisions are made, (ii) the flexibility to tackle various kinds of circuits, (iii) the coverage of the overall design space defined by the possible topologies, and (iv) the capability to optimize the parameters of a circuit topology. Based on the four criteria, the authors suggest that current synthesis tools can be categorized into four main classes: (i) tools that select a topology based on designer experience or input from knowledge-based expert system, (ii) methods that decide the topology in parallel with parameter sizing based on circuits and sub-circuits stored in a library, (iii) tools that produce a topology through top-down synthesis starting from high-level descriptions, and (iv) bottom-up techniques that create architectures through systematic rules or stochastic evolution to connect devices into structures. This section reviews some of the existing synthesis tools in every category and then explains the need for novel approaches in circuit synthesis.

Synthesis tools that mimic knowledge-based expert systems utilize a static library of design rules that can tackle a certain family of circuits. IDAC [5] is an interactive design tool for a range of circuits, like OTAs, OpAmps, voltage and current amplifiers, comparators, and oversampling ADCs. The design methodology utilizes three types of knowledge, specific to the schematic, general circuit theory, and related to the circuit family. For instance, for OTA design, the first step performs worst-case distortion analysis due to variations of temperature and bias currents to derive the acceptable, nominal gain-bandwidth product, gain, slew rate, noise and phase margins. The second step sizes the devices based on a pre-specified design plan. Finally, the third step evaluates the correctness of the design, such as the resulting phase margin. Another expert-system for analog circuit design, BLADES [9], implements a given set of inference steps in which specification requirements and design knowledge is used to reason out the circuit topology and device dimensions. The topology selection algorithm in FASY [26] uses fuzzy rules based on specification requirements. OPASYN [15] utilizes a decision tree to select circuit topologies. A decision tree distinguishes the common topologies from special circuits based on requirements like area, open-loop gain, power supply rejection ratio, and fully differential structure.

Template-based analog circuit synthesis can be viewed as an extension of knowledge-based design methods. They use a set of invariant, circuit-specific constraints (e.g., design templates) to find performance optimized AMS topologies, thus perform simultaneously topology selection and parameter sizing. Design templates usually describe implicitly an entire family of solutions, out of which the synthesis method identifies the best solution for the current specification. OASYS [11] utilizes a hierarchical set of templates, in which the top level corresponds to a certain type of circuits, like successive approximation ADC, the next level defines the structure of the circuits out of building blocks, i.e. comparator, sample-and-hold, D/A converter, the third level defines the structure of each building block including OpAmps and their RC networks, and the fourth level presents possible structures of OpAmps. A different kind of templates has been used to synthesize $\Delta\Sigma$ ADCs [24] and reconfigurable $\Delta\Sigma$ ADCs [28]. In this work, a template is a set of mixed integer nonlinear programming (MINLP) equations that express the alternative structures and parameters of the converters as well as the impact of circuit nonidealities on performance, e.g., finite gain, bandwidth, and noise. The MINLP equations of a template are solved to find the ADC structure, including its order, feedback and feedforward loops, and the coefficients of the structure.

Top-down synthesis flows transform high-level descriptions of circuits or systems into implementations optimized for the specification requirements. The transformation process includes steps for selecting the topology and sizing its parameters in addition to placement and routing the design. Various high-level descriptions have been explored, including state-space model [1] and signal-flow graphs [7, 13]. Using VHDL-AMS as a specification language for AMS synthesis is discussed in [8]. The transformation steps are realized based on predefined rules [1], exploration [8], or constraint transformation [3].

More recently, stochastic evolution has been proposed to synthesize AMS circuits and systems. The method in [22] utilizes genetic programming to evolve CMOS OpAmps utilizing basic devices and building blocks. Current flow analysis is employed to verify basic electrical requirements of an evolved topology, such as having current flow through the component lists of a circuit, operating all transistors in their correct regions, and avoiding floating and isolated devices. The approach in [17] performs variation-aware structural synthesis of analog circuits to produce decision trees that indicate the topologies that implement better a set of requirements. Decision trees partition the performance space such that every internal node describes a performance constraints, like power consumption, area, gain, dynamic range, etc., and every leaf node is a circuit topology. A path through the decision tree describes a circuit (leaf) and the set of requirements that are met by the circuit (the reunion of the leaf node constraints). The synthesis approach utilizes hierarchical specified analog building blocks representing one and two stage OpAmps, various inputs (e.g., single-ended, differential, stacked, folded cascode, etc.) and loads (i.e. cascode, noncascode, resistor), different current mirror circuits, and outputs. Structural homotopy avoids early elimination of attempts to create new topologies by attaching a maximum age for the individuals of each layer of the evolutionary synthesis algorithm. The technique proposed in [4] creates topologies bottom-up starting from devices and gradually creating, based on rules, blocks of more complex functionality. Blocks are stored for re-use in a library together with information about their neighbors, which are blocks placed adjacently.

In addition to the criteria enumerated at the start of the section, we consider that circuit synthesis tools can be also distinguished based on their capability to "understand" and "learn" from the features of previous solutions, such as to use the explored design space to extract new knowledge on effective design decisions (e.g., to use some structures together and to size parameters according to specific constraints), to identify design space regions that have not been explored yet, and to produce design strategies, i.e. parameter optimization plans, of superior quality, shorter design closure, and higher transparency with respect to the selected decisions. Higher transparency helps designers understand easier the purpose of the selected design decisions, thus aids in design validation. Traditional reuse-based methodologies rely exclusively on static libraries of circuits, design rules and equations, circuit models, and layout templates [2]. This chapter argues that understanding and learning new design features produced dynamically during synthesis requires incorporating into synthesis flows the capability to automatically identify, characterize, and reuse design features created dynamically during synthesis. [25] indicates that reusing automatically mined constraints on design parameters significantly improves convergence of synthesis and reduces synthesis time.

Automatically understanding and learning the useful features in circuits requires to find the similarities and differences between the electrical behavior and performance of the circuits. Simulating and/or modeling individually each circuit gives some insight but understanding in detail the links between topological changes and performance differences requires complicated manual analysis. Previous work suggests that two circuits can be compared using their performance space descriptions [6, 16, 23]. Performance space descriptions specify the performance trade-offs of a circuit, like DC gain, 3db frequency, and slew rate [23], by abstracting away all design variables. However, they give little insight into how topological differences in two circuits introduce new design variables that create novel trade-offs and new opportunities (flexibility) to improve performance. Also, there is no indication about how design-specific constraints can improve performance, such as certain pole-zero placements, or constraints on device parameters, like transconductances. This insight is important for circuit topology selection and design reuse, including circuit resizing for new requirements, incremental topology changes for novel

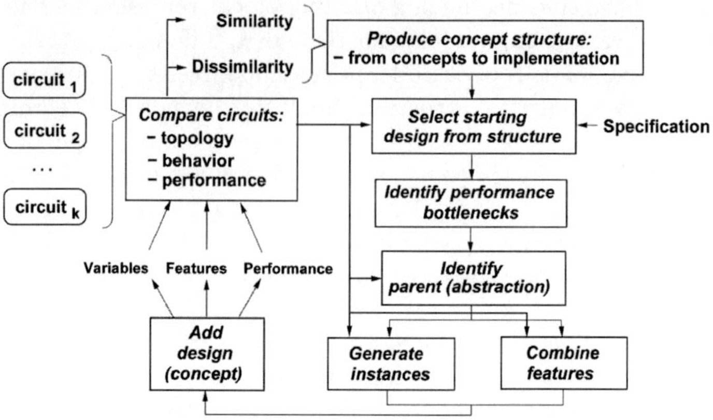

Fig. 4.1 Synthesis flow based on design concept comparison, combination, learning, and re-using

applications, and design migration to different fabrication processes. The next section presents a synthesis method based on design feature comparison and combination.

4.3 Circuit Synthesis Based on Concept Comparison and Combination

This section presents a novel circuit synthesis flow based on design concept comparison, learning, combining, and re-using. The main premise of the synthesis flow is that solving a circuit design problem requires to identify a set of design steps, so that every step is justified by the fact that it improves performance (i.e. at least one performance attribute) or relaxes design constraints (e.g., at least one constraint). Every synthesis step attempts to address the performance bottlenecks of a circuit topology by changing the relations between the design variables of the bottlenecks. Relations are changed by (i) searching for previous designs with related bottlenecks and then combining their features with the current solution, or by (ii) exploring orthogonal ways of relating the variables of the bottlenecks through new ways of interconnecting circuit nodes. Circuit comparison is a main step of this synthesis flow as it identifies the commonalities and differences among circuit features and their importance.

The proposed synthesis flow is shown in Figure 4.1. The concept structure used in design is constructed based on a set of existing circuits, e.g., $circuit_1$ to $circuit_k$ in the figure. The circuits are compared pairwise with regard to their topology, electrical behavior, and performance. Section 4.4 details the comparison procedure. The information about the similar and dissimilar features of the circuits is used to construct a concept structure in which leaf nodes represent circuit designs, and intermediate nodes express the common design features of the children. Hence, the children of an arbitrary node n describe a sampling of the conceptual space represented by node n. Similar to [10], a classification method based on entropy can be used to create the concept structure that maximizes the amount of common design features of the intermediate nodes and minimizes the number of common features across different intermediate nodes. This concept structure optimizes the partitioning of the different circuit features depending on their impact on performance. Note that the lower level concepts of the structure represent more complex and detailed solutions while the higher level concepts correspond to more abstract (conceptual) designs.

The synthesis flow starts by selecting a concept of the design structure, such that its performance attributes are closest to the problem description and its bottlenecks do not conflict with the description. This design is likely to be efficiently refined and modified to accommodate the requirements of the specification. Next, the iterations of the synthesis flow attempt to minimize the miss-matching between the design performance and the specification requirements by conducting the following steps. First, it analyzes the nature of the performance bottlenecks of the current solution and then finds bottom-up in the design structure the first parent node that does not have the bottlenecks. The parent node is found by comparing the current node with the nodes placed bottom-up in the structure. Then, a child without the features that cause the bottleneck is used to further synthesize the solution. Alternatively, the method attempts to remove the current bottlenecks by adding (combining) features that are present in other designs (of the concept structure) that do not have the bottlenecks. The new features are structural connections between the circuit nodes that create the current bottlenecks. This changes the relations among the nodal currents and voltages. The produced performance modifications are evaluated by comparing the modified and original circuits. The third option is to exhaustively create new ways of relating the nodal variables of the bottlenecks, so the bottleneck is changed and a better matching to the problem description is possible. The performance improvement of the explored structures is characterized by comparing the original and the modified circuits. Any new solutions created during the steps of concept combination or new feature generation are added to the concept structure.

A case study example for designing differential MOS transconductor cells based on the concept structure is detailed in [31]. The problem requirements are to improve linearity and input range. The simplified concept structure is shown in Figure 4.2. The top level concept, common to all circuits, guarantees that the transconductor functionality is implemented, $I_o = f(V_i)$, where I_o and V_i are the differential output current and input voltage, respectively. Child concepts add details to the structure. One alternative is to use the same source voltage for both input MOS devices ($V_{S1} = V_{S2}$) and a direct

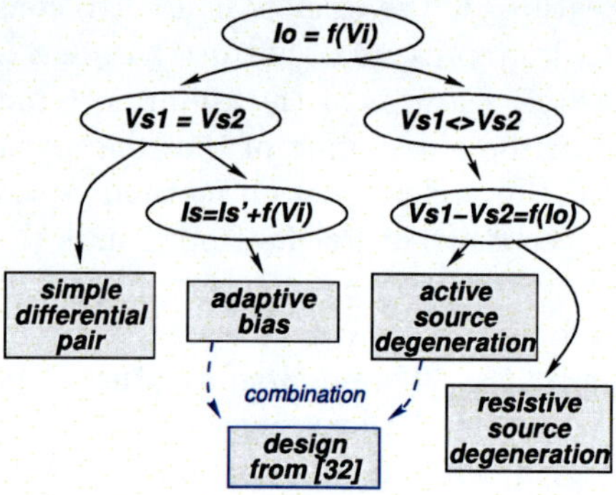

Fig. 4.2 Simplified concept structure for transconductor designs

implementation is the simple differential pair transconductor. If linearity and/or input range of this solution are insufficient for the given specification, another child concept of the parent node can be selected for synthesis. This alternative concept maintains the same source voltage, but adds an input dependence to the tail biasing current, $I_S = I'_S + f(V_i)$. This improves linearity and a specific implementation is the adaptive bias transconductor. Another option is to explore the concept utilizing different source voltages for the input transistors ($V_{S1} <> V_{S2}$). One possible child of this concept correlates the difference in source voltages to I_o and can be specifically implemented through either resistive or active source degeneration topologies. If none of the designs represented in the concept structure offer adequate performance, concept combination can be attempted to relax constraints. For example, combining the features of the adaptive bias and source degeneration transconductors can produce a highly linear design [32].

The concept structure of the flow is the main data structure used to learn, combine, and re-use design features. It is produced based on comparing structural, behavioral, and performance attributes of circuits, as presented in the next section.

4.4 Systematic Comparison of Analog Circuits

Comparing the nodal behavior and performance of two analog circuits comprises of two activities: (i) relating the electrical behavior of the circuit nodes and sub-circuits in the two circuits, and (ii) understanding how commonalities and differences in the nodal behavior impact performance. The two activities require a dual topological-symbolic matching to find nodes connected in similar structures and with similar electrical behavior. Topological matching alone is insufficient as it only identifies the circuit nodes with similar connectivity (to other devices) and the node groups (clusters) with same structure, but does not perform any analysis of the electrical behavior. It is known that similar topological structures can produce different signal flows depending on biasing, output sensing, and connectivity with other structures. The literature presents several topological matching approaches, including clan-based matching [12] and string-based matching [21].

Two circuit nodes (in different circuits) have similar electrical behavior, if there are conditions under which the transfer function (TFs) between the nodes and inputs can be matched, such that the two TFs represent the same mathematical expressions. Examples of such conditions include requirements that certain device parameters are equal (matching), some device values are much larger (smaller) than others, certain device parameters can be neglected, and so on. Characterizing the performance differences of two circuits must capture how topological and behavioral changes modify performance attributes with respect to the following key aspects of a design: (i) new trade-offs, (ii) availability of free (orthogonal) variables to control specific performance attributes, (iii) achievable performance values, and (iv) hardness to find the parameter values that set a desired performance value.

The circuit comparison procedure performs the following four steps to quantify the above key aspects. The first step, *topological matching*, relates the structural features of the two circuits, e.g., it identifies maximal sets of nodes with similar poles and connectivity to other nodes. The second step, *symbolic matching*, matches the electrical behavior of the circuit nodes using the topologically matched nodes as a reference. It computes transfer functions H_{comm} and H_{diff} expressing the similarities and differences in the electrical behavior of the circuit nodes. *Constraint generation*, the third step, creates constraints defining how functions H_{comm} and H_{diff} impact performance, such as the resulting ΔDC gain, Δbandwidth, Δnoise, ΔCMRR, and Δsensitivity. Finally, the *performance characterization* step describes the capability of a design to meet the generated constraints, and thus achieve certain performance and trade-off values. The four steps are detailed next.

4.4.1 Topological Matching

We define that two nodes are topologically matched if they have the same number and kind of devices connected to the nodes. Two clusters of connected nodes (pertaining to two circuits) are topologically matched if all nodes in the clusters are also topologically matched.

Topological matching finds the maximal node clusters with similar structure. Given two circuits C_1 and C_2, the topological matching algorithm starts by identifying for every node i in circuit C_1 the list of nodes in C_2 that are candidates to be topologically matched with node i. Then, it exhaustively considers for every node in C_1 every matching candidate in C_2 while maximizing the size of the matched clusters that are found using the candidate.

Figure 4.3(a) shows the superimposed schematics of a compensated two-stage amplifier and a class AB two-stage amplifier [19]. The two topologies

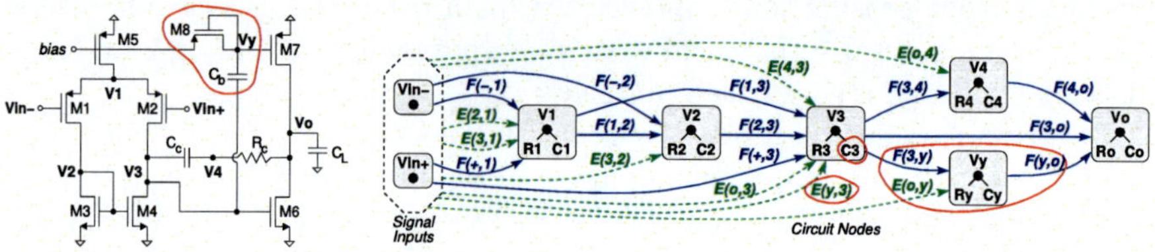

Fig. 4.3 Schematic and macromodel for simple OpAmp and class AB two-stage OpAmp [19]

have all nodes matched with the exception of nodes V_3, and V_y. The matched clusters of the two circuits include all nodes but these two nodes.

4.4.2 Symbolic Matching

Given two circuits, symbolic matching finds the nodes with similar electrical behavior, including any constraints on the device parameters that make the behavior similar. The proposed method identifies the similarities and differences in the behavior of two nodes by computing the following TFs: (i) H_{comm} defines the common symbolic parts of the TFs of the compared nodal voltage, and (ii) H_{diff} expresses the distinguishing symbolic terms of the two TFs. Figure 4.4(a) illustrates the TFs of the two structures shown as signal-flow graphs. The topologically matched nodes are highlighted.

The electrical behavior is described in the proposed method through macromodels, which are built using the technique in [29]. The uncoupled models are structural and represent all circuit nodes. Based on the nodal formulation, frequency domain behavior is expressed at the circuit nodes using parameters of connected devices. The voltage at each node V_i is characterized by a pole expression (R_i and C_i) and the couplings from other nodal voltages in the circuit in the form of voltage-controlled current sources (VCCS). For MOS transistors, a decoupled model is used, obtained from an error free transformation of the hybrid-π representation. Considered device parameters include terminal transconductances (g_{mg}, g_{md}, and g_{ms}) and capacitances (C_{gd}, C_{gs}, C_{gb}, C_{db}, and C_{sb}). For example, the drain terminal voltage V_D in the decoupled MOS model is determined by pole components $R_D = 1/g_{md}$ in parallel with $C_D = C_{gd} + C_{db}$ and the parallel coupling (VCCS) from the gate $(sC_{gd} - g_{mg})V_G$ and source $g_{ms}V_S$ terminals. Similar forms are expressed for the gate and source voltages in the decoupled MOS model [29].

Figure 4.3(b) presents the superimposed macromodels of the two amplifiers. The encircled nodes and edges correspond to the unmatched nodes. All circuit nodes V_i are represented in the model. Coupling between nodes can be direct (expressions $F(i, j)$), or equivalent, after removing feedback connections (expressions $E(k, j)$). The voltage at any node V_j is expressed as follows:

$$V_j = \frac{R_j}{1 + sR_jC_j} \times \left(\sum_i F(i,j)V_i + \sum_k E(k,j) \right); \ i,k \neq j. \tag{4.1}$$

Terms R_j and C_j are the resistive and capacitive components of the pole at node j, and depend on the transconductances and capacitances of the devices connected at node j. For example, for V_1 in Figure 4.3, $R_1 = 1/(g_{ms1}+g_{ms2}+g_{md5})$ and $C_1 = C_{gs1} + C_{sb1} + C_{gs2} + C_{sb2} + C_{gd5} + C_{db5}$.

Direct coupling has the general form:

$$F(i,j) = sC_m \pm G_m. \tag{4.2}$$

C_m and G_m are the junction capacitances and terminal transconductances of the MOS devices connected to nodes i and j. For example, at node V_1 in Figure 4.3, $F(-,1) = sC_{gs1} + g_{mg1}$.

The equivalent coupling results by decoupling and eliminating all feedback relations from the initial coupled model and replacing them with equivalent edges:

$$E(k,j) = (sC_m \pm G_m) \times V_{k\ eq}. \tag{4.3}$$

k denotes the circuit node where the (eliminated) feedback originated and j is the influenced node. C_m and G_m are the junction capacitances and terminal transconductances of the devices connected to nodes k and j. $V_{k\ eq}$ is the equivalent voltage of node k after removing the feedback to node j.

The decoupling sequence is performed following the input-output signal path to order the nodes of the coupled model and identify which $V_{k\ eq}$ needs to be solved first. Each equivalent voltage is then expressed in the general s-polynomial form, in terms of the circuit inputs. The symbolic expression is explored from lower to higher order with controllable accuracy. The algorithm exploits the pattern of each coefficient of s^k in the expression, $a_k = \sum(\pm \prod^k C_m \prod^{N-k} G_m)$, where N is the total number of nodes. An optimization method identifies the combination of model parameters for each coefficient a_k such that the cumulative error between the original coupled and current uncoupled models is minimized across the frequency range and different operating points [29].

The symbolic matching step analyzes separately each set of topologically matched clusters, so that the effect of the unmatched nodes is kept locally to the cluster. It computes for every topologically matched pair of nodes in the clusters the TFs H_{comm} of the common part, and $H_{1,diff}$ and $H_{2,diff}$ of the unmatched nodes in each of the two circuits. TFs H_{comm} and H_{diff} are using the following expression:

$$H = \sum_p \left(\frac{\prod_t F_t(s)}{\prod_j \frac{(1+sR_jC_j)}{R_j}} \right). \tag{4.4}$$

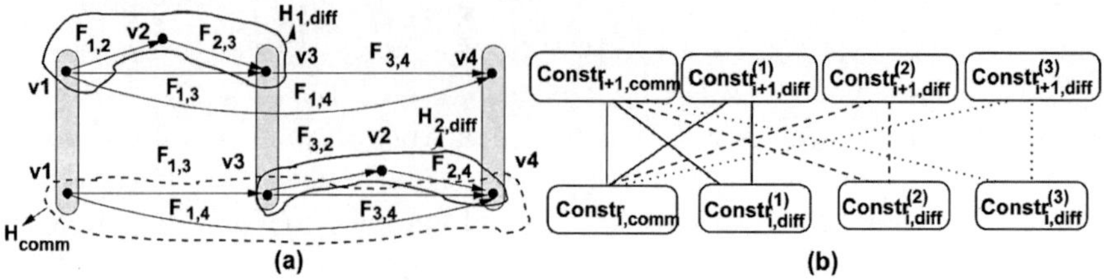

Fig. 4.4 (a) Symbolic matching and (b) constraint generation

p are the signal paths of a structure (including matched and unmatched nodes), j represents all nodes of a path with pole components R_j and C_j, and F_t are the edge labels in the signal flow graphs of the macromodel.

Example: The following TFs exist for the two structures in Figure 4.4(a). TF $H_{comm} = \frac{F_{1,3}F_{3,4}}{P_1P_3P_4} + \frac{F_{1,4}}{P_1P_4}$, where P_i are the symbolic expressions of the pole in node i. $F_{i,j}$ is the symbolic expression defining the connection between nodes i and j. TF $H_{1,diff} = \frac{F_{1,2}F_{2,3}F_{3,4}}{P_1P_2P_3P_4}$ is the difference between the top and bottom structure, and TF $H_{2,diff} = \frac{F_{1,3}F_{3,2}F_{2,4}}{P_1P_2P_3P_4}$ distinguishes the bottom from the top structure.

4.4.3 Constraint Generation

The impact on performance attribute $\mathcal{P}_i$ (e.g., DC gain, bandwidth, noise, CMRR, and sensitivity) due to topological and TF differences are estimated using H_{comm} and H_{diff} for every matched cluster of the circuits. Each TF generates a set of constraints that must be met to satisfy performance $\mathcal{P}_i$. Figure 4.4(b) shows the constraint sets corresponding to the TFs in Figure 4.4(a).

1. DC gain. Series connection of TFs H_{comm} and H_{diff} produces a DC gain of $DCgain_{H_{comm}} \times DCgain_{H_{diff}}$. Parallel connection creates a DC gain of $DCgain_{H_{comm}} + DCgain_{H_{diff}}$.

2. Bandwidth. The relation between the TFs H_{comm} and H_{diff} and bandwidth can be estimated based on Loop-Gain-Poles product (LP) [27]:

$$\omega_{max} \approx (|1 - DC\ gain| \prod_{i=1}^{n} p_i)^{\frac{1}{n}}. \tag{4.5}$$

ω_{max} is the estimated maximum bandwidth, and n is the number of dominant poles p_i. The set of considered dominant poles (usually $n \leq 2$) is selected from the total number of poles, given by the number of circuit nodes in the matched cluster. Then the bandwidth corresponding to TF H_{comm} has the following upper bound:

$$\omega_{max}^{H_{comm}} \approx (|1 - DC\ gain_{H_{comm}}| \prod_{k=1}^{m} p_k)^{\frac{1}{m}}, \tag{4.6}$$

where p_k are the m dominant poles of the common TF.

The difference TF, H_{diff}, modifies the bandwidth expression ω_{max} depending on how H_{comm} and H_{diff} are connected with each other, e.g., series or parallel.

For series connection, the bandwidth corresponding to TF $H_{comm} \times H_{diff}$ is equal to the following value:

$$\omega_{max}^{H_{comm}H_{diff}} \approx (|1 - DC\ gain_{H_{comm}} DC\ gain_{H_{diff}}| \prod_{k=1}^{m+n} p_i)^{\frac{1}{m+n}}, \tag{4.7}$$

where TF H_{diff} has n dominant poles that are not among the m dominant poles of TF H_{comm}. The change in bandwidth due to TF H_{diff} is equal to the expression:

$$\frac{\omega_{max}^{H_{comm}H_{diff}}}{\omega_{max}^{H_{comm}}} \approx \left[\frac{(DC\ gain_{H_{diff}} \prod_n p_i)^m}{(DC\ gain_{H_{comm}} \prod_m p_i)^n}\right]^{\frac{1}{m(m+n)}}. \tag{4.8}$$

For $m = n$, a sufficient condition for increasing the resulting bandwidth is if the DC gain of H_{diff} is higher than that of H_{comm} and the distance of the dominant poles to the origin is higher for TF H_{diff} than for TF H_{comm}.

For parallel connection, the bandwidth change due to $H_{comm} + H_{diff}$ can be estimated when assuming that each TF is expressed as $H_i = \frac{\prod_j z_j}{\prod_i p_i}$, where z_j are zeros and p_i are poles. The two bandwidths can then be related as in the next expression:

$$\frac{\omega_{max}^{H_{comm}+H_{diff}}}{\omega_{max}^{H_{comm}}} \approx \left[\frac{[(1 + \frac{DCgain_{H_{diff}}}{DCgain_{H_{comm}}}) \prod_n p_i)]^m}{(DC\ gain_{H_{comm}} \prod_m p_i)^n}\right]^{\frac{1}{m(m+n)}}. \tag{4.9}$$

Expressions (4.8) and (4.9) are used repeatedly for generalized products and sums of TFs.

Similar constraints are formulated for CMRR and noise.

4.4.4 Performance Characterization

Let $\mathcal{R}_i(H, p_1, p_2, ..., p_k)$ be the constraint introduced on performance attribute i through TF H with parameters p_i. The difficulty in meeting constraint $\mathcal{R}_i$ can be approximated by the value $Rate(i, H) = \max|\frac{\sum_{i,j} \frac{\partial^2 \mathcal{R}_i}{\partial p_i \partial p_j}}{\sum_i \frac{\partial \mathcal{R}_i}{\partial p_i}}|$, an estimation of the minimum rate of convergence when optimizing constraint

```
PROCEDURE (Set of constraints) IS
FOR every equation ej DO
   FOR every variable xi DO
     compute sensitivity of ej with respect to variable xi;
FOR every equation ej DO
   find sets FVj of free variables and TVj of trade-off variables;
   bind variables in sets TVj and FVj;
FOR every equation ej DO
   Fl = Fl + cardinality of set FVj;
   FOR every variable xi in FVj DO
      Diffj = Diffj * Ratei;
   Diff = Diff + Diffj;
END PROCEDURE;
```

Fig. 4.5 Procedure to estimate the flexibility and convergence of a design

$\mathcal{R}_i(H, p_1, p_2, ..., p_k)$ over the domains of parameters p_i. If circuit C_i is defined by transfer functions H_{comm} and $H_{i,diff}$ then the difficulty in optimizing a set of performance attributes m is approximated as $\sum_m Rate(m, H_{comm}) + \sum_m Rate(m, H_{i,diff})$.

The flexibility in meeting a set of constraints depends on the maximum number of free variables that intervene in the constraints. The difficulty in meeting the constraints can be estimated depending on the value *Rate*. Figure 4.5 presents the algorithm for estimating the flexibility and difficulty of optimizing a design. First, the algorithm computes for every equation e_j the sets FV_j and TV_j of free and trade-off variables. A variable x_i is free, if it has the same kind of sensitivity for all equations, e.g., only positive or only negative. A variable x_i is a trade-off variable if it has both positive and negative sensitivities. Some trade-off variables are removed from sets TV_j by binding free variables in sets FV_j, such that they cancel out the effect of the trade-off variables. The number of remaining free variables describes the available flexibility in deciding the performance values described through constraints. The product of the variable rates indicates the difficulty of a gradient-based optimization algorithm to find the solution.

4.5 Experiments

This section presents the symbolic comparison of two circuits to understand the design trade-offs and performance changes introduced by the dissimilar topological features of the circuits.

Figure 4.6 shows two low-voltage amplifier circuits, named AMP_1 and AMP_2. The first design is a two-stage class-AB topology [19]. The second circuit is a three-stage amplifier with positive feedback compensation [20].

First, topological matching found the similar and distinct nodes of the two circuits with respect to their structure and electrical behavior. Figure 4.7 illustrates the nodes and couplings of the two amplifiers. Nodes V_{in+}, V_{in-}

(signal inputs), V_1, and V_o (output) have identical symbolic pole expressions in both designs. Similarly, the couplings between nodes, Fc_i $(i = \overline{1,9})$, are the same in both circuits. Nodes V_3 and V_7 are only partially-matched due to small differences in their symbolic pole expressions. Enforcing that nodes V_3 have comparable pole components in both circuits results in the following two constraints: $g_{mg3}|_{AMP_1} \equiv g_{ms8}|_{AMP_2}$ and $(C_{gs3} + C_{gb3})|_{AMP_1} \equiv (C_{gs8} + C_{sb8})|_{AMP_2}$, when device M_1 parameters are matched between the two circuits, and transistors M_3 and M_4 are functionally matched in AMP_1 (current mirror). The constraints imply that device parameters of M_3 in circuit AMP_1 are paired with those of device M_8 in circuit AMP_2.

Similar matching constraints exist for the pole at node V_7: $(g_{mdR} + g_{mgR})|_{AMP_1} \equiv (g_{md12} + g_{md13})|_{AMP_2}$ and $(C_{dbR} + C_{gsR} + C_{gbR})|_{AMP_1} \equiv (C_{gd12} + C_{db12} + C_{gd13} + C_{db13})|_{AMP_2}$, when devices M_7 and capacitances $C_b \equiv C_{m2}$ are matched between the two designs. The two constraints link the parameters of device M_R in AMP_1 to the combined parameters of two devices, M_{12} and M_{13}, in AMP_2. The second circuit has a greater flexibility in meeting the matching of the above conditions.

In Figure 4.7, the input and output blocks of circuits AMP_1 and AMP_2 are matched and express similar electrical behavior. The distinguishing substructures are at nodes V_2, V_4, V_5, and V_6 with couplings between the nodes given by symbolic expressions Fd_j, $j = \overline{1,5}$. For example, the additional nodes V_4 and V_6 in AMP_2 impose different model graph edges: Fd_1 and Fd_2 for the input block, and Fd_4 and Fd_5 for the output block. The extra block in circuit AMP_2 has no equivalent in AMP_1.

Finally, the set of symbolic transfer functions for each circuit block are generated: (i) H_{comm} defines the common symbolic parts of the blocks in both circuits, and (ii) H_{diff} expressing the distinguishing symbolic terms for the blocks in each circuit.

The transfer functions of the input blocks of circuits AMP_1 and AMP_2, respectively, to node V_2 are as follows:

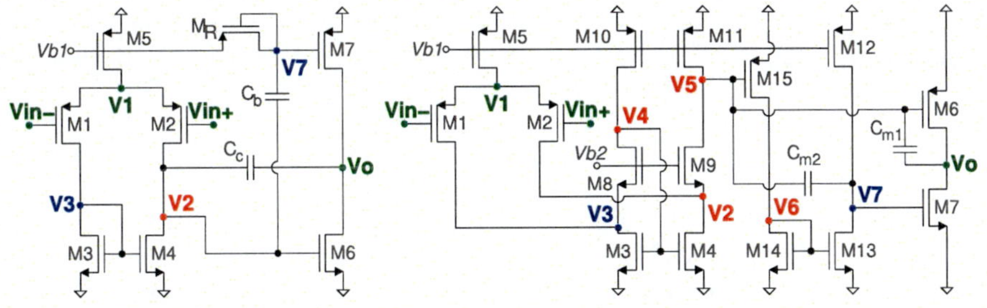

Fig. 4.6 Two low-voltage differential amplifiers: (a) class-AB 2-stage AMP_1 [19] and (b) positive feedback compensation 3-stage AMP_2 [20]

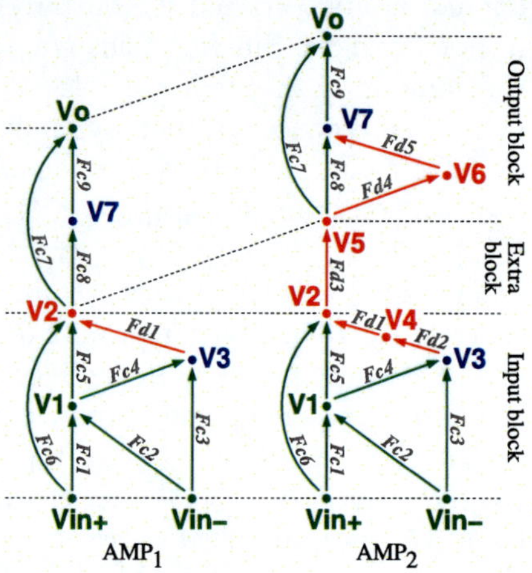

Fig. 4.7 Model graphs and sub-blocks for circuits AMP_1 and AMP_2

$$AMP_1 : H_2 = H_{comm_1} \times H_{diff_1} + H_{comm_2} \times H_{diff_2}, \quad (4.10)$$

$$AMP_2 : H_2 = H_{comm_1} \times H_{diff_3} + H_{comm_2} \times H_{diff_4}, \quad (4.11)$$

where H_{comm_1} and H_{comm_2} are the common signal paths in both designs and H_{diff_i} $(i = \overline{1,4})$ define the differences of the two input blocks:

$$H_{diff_1} = \frac{R_2}{1 + sR_2C_2}, H_{diff_2} = \frac{R_2Fd_1}{1 + sR_2C_2}, \quad (4.12)$$

$$H_{diff_3} = \frac{R_2}{1 + sR_2C_2}, H_{diff_4} = \frac{R_2R_4Fd_1Fd_2}{(1 + sR_2C_2)(1 + sR_4C_4)}. \quad (4.13)$$

Expressions (4.10)-(4.13) indicate that the input blocks of circuits AMP_1 and AMP_2 differ because of the poles at nodes V_2 and V_4, and the coupling between nodes $V_3 \rightarrow V_2$ and $V_3 \rightarrow V_4 \rightarrow V_2$.

The extra block of design AMP_2 has no equivalent in circuit AMP_1 (see Figure 4.7). Its transfer function to circuit node V_5 is defined only by unmatched components:

$$H_5 = H_{diff_5} = \frac{R_5Fd_3}{1 + sR_5C_5}. \quad (4.14)$$

The matching illustrated in Figure 4.7 shows that the output block of AMP_1 is composed of only matched nodes. Only AMP_2 exhibits differences in this circuit block with a transfer function to node V_o defined as follows:

$$H_o = H_{comm_3} + H_{comm_4} \times H_{diff_6}, \quad (4.15)$$

where $H_{comm_{3,4}}$ represent the common structures (also present in AMP_1). H_{diff_6} is defined by the unmatched pole and edges related to node V_6, $H_{diff_6} = \frac{R_6 Fd_4 Fd_5}{1+sR_6C_6}$.

For constraint extraction, the terms of the common (H_{comm}) and different (H_{diff}) transfer functions are instantiated with their respective macromodel symbolic expressions based on device parameters. Considering the corresponding matched parameters constant, constraints are expressed such that unmatched parameters of H_{diff} can be used to improve performance.

For example, in the output block of AMP_2, the node coupling terms of H_{diff_6} from equation (4.15) are expressed as $Fd_4 = sC_{gd15} - g_{mg15}$ and $Fd_5 = sC_{gd13} - g_{mg13}$. The pole components for node V_6 are $R_6 = \frac{1}{g_{md15}+g_{md14}+g_{mg14}}$ and $C_6 = C_{gd15} + C_{db15} + C_{db14} + C_{gs14} + C_{gb14} + C_{gs13} + C_{gd13} + C_{gb13}$.

The DC-gain of each circuit sub-structure is characterized by transfer functions $H_{comm_i}(0)$ and $H_{diff_i}(0)$. For analysis, considering $H_{comm_i}(0)$ constant, the constraints on $H_{diff_i}(0)$'s design variables are identified, such that the overall gain is improved. For the output block of AMP_2, the constraint on different parameters is given by the following expression:

$$K_1 + K_2 \frac{g_{mg15}}{g_{md15} + g_{md14} + g_{mg14}} \nearrow . \tag{4.16}$$

Constants $K_1 = \frac{g_{mg6}}{g_{md6}+g_{md7}}$ and $K_2 = \frac{g_{mg7}g_{mg13}}{(g_{md6}+g_{md7})(g_{md12}+g_{md13})}$ correspond to the parameters of common transfers $H_{comm_{3,4}}$ from equation (4.15). Note that we also consider constant parameters for M_{13} in the expression of Fd_5 as it contains device parameters topologically matched for the pole at V_7.

The gain-poles product (GPP) is used to estimate bandwidth [27] in each circuit block by selecting at most two dominant poles. Any remaining poles of the block are considered non-dominant and their constraints are expressed. For the output block of AMP_2, the only valid dominant pole set is P_o and P_7, introducing two constraints on distinguishing parameters, $\frac{1}{g_{md15}+g_{md14}+g_{mg14}} \searrow$ and $K_3 + C_{gd15} + C_{db15} + C_{db14} + C_{gs14} + C_{gb14} \searrow$. Then the gain-pole product increases when:

$$K_4 \frac{g_{mg15}}{g_{md15} + g_{md14} + g_{mg14}} \nearrow . \tag{4.17}$$

The constants for this case are expressed as $K_3 = C_{gs13} + C_{gd13} + C_{gb13}$ and $K_4 = \frac{g_{mg7}g_{mg13}}{g_{md12}+g_{md13}}$.

Noise is modeled for each circuit block of the two circuits using the method in [29]. Considering the common path noise fixed, constraints are extracted such that the noise contribution of the differences is diminished. The relevant constraints for the output block of AMP_2 are $\frac{g_{mg15}}{g_{md15}+g_{md14}+g_{mg14}} \searrow$ and $K_3 + C_{gd15} + C_{db15} + C_{db14} + C_{gs14} + C_{gb14} \nearrow$; or $g_{m15} \approx g_{m14}$.

With respect to CMRR, both circuits include only matched design parameters. The relation between the parameters of devices M_5, M_1, and M_2 controls this performance. The differences impact DC-gain and CM-gain in

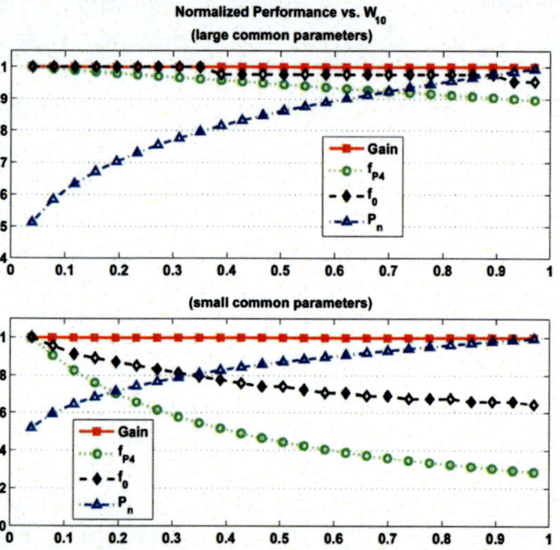

Fig. 4.8 Input block performance trade-offs for comparing circuits AMP_2 and AMP_1 with respect to W_{10}

the same way. This implies that similar CM performance can be expected in AMP_1 and AMP_2 when common signal path variables are precisely matched.

Constraints are extracted in the same manner for all circuit blocks [30]. Table 4.1 summarizes the required trends (except equality relations) for the distinguishing design variables of circuit AMP_2 across the entire amplifier, including input, extra, and output blocks. The trends were determined using the extracted constraints. For example, increasing the output block's DC-gain in AMP_2 by satisfying constraint (4.16) can only be performed by deteriorating noise performance. Parameters of device M_{10} have two variations for dominant poles and bandwidth, with respect to each feasible set. Design variables g_{mg15} and g_{md15} are bound by conflicting trade-offs. However, both parameters are determined by device M_{15}, as is the case of $C_{gd15} + C_{db15}$, and the common dependence of both transconductance and capacitance on W_{15} correlates these parameters.

Table 4.1 Desired variable trends with respect to performance in circuit AMP_2

Variables	Gain	CM	Noise	Pole	GPP
g_{md10}	$\searrow$	-	$\nearrow$	$\nearrow,\searrow$	$\searrow$,-
$C_{gd10} + C_{db10}$	-	-	$\nearrow$	$\searrow,\nearrow$	-,$\searrow$
g_{mg15}	$\nearrow$	-	$\searrow$	-	$\nearrow$
g_{md15}	$\searrow$	-	$\nearrow$	$\nearrow$	$\nearrow$
$C_{gd15} + C_{db15}$	-	-	$\nearrow$	$\searrow$	-
$C_{gs15} + C_{gd15} + C_{gb15}$	-	-	$\nearrow$	$\searrow$	-
$g_{md14} + g_{mg14}$	$\searrow$	-	$\nearrow$	$\nearrow$	$\searrow$
$C_{db14} + C_{gs14} + C_{gb14}$	-	-	$\nearrow$	$\searrow$	-

The procedure further investigates these parameter correlations and performance trade-offs by evaluating the model of each individual circuit block. We vary the widths of devices identified by parameters in Table 4.1 across a predefined range and characterize the normalized performance trends. Common parameters and branch bias currents through matched devices are considered constant. This allows us to estimate the performance sensitivities to distinguishing circuit parameters and provides insight about the most appropriate relative sizing strategy that can improve performance.

For the input block of circuit AMP_2, design parameters related to device M_{10}, i.e. g_{md10} and $C_{gd10} + C_{db10}$, influence the resistive (R_4) and capacitive (C_4) components of the pole at node V_4. Figure 4.8 shows the correlations between W_{10} and gain, pole frequency (f_{P4}), unity-gain frequency (f_0), and total block output noise (P_n). The normalized performances are presented for two scenarios: when the common path parameters are (i) large or (ii) small with respect to those of M_{10}. For either case, P_4 is the first dominant pole. We observe that parameters of device M_{10} have virtually no impact on the circuit block's gain. This is due to the constant branch current imposed by the matched devices. The total noise exhibits the same sensitivity to M_{10} parameters regardless of the constant parameter values and is minimized for relatively small transistor widths. The variation of 50% across the investigated range suggests that when the common parameters are matched between the two designs, M_{10}'s width should be kept low. This would be required to attempt comparable noise performance in both AMP_1 and AMP_2. For pole and unity-gain frequency, W_{10} is again best kept low especially in the case of small common parameters. There is a deterioration in first dominant pole frequency (equivalent to the -3dB point) of up to $\approx$ 70% from the maximum. Impact on unity-gain frequency is reduced, but can still decrease with up to $\approx$ 35% as the size increases. The opposing trends between total noise and pole frequencies suggests that g_{m10} is a dominant parameter for noise. $C_{gd10} + C_{db10}$ best controls bandwidth performance and has limited impact on total noise. In terms of sensitivity, all impacted performances provide a more pronounced variation within the first third of the width range. As W_{10} is further increased, the impact on performance is reduced.

Considering the output block of circuit AMP_2, Figure 4.9 depicts the normalized performance plots based on the width of device M_{15} when M_{14} is kept constant. The analyzed parameters are g_{md15} and $C_{gd15} + C_{db15}$ controlling the non-dominant pole at node V_6 (P_6) and $sC_{gd15} - g_{mg15}$ defining the coupling between nodes V_5 and V_6. Gain follows the same increasing trend for both common parameters cases, showing that the impact of W_{15} on this performance is dominant. Furthermore, the increase in gain across the entire analyzed range suggests that g_{mg15} is the dominant parameter and compensates for the smaller increase in g_{md15}. Unity-gain frequency also follows similar trends in either case. While f_0 is dominated by the common path attributes, it can still be increased by up to 20% for the maximum analyzed

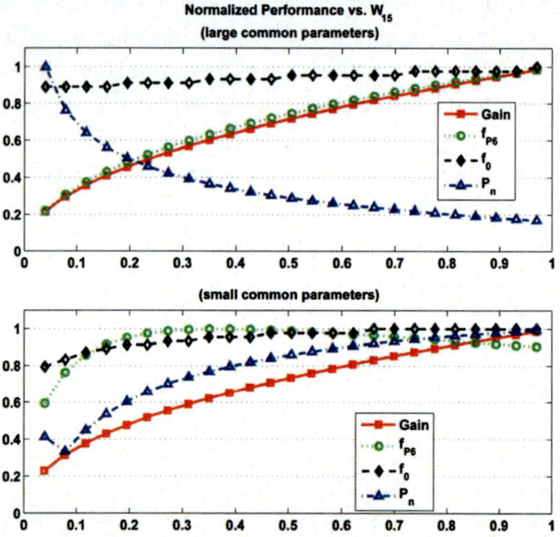

Fig. 4.9 Output block performance trade-offs for comparing circuits AMP_2 and AMP_1 with respect to W_{15} when W_{14} is constant

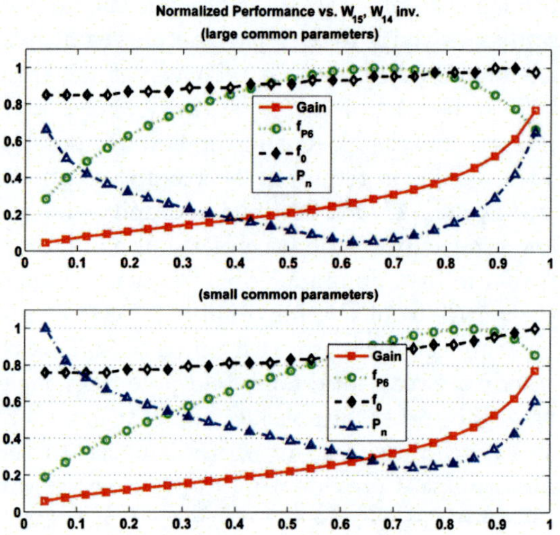

Fig. 4.10 Output block performance trade-offs for comparing circuits AMP_2 and AMP_1 with respect to W_{15} and W_{14} with opposing variation

value of W_{15}. Factoring in the total noise for the output block of AMP_2, the scenario of large common parameters becomes favorable. This allows for increasing gain and f_0, while reducing noise. An optimal point is reached when W_{15} is maximum. For small common parameters, this advantage is lost after the first tenth of the analyzed range, when gain is $\approx 70\%$ less than the maximum. In addition, beyond this point, noise exhibits a more pronounced variation and deteriorates faster than gain increases.

An interesting situation occurs when W_{14} is also varied, but in opposition to W_{15}, illustrated in Figure 4.10. The added parameters of device M_{14} are $g_{md14} + g_{mg14}$ and $C_{db14} + C_{gs14} + C_{gb14}$ influencing the pole at node V_6 (P_6).

As in the previous scenario, a limited variation in unity gain is observed, which increases as W_{15} increases and W_{14} decreases. Similarly, gain increases across the range. However, the impact is now more pronounced in the last third of the interval, when W_{15} is significantly larger than W_{14}. For noise performance, it is now possible to minimize the impact for both large and small common path parameters. However, this minimum is no longer achieved when the gain and f_0 are maximized. This suggests that in this scenario gain is sacrificed in the output block of AMP_2 in order to reduce the noise impact of distinguishing features. Closer inspection shows that the achievable noise minimum is still better for the case of large common parameters.

Overall, the analysis of AMP_2 suggests that compared to the structures of AMP_1 this topology can exploit the distinguishing attributes through different sizing schemes and offer performance improvements.

4.6 Conclusion

This chapter presents a novel circuit synthesis flow based on concept comparison, combination, learning, and re-use. A technique for systematically producing comparison data between two analog circuits is introduced. The comparison data refers to DC-gain, bandwidth, noise, CMRR, and sensitivity. The nodes with similar electric behavior in the two circuits are found through a dual matching approach of circuit topologies and symbolic expressions. Dissimilarities are also identified in the process. Next, the method computes the constraints that relate the electrical behavior to changes of the performance attributes. Using the constraints, the final step produces the comparison data, which includes modification of design trade-offs, availability of free design variables, achievable performance values, and hardness to find optimized design parameters.

Acknowledgements. This work was supported by major collaborative research National Science Foundation CreativeIT Grant no. IIS 0856038.

References

1. Antao, B., Brodersen, A.: ARCHGEN: automated synthesis of analog systems. IEEE Transactions on VLSI Systems 3(2), 231–244 (1995)
2. Castro-Lopez, R., Fernandez, F., Guerra-Vinuesa, O., Rodriguez-Vazquez, A.: Reuse-Based Methodologies and Tools in the Design of Analog and Mixed-Signal Integrated Circuits. Kluwer (2006)
3. Chang, H., Charbon, E., et al.: A Top-Down Constraint Driven Design Methodology for Analog Integrated Circuits. Kluwer (1996)

4. Das, A., Vemuri, R.: Topology Synthesis of Analog Circuits based on Adaptively Generated Building Blocks. In: Proceedings of Design Automation Conference, pp. 44–49 (2008)
5. Degrauwe, M., Nys, O., Dijkstra, E., Rijmenants, J., Bitz, S., Goffart, B., Vittoz, E., Cserveny, S., Meixenberger, C., Van Der Stappen, G., Oguey, H.: IDAC: An Interactive Design Tool for Analog CMOS Circuits. IEEE Journal of Solid-State Circuits SC 22(6), 1106–1116 (1987)
6. De Smedt, B., Gielen, G.: WATSON: Design Space Boundary Exploration and Model Generation for Analog and RF IC Design. IEEE Transactions on CAD of Integrated Circuits and Systems 22(2), 213–224 (2003)
7. Doboli, A., Vemuri, R.: Exploration-Based High-Level Synthesis of Linear Analog Systems Operating at Low/Medium Frequencies. IEEE Transactions on CAD of Integrated Circuits and Systems 22(11), 1556–1568 (2003)
8. Doboli, A., Dhanwada, N., Nunez, A., Vemuri, R.: A Library-Based Approach to Synthesis of Analog Systems from VHDL-AMS Specifications. ACM Transactions on Design Automation 9(2), 238–271 (2004)
9. El-Turky, F., Perry, E.: BLADES: An Artificial Intelligence Approach to Analog Circuit Design. IEEE Transactions on CAD of Integrated Circuits and Systems 8(6), 680–692 (1989)
10. Ferent, C., Doboli, A.: A Symbolic Technique for Automated Characterization of the Uniqueness and Similarity of Analog Circuit Design Features. In: Proceedings of Design, Automation and Test in Europe Conference (2011)
11. Harjani, R., Rutenbar, R., Carley, L.: OASYS: a framework for analog circuit synthesis. IEEE Transactions on CAD of Integrated Circuits and Systems 8(12), 1247–1266 (1989)
12. Hassoun, S., McCreary, C.: Regularity Extraction via Clan-based Structural Circuit Decomposition. In: Proceedings of ICCAD, pp. 414–419 (1999)
13. Horta, N., Franca, J.: Algorithm-driven synthesis of data conversion architectures. IEEE Transactions on CAD of Integrated Circuits and Systems 16(10), 1116–1135 (1997)
14. Hum, R.: Where are the dragons? Presentation at Frontiers in Analog Circuit Synthesis and Verification Workshop (2011)
15. Koh, H., Sequin, C., Gray, P.: OPASYN: a compiler for CMOS operational amplifiers. IEEE Transactions on CAD of Integrated Circuits and Systems 9(2), 113–125 (1990)
16. Li, X., Wang, J., Pileggi, L., Chen, T.-S., Chiang, W.: Performance-Centering Optimization for System-Level Analog Design Exploration. In: Proceedings Design Automation Conference, pp. 422–429 (2005)
17. McConaghy, T., Palmers, P., Steyaert, M., Gielen, G.: Variation-Aware Structural Synthesis of Analog Circuits via Hierarchical Building Blocks and Structural Homotopy. IEEE Transactions on CAD of Integrated Circuits and Systems 28(9), 1281–1294 (2009)
18. Martens, E., Gielen, G.: Classification of Analog Synthesis Tools based on their Architecture Selection Mechanisms. Integration, the VLSI Journal 41, 238–252 (2008)
19. Ramirez-Angulo, J., Carvajal, R., Galan, J., Lopez-Martin, A.: A Free But Efficient Low-Voltage Class-AB Two-Stage Operational Amplifier. IEEE Transactions on Circuits & Systems - II 53, 568–571 (2006)

20. Ramos, J., Steyaert, M.: Positive Feedback Frequency Compensation for Low-Voltage Low-Power Three-Stage Amplifier. IEEE Transactions on Circuits & Systems - II 51, 1967–1974 (2004)
21. Rao, D., Kurdahi, F.: On Clustering for Maximal Regularity Extraction. IEEE Transactions on CAD of Integrated Circuits and Systems 12(8), 1198–1208 (1993)
22. Sripramong, T., Toumazou, C.: The Invention of CMOS Amplifiers Using Genetic Programming and Current-Flow Analysis. IEEE Transactions on CAD of Integrated Circuits and Systems 21(11), 1237–1252 (2002)
23. Stehr, G., Graeb, H., Antreich, K.: Analog Performance Space Exploration by Normal-Boundary Intersection and by Fourier-Motzkin Elimination. IEEE Transactions on CAD of Integrated Circuits and Systems 26(10), 1733–1748 (2007)
24. Tang, H., Doboli, A.: High-Level Synthesis of Delta-Sigma Modulators Optimized for Complexity, Sensitivity and Power Consumption. IEEE Transactions on CAD of Integrated Circuits and Systems 25(3), 597–607 (2006)
25. Tang, H., Zhang, H., Doboli, A.: Refinement based Synthesis of Continuous-Time Analog Filters Through Successive Domain Pruning, Plateau Search and Adaptive Sampling. IEEE Transactions on CAD of Integrated Circuits and Systems 25(8), 1421–1440 (2006)
26. Torralba, A., Chavez, J., Franquelo, L.: FASY: a fuzzy-logic based tool for analog synthesis. IEEE Transactions on CAD of Integrated Circuits and Systems 15(7), 705–715 (1996)
27. Verhoeven, C., et al.: Structured Electronic Design - Negative-Feedback Amplifiers. Kluwer (2006)
28. Wei, Y., Tang, H., Doboli, A.: Systematic Methodology for Designing Reconfigurable Delta Sigma Modulator Topologies for Multimode Communication Systems. IEEE Transactions on CAD of Integrated Circuits and Systems 26(3), 480–496 (2007)
29. Wei, Y., Doboli, A.: Structural Macromodeling of Analog Circuits Through Model Decoupling and Transformation. IEEE Transactions on CAD of Integrated Circuits and Systems 27, 712–725 (2008)
30. Ferent, C., Doboli, A.: Systematic Comparison of Two Low-Voltage Amplifiers using Topology Matching and Performance Constraints. IEEE International New Circuits And Systems Conference, NEWCAS (2012)
31. Ferent, C., Doboli, A.: A Prototype Framework for Conceptual Design of Novel Analog Circuits. In: International Conference on Synthesis, Modeling, Analysis and Simulation Methods and Applications to Circuit Design, SMACD (2012)
32. Kuo, K.-C., Leuciuc, A.: A Linear MOS Transconductor using Source Degeneration and Adaptive Biasing. IEEE Transactions on Circuits & Systems - II 48, 937–943 (2001)

Chapter 5
Graph-Based Symbolic and Symbolic Sensitivity Analysis of Analog Integrated Circuits

S. Rodriguez-Chávez, A.A. Palma-Rodriguez,
E. Tlelo-Cuautle, and S.X.-D. Tan

Abstract. A graph-based symbolic technique is introduced for the computation of fully-symbolic expressions of analog integrated circuits. The system of equations is formulated by using nullor-equivalents of the SPICE-like circuit elements, and then by applying just nodal analysis (NA). The proposed technique is extended to perform symbolic sensitivity analysis not only with respect to each symbolic-element but also with respect to multi-parameters. Several examples are presented in order to compare the performance of the proposed graph-based symbolic technique with the already known determinant decision diagram (DDD) method and with HSPICE^{TM}. As a result, it is shown that as for the DDD method, the graph-based symbolic technique is compact, unique and the complexity to obtain the symbolic expression depends on the size of the graph.

5.1 Introduction

Symbolic analysis has demonstrated its usefulness to capture the dominant behavior or derive an exact characteristic of a circuit with its variables (dependent and independent) and with all or some of its circuit elements represented by symbols [4, 10, 17, 19, 23–25, 27, 30, 31, 33, 34]. Symbolic techniques had a growing interest between 1960's and 1980's because of the increasing computing power and that many computer analysis techniques were proposed. Those contributions were adopted by the integrated circuit (IC) design

S. Rodriguez-Chávez · A.A. Palma-Rodriguez · E. Tlelo-Cuautle
INAOE (Instituto Nacional de Astrofisica, Optica y Electrónica), México
e-mail: {srodriguez,adairpalma,etlelo}@inaoep.mx

S.X.-D. Tan
University of California at Riverside, CA, USA
e-mail: stan@ee.ucr.edu

M. Fakhfakh et al. (Eds.): *Analog/RF & Mixed-Signal Circuit Sys. Design*, LNEE 233, pp. 101–122.
DOI: 10.1007/978-3-642-36329-0_5

community from late 1980's. Further, that interest resulted in the successful development of symbolic analyzers described in [10, 11, 27, 35].

Knowledge on the behavior of a circuit is very important in the design and optimization processes for analog ICs [7, 12, 19, 21, 23, 25, 33]. For this reason, symbolic simulators become a very useful design automation tool as they give as the result the analytic expression in a closed form of the circuit presenting the relationships between its parameters, this being the main advantage over repetitive numerical simulations [10, 30]. In addition, symbolic analysis is useful when many numerical cancellations leads us to large round off error, because one can verify accurately the behavior of a circuit through evaluating the analytical expression from a numerical simulation, i.e. the symbolic expression is always valid even if the parameter values change, as long as the circuit topology remains the same and the parameters have values under the allowed variation ranges [7, 19].

Up to day, symbolic analysis techniques can be categorized in the following approaches: Tree enumeration methods, signal flow graph methods and determinant based methods [10]. For instance, Tree enumeration methods were proposed by Maxwell and Kirchhoff and first used for analysis of RCL networks represented as weighted undirected graphs. The advantage of that method is that the expressions are irreducible. However, for extended networks like RLC-g_m ones, the tree enumeration cannot be applied directly and tradeoffs between sign calculation and obtaining cancellation-free product terms are encountered [29]. Signal flow graph approaches work with weighted directed graphs, but the product terms obtained by this method are still not irreducible or cancellation-free [10]. Determinant-based methods are also not cancellation-free but it has been proven that topological methods (such as tree enumeration and signal flow graph ones) do not offer any advantage over these methods. On the other hand, from 2000's the symbolic technique based on Determinant Decision Diagrams (DDDs) has been the best way to derive exact and simplified symbolic expressions [10]. Besides, it is well suited just for sparse matrices [16, 26].

Exploiting the success of DDDs, this chapter introduces a graph-based symbolic approach that computes the solution of a system of equations by graph manipulations and applying some DDD properties. In this manner, as for DDD-based approaches, the proposed graph representation was developed to work with the nodal admittance matrix [24], which should be sparse to highlight its usefulness in deriving analytical expressions that very often shares many sub-expressions. Henceforth, the determinant representation by applying our proposed graph-based symbolic technique is compact, unique and the complexity to obtain the symbolic expression depends on the size of the graph. Additionally, this chapter highlights its suitability to compute sensitivities with respect to one or many parameters, provided that a symbolic transfer function exists [2, 5, 8, 13, 15]. Indeed, sensitivity analysis is very important in IC design since it helps us to optimize the behavior of a given circuit by showing us which components of the entire systems are more sensitive [7, 20]. That way, it can help to reduce costs of production given that an IC designer can replace the less sensitive components with cheaper ones and critical components with high quality components.

5.2 Nullor Circuit Equivalents

The Nullator and Norator are abstract elements which together not only model the ideal characteristics of an operational amplifier [10], but also by manipulating them along with linear passive elements it is possible to synthesize a wide variety of analog ICs [9]. More recently, other pathological elements called current mirror and voltage mirror have been applied to formulate compact systems of equations of analog ICs, as shown in [14,24,28,32]. Those formulation methods are used herein, and the solution is performed by applying the proposed graph-based symbolic technique.

The properties of the nullator are that the voltage across its terminals is zero and does not allow current flowing through it. Even though it has some properties of an open circuit, and it does not have an immittance or a scattering representation. For the norator, the voltage between its terminals is arbitrary and an arbitrary current can flow through it. The circuit representations of the nullator and norator are shown in Fig. 5.1, where V1=V2 and Ix=0 for the nullator, and V1=V2=Ix=undefined or arbitrary for the norator.

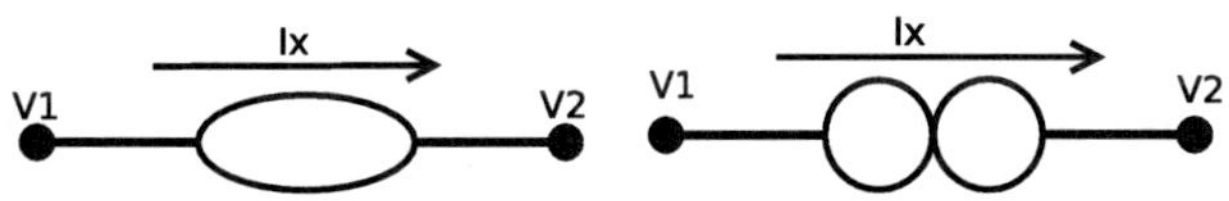

Fig. 5.1 Nullator (left), and norator (right) representations

One can combine these two elements in order to obtain a four-terminals element called Nullor. The nullor can also be described as a two port element, where its input port is the Nullator and the output port the Norator. That is why the Nullator-Norator pair models the ideal behavior of an operational amplifier, and they receive the name of Nullor, as introduced by Carlin [6]. The nullor element has proven its usefulness in areas like symbolic analysis [10], and circuit synthesis [9].

One of the main advantages of the Nullor element in symbolic analysis is its suitability to performing pure nodal analysis (NA) for an analog IC consisting of any kind of active devices [24]. In the same way, the behavior of the MOSFET can not only be modeled using nullors but also one can include parasitics as the gate-source and gate-drain capacitances, as shown in Fig. 5.2. From this model, one can derive the Nullor equivalents for a wide variety

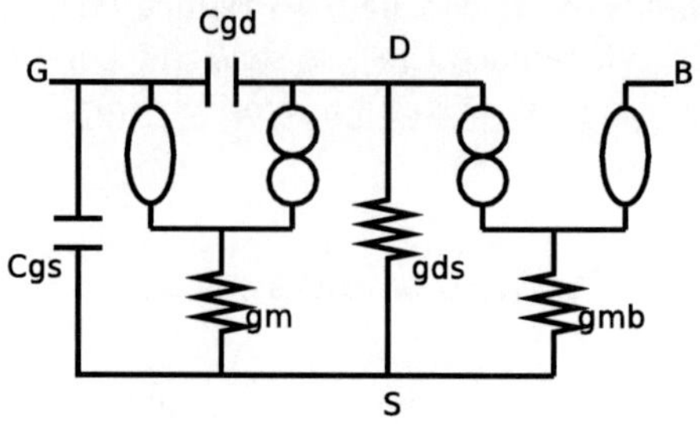

Fig. 5.2 Nullor representation of the MOSFET

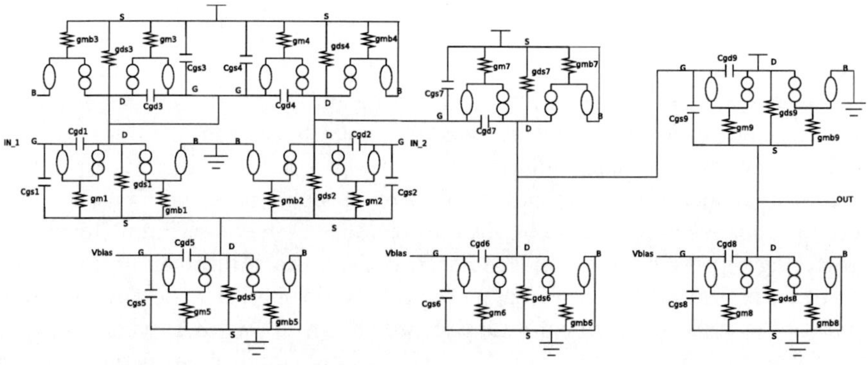

Fig. 5.3 Nullor equivalent of a three stages OTA

of amplifiers. For example, the three stages operational transconductance amplifier (OTA) has the Nullor equivalent shown in Fig. 5.3 [22].

Using nullor equivalents to describe analog ICs, the guidelines for obtaining the nodal admittance matrix by applying nodal analysis is summarized below [22, 24].

The nullator properties are associated to manipulating the ROWs of the admittance matrix, and the norator properties to its COLs. Therefore, two sets of pairs of nodes are formed, one for ROWs and one for COLs. These two sets are then used to formulate the nodal admittance matrix by performing the Cartesian product of every subset. In the ROW group a subset is formed for every node with no Norator(s) connected to it, and a different subset for every group of nodes connected by floating Norators. In the same way, in the COL group a subset is formed for every node with no Nullator(s) connected to it, and a different subset for every group of nodes connected by floating Nullators. Two groups of admittances are formed, the first (group A) containing a subset for every node listing all the admittances connected to it, and the other (group B) listing the floating admittances with their corresponding pair of nodes. If a node is present in a subset in ROWs and a subset in COLs then the corresponding subset of admittances (from group A) is summed at the matrix position (ROW index, COL index). If a pair of nodes is present, one in a subset of ROWs and the other in a subset of COLs, the corresponding admittance (from group B) is summed with negative sign at the matrix position (ROW index, COL index). Further, in performing symbolic nodal analysis (NA) of nullor circuits [24,33], the non-NA compatible elements are replaced by their nullor equivalents. For example: the independent voltage source is transformed to a current source equivalent circuit using one nullor, as shown in Fig. 5.4, as well as the voltage-controlled voltage source.

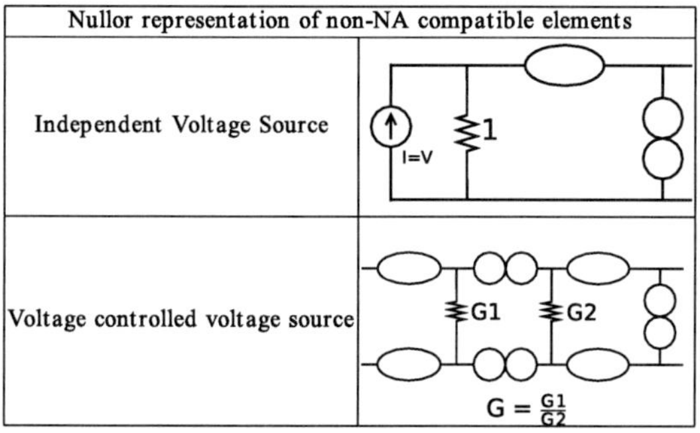

Fig. 5.4 Nullor equivalents of non-NA compatible elements

5.3 Graph-Based Determinant Representation

The circuit size is a challenge in performing symbolic analysis because a large number of symbolic terms are manipulated [10]. Fortunately, this problem is mitigated when applying DDDs [29], and variants of it. Furthermore, the graph-based symbolic technique has a special representation of the admittance matrix as shown in this section. That graph representation is unique and compact for a large class of analog ICs. In this sense, every determinant has a unique representation, and is liable to symbolic manipulations. To understand how this graph approach works, lets us consider the following determinant [10] [29]:

$$det(M) = \begin{vmatrix} a & b & 0 & 0 \\ c & d & e & 0 \\ 0 & f & g & h \\ 0 & 0 & i & j \end{vmatrix} = adgj - adhi - aefj - bcgj + bchi \qquad (5.1)$$

If the determinant's size is $n \times n$, we expect to have paths of $n+1$ levels. So if there is a path that is not complete, i.e. that does not have $n+1$ levels or that has a zero in any element of the path, it will be eliminated completely since this means that this expression is multiplied by zero. This structure is built in a depth-first search (DFS) fashion. Every element of the graph structure corresponds to a position of the admittance matrix. As a result, for this particular example one obtains the graph shown in Fig. 5.5, where all multiplications by zero were already omitted.

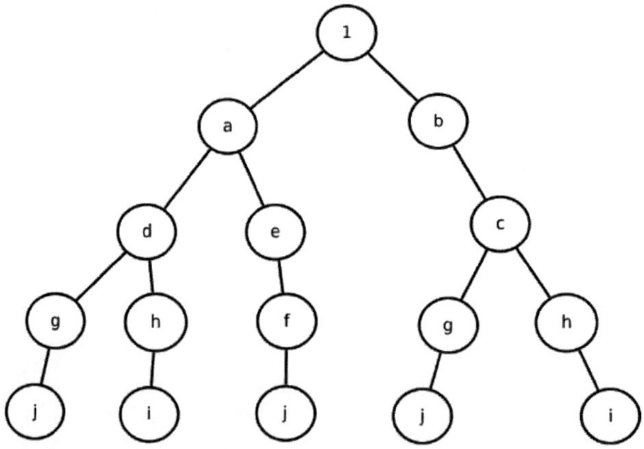

Fig. 5.5 Graph representation of (5.1)

Sending a prune signal if a zero is found inside the path performs path eliminations. This prune signal is propagated all the way up until a summing point is reached so the whole branch does not form part of the final structure and the path is terminated to zero instead. As one sees, the implementation of this graph structure is by a tree in which the arithmetic operations are encoded in the depth of the tree node, that is, different depth implies multiplication while equal depth implies addition. This leads us to get the expression:

$$det(M) = a\left[d\left(gj - hi\right) + e\left(-fj\right)\right] + b\left[c\left(-gj + hi\right)\right] \tag{5.2}$$

A key point is related to the assignation of signs to each node in the expanded graph. They are established by applying the rule of signs from Cramer's rule. When applying graph methods to evaluate a determinant, not only one can obtain a factorized exact symbolic expression, but also derive all transfer relationships with respect to each node, and in a post-processing step to each branch circuit variable.

5.4 Sensitivity Analysis

Circuit sensitivity can be defined as the influence of a change in the circuit characteristics, i.e. how much a particular circuit characteristic changes as a particular circuit-component value varies. This gives us an insight on how certain parameters influence in the response of a specific circuit. In this chapter, small-signal or AC sensitivities will be treated. For instance, one of the more popular notions used in analog IC design is the adjoint network analysis [10], also implemented in the circuit simulator SPICE. However, the drawback of using the numerical circuit simulator SPICE to obtain the sensitivity of an analog IC with respect to a given circuit element, is that one has to execute AC sensitivity command, then calculate the finite difference and finally one should apply normalization to get the numerical sensitivity. Subsequently, one needs to plot the real and imaginary parts versus the frequency ω to generate the sensitivity curves. This is the main reason why we propose to apply our proposed graph-based symbolic technique to compute symbolic sensitivities of analog ICs [18].

For instance, given the transfer function $H(s)$ seen as:

$$H(s) = \frac{N(s)}{D(s)} \tag{5.3}$$

Both $N(s)$ and $D(s)$ can be represented by a different graph. The former by using the solution vector y from the formulation $Ax = y$ and by applying Cramer's rule. The graph for computing $D(s)$ is simply matrix A or the determinant of the nodal admittance matrix.

In general, the AC-sensitivity is given by the following equation:

$$Sens(H(s), W) = \frac{W}{H(s)} \frac{\partial H(s)}{\partial W} \tag{5.4}$$

Substituting (5.3) into (5.4) and applying the chain rule we have:

$$Sens\,(H(s), W) = \left(\frac{WD(s)}{N(s)}\right)\left(\frac{\frac{\partial N(s)}{\partial W}D(s) - N(s)\frac{\partial D(s)}{\partial W}}{D^2(s)}\right) \tag{5.5}$$

Regrouping similar terms in (5.5) we get to the expression

$$Sens(H(s), W) = W\left(\frac{1}{N(s)}\frac{\partial N(s)}{\partial W} - \frac{1}{D(s)}\frac{\partial D(s)}{\partial W}\right) \tag{5.6}$$

Equation (5.6) is quite suitable to perform symbolic sensitivity computation of analog ICs by applying the graph-based symbolic technique. For instance, one advantage is that in the resulting sum of symbolic product-terms, one can derive each product with respect to the desired variable, directly. Moreover, that desired symbolic variable in the generated tree can be replaced by 1 in the paths it is contained, while eliminating those paths that does not contain the symbolic variable, as introduced in [20].

For example, performing the sensitivity analysis with respect to h in (5.2), the resultant graph is shown in Fig. 5.6. It is clear that from Fig. 5.5, those paths including h survive, and 1 replaces the nodes containing h.

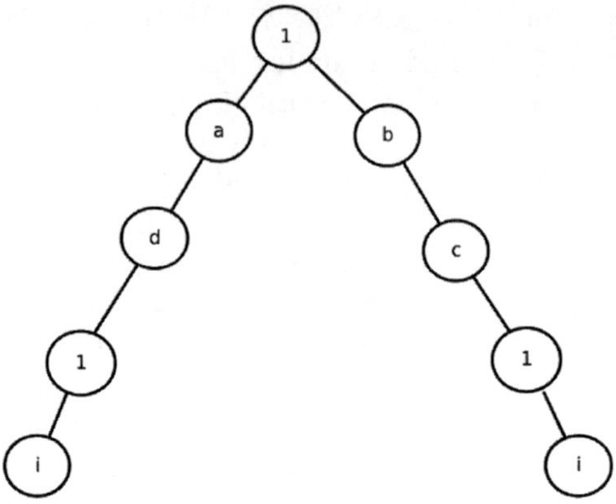

Fig. 5.6 Resulting graph by applying the derivative

5.5 Graph-Based Symbolic Technique

The graph-based symbolic technique is applied to compute the solution of a determinant that is associated with the nodal admittance matrix. The technique starts off with a SPICE netlist as input, from which the nodal admittance matrix is formulated as described in Sect. 5.2. The elements implemented are R, C, L, V, I, E, G and M, being resistor, capacitor, inductor, independent voltage source, independent current source, voltage-controlled voltage source, voltage-controlled current source and MOSFET. Both independent voltage and current sources can be DC, AC or both.

The netlist is used as a way to input the circuit topology along with the circuit elements and numerical values. If the numerical evaluation of the resulting symbolic expression as the transfer function is required, the small signal values for the parasitic elements and the operating point conditions are taken from the output listing in HSPICE. Those numerical values are used for evaluating the symbolic expressions to verify correctness, as well as a way to rank the sensitivity with respect to each symbolic-element. In general, the flow of the graph-based symbolic technique is shown in Fig. 5.7, where one can appreciate the module called *Derivative Module*. This module computes the derivative required for performing sensitivity analysis. Obviously, there is an algorithm to evaluate the required symbolic derivatives with respect to different variables. The details of the module *Derivative Module* in Fig. 5.7, are shown in Fig. 5.8, where DFS stands for depth first search, as described in Sect. 5.3.

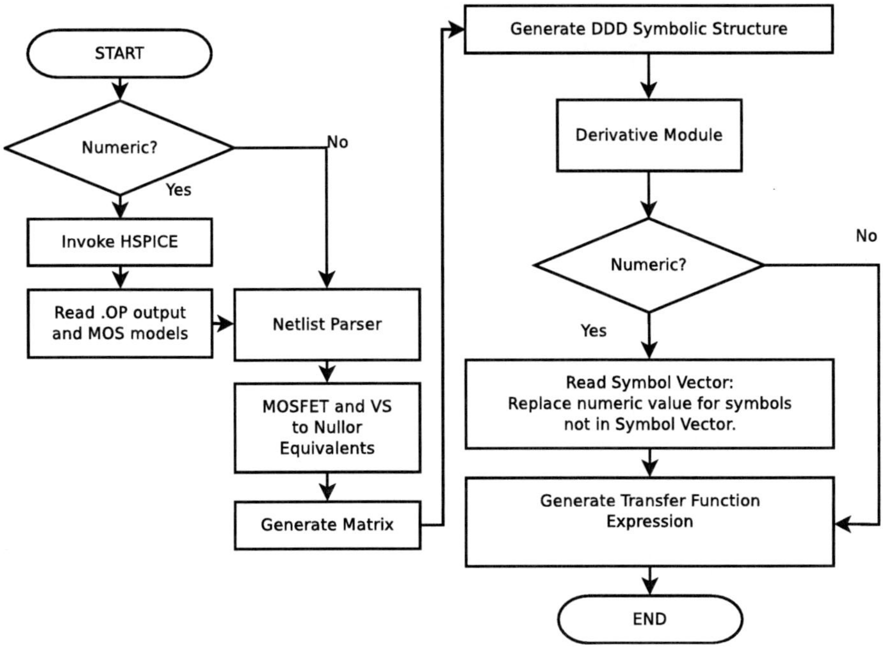

Fig. 5.7 Graph-based symbolic technique flow

5.5.1 Parsing the Netlist

The first step in formulating the nodal admittance matrix is devoted to parsing the netlist (from HSPICE), and building suitable data structures for each group of circuit elements. The symbolic name given to every composed circuit element like the MOSFET, is tracked to the one associated to its name. In this manner, to keep consistency, the symbol name is taken exactly as specified in the netlist (ex. R_name, C_name, M_name, etc) input, it is not case sensitive and is parsed as lowercase. For instance, the circuit elements are first grouped into one of four different tables: conductances, independent sources, controlled sources and Mosfet, as shown in Table 5.1.

Table 5.1 Elements Tables

Table Type	Fields
Conductances	Name, Node 1, Node 2, Value
Independent Sources	Name, Node 1, Node 2, DC, AC
Controlled Sources	Name, Node 1, Node 2, Node 3, Node 4, Gain
MOSFET	Name, Drain, Gate, Source, Bulk, Width, Length, ModelName

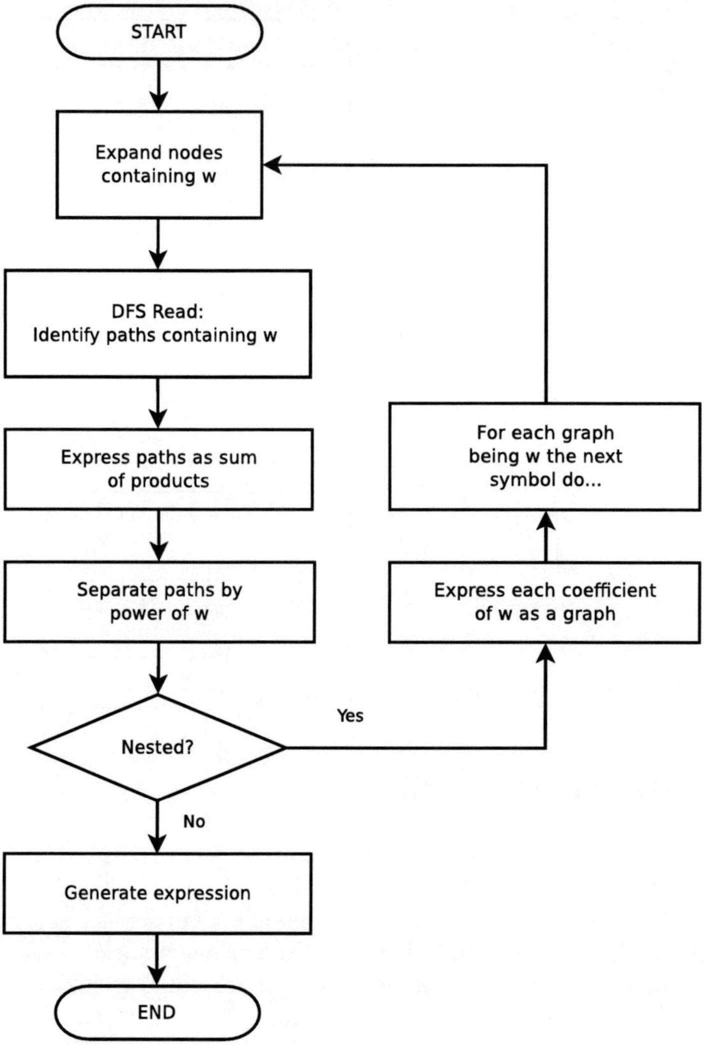

Fig. 5.8 Details of the module *Derivative Module* in Fig. 5.7

Values for the circuit elements and independent sources are considered strings of characters, so that the graph-based symbolic tool is agnostic of whether they are actually numbers or functions until the numerical evaluation of the output is performed.

The network relationships of the different elements are codified in their node connections. Because it is a lot easier to treat nodes as indexes, a mapping is created first by building an array with all the nodes as they appear in the netlist and any duplicate occurrences are removed. The tool is intended to perform AC analysis so it is of no use to keep independent sources with only

DC value, as they are only valid in a DC or transient context. When removing DC voltage sources their nodes are collapsed, while for DC current sources their nodes are left floating. Collapsing two nodes in a circuit is equivalent to assigning to both nodes the same mapping and so is done in this step. If the numerical evaluation of the resulting symbolic expression or transfer function is to be performed, the output listing (.lis file for $HSPICE^{TM}$) is read and the values from the operating point analysis are parsed and stored in an array with the same index correspondence as the element symbol array so that the mapping between the symbols and their numerical values is direct.

Up to this point the elements are separated in groups according to their type, nodes are collected in a single array and are mapped to indexes to facilitate their treatment, and DC sources are removed.

5.5.2 Small Signal Models and NullOr Equivalents

In a previous section it was explained why symbolic modeling is intended to give an insight into certain behaviors and tendencies of the circuit under design. With this in mind, it is now evident that the more complex the small signal model of a device the more accurate is the resulting simulation but at the cost of increasing machine operation time. A compromise can be reached where the qualitative behavior is roughly preserved at the expense of numerical accuracy.

In the introduced graph-based symbolic technique we have implemented three levels of parasitic elements for the small signal model for the MOSFET:

- Level 0 has no parasitic elements and models only the voltage-controlled current source (VCCS) with gate-source as the controlling branch voltage and transconductance g_m.
- Level 1 accounts for level 0 plus C_{gs}, C_{gd} and g_{ds}.
- Level 2 accounts for level 1 plus the voltage-controlled current source whose controlling branch voltage is bulk-source with transconductance g_{mb}.

From this first stage the elements are classified as *NA compatible* or *NA incompatible* ones, according to their small signal models. Incompatible elements are then treated as their respective Nullator/Norator (Nullor) equivalents as shown before in Fig. 5.4.

There are two reasons why elements can be incompatible. Firstly, if the dependent variable of the function for a given element is voltage, as is the case for voltage sources and secondly, if the independent variable for the element function is current. The reason behind this as the reader may already be aware is that in symbolic nodal analysis [24], the unknown vector is composed of voltages (node voltages) while the independent vector is formed just by current sources. Then, the need arises to apply a manipulation such that those conditions are preserved while maintaining a physical equivalence.

On the other hand, a fully differential OTA can be modeled as a single VCCS, which in turn can be viewed as a pair of MOSFET transistors in differential input configuration, as shown in Fig. 5.9.

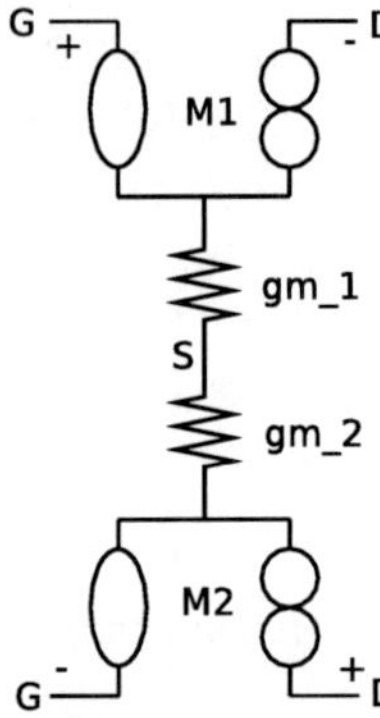

Fig. 5.9 Nullor Equivalent of the MOSFET Differential Pair

The ideal conditions for each of the two MOSFETs are kept: infinite input impedance (zero branch current in input Nullator), zero output impedance (arbitrary voltage) and ideal g_m. Parasitic elements can then be attached around this basic or generic nullor equivalent block.

As explained before, pairs of Nullator-Norator equivalents are arranged to account for the intrinsic VCCS of the small signal model of the MOSFET shown in Fig. 5.2, voltage sources and controlled sources shown in Fig. 5.4.

In most cases, the basic building blocks for the nullor equivalents are the Voltage Follower shown in Fig. 5.10a and the Current Follower shown in Fig. 5.10b [9].

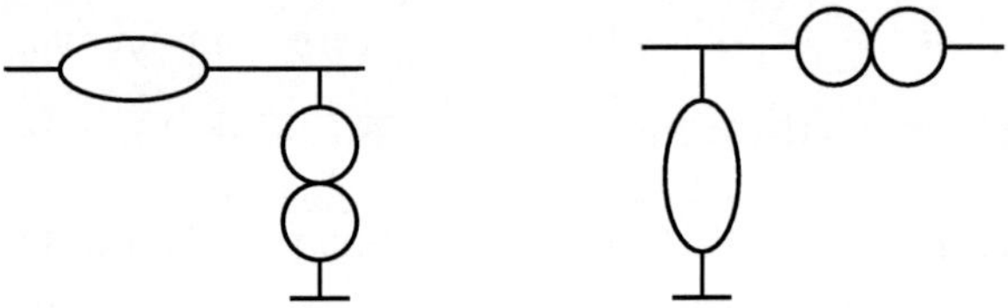

(a) Voltage Follower (b) Current Follower

Fig. 5.10 Nullor based (a) Voltage and (b) Current Followers

When replacing a certain active device with its NullOr equivalent there are three basic operations that have to be performed:

1. Add an element to the proper array (nullator, norator, conductance or independent source)
2. Add extra nodes if required (MOSFET, Voltage Source, etc.)

3. Assign a unique name to the element and the extra nodes (as required)

Adding a new element is as easy as appending the new entry to the corresponding structure (conductance, independent source, nullator or norator) and updating the number of elements for the given structure. As a list of the nodes originally read from the netlist is kept, it is easier to keep track of the nodes added afterwards if any new node is appended at the end of the structure containing the numerical mapping. When adding an element it is useful if the name is associated to the name of the original element it is bound to. For instance, if we are including C_{gs} to a MOSFET named M_1, the name of the new capacitor can be Cgs_1. In this way, it becomes easier to know to which netlist element a given symbol belongs.

Now that we have a mapping of unique node elements, we can proceed to replace elements for their corresponding Nullor equivalents. For the case of the MOSFET we assign the desired small signal model and corresponding parasitic elements.

5.5.3 Matrix Equations

Now that we know the node mappings we can create an adjacency matrix for each of the circuit elements. From this adjacency matrix the NA formulation is performed by following the method introduced in [24], which is reproduced here in a short form for convenience in Algorithm 1.

This procedure is shown for a simple common source circuit as the one shown in Fig. 5.11.

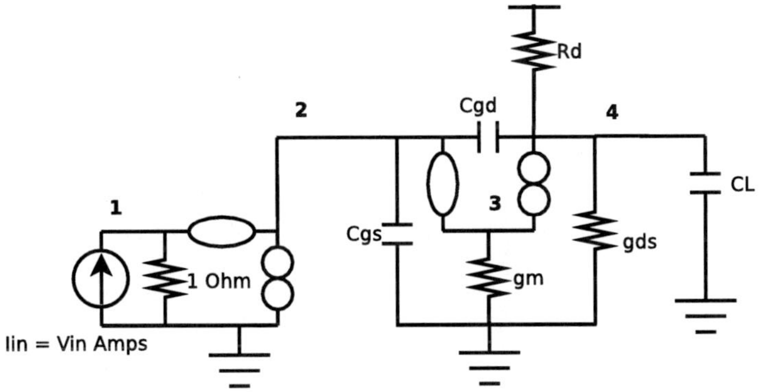

Fig. 5.11 Nullor equivalent of the common source amplifier

The formulated groups are shown in Table 5.2. The resulting matrix is a 2 x 2 system given by:

Algorithm 1. NA Formulation with NullOr elements

1. Two groups of nodes are formulated, one named ROW and another named COL. The nodal admittance matrix is formulated by performing the Cartesian product of these two groups.
2. Formulate ROW group of nodes:
   - Add a subset for each node with no Norator connected to it.
   - Add a subset for every chain of nodes connected by floating Norators.
3. Formulate COL group of nodes:
   - Add a subset for each node with no Nullator connected to it.
   - Add a subset for every chain of nodes connected by floating Nullators.
4. Group A: Admittances seen in a node.
   - A set for each node containing all admittances conected to it.
5. Group B: Floating Admittances.
   - A set for each floating admittance containing its pair of nodes.
6. Formulate Matrix
   - If the same node is present in a subset in ROWs and a subset in COLs then the corresponding subset of admittances (from group A) is summed at the admittance matrix position (ROW index, COL index).
   - If different nodes are present, one in a subset of ROWs and the other in a subset of COLs, the corresponding admittance (from group B) is summed with negative sign at the admittance matrix position (ROW index, COL index).

Table 5.2 NA Formulation

ROWS	COLS	FLOATING Y	GROUNDED Y
(1)	(1,2,3)	(2,4) : sC_{gd}	$1 : 1Ohm$
(3,4)	(4)		$2 : sC_{gs}, sC_{gd}$
			$3 : gm$
			$4 : sC_{gd}, g_{ds}, \frac{1}{R_d}, sC_L$

$$\begin{array}{c} \\ (1) \\ (3,4) \end{array} \begin{array}{c} \begin{array}{cc} (1,2,3) & (4) \end{array} \\ \begin{bmatrix} 1 & 0 \\ g_m - sC_{gd} & -(sC_{gd} + g_{ds} + \frac{1}{R_d} + sC_L) \end{bmatrix} \end{array} \begin{bmatrix} v_{(1,2,3)} \\ v_{(2,4)} \end{bmatrix} = \begin{bmatrix} I_{in} \\ 0 \end{bmatrix} \quad (5.7)$$

The solution for this system of equations becomes an application for the graph-based symbolic tool introduced in this chapter. The system of equations is solved by using Cramer's rule by computing $n+1$ determinants for $[v_1, v_2, \cdots, v_n]^T$ node voltages in order to completely define the circuit.

For the system of equations $\boldsymbol{A}x = b$ where $\boldsymbol{A}$ is an n x n matrix, the solution by Cramer's Rule is given as $x_i = \frac{|\boldsymbol{b}\rightarrow\boldsymbol{A}_i|}{|\boldsymbol{A}|}$.

5.5.4 Symbolic Determinant Formulation by Graph Operations

The symbolic manipulations performed are agnostic of the origin of the nodal admittance matrix. That is, the only conditions are that the matrix is square. Each node in the graph structure corresponds to a matrix non-zero entry, which in turn encapsulates the summation of one or more circuit elements. For this purpose, three different data structures are required. The first and most obvious is the node structure that contains the following fields:

- Node Name: A unique name for the given node. It is assigned as an index number of type int.
- Terms: An array containing the index and sign of the element (mapping it to the element table described in Sect. 5.5.1).
- Row and Column: The row and column of the non-zero entry that this node represents.
- Descendants: An array of node pointers linking to the descendants of the current node in the graph structure.

The graph is built starting with a trivial node named 0 with term value of 1. One shall remember that the multiplication is codified as the depth in the graph. The different nodes are linked accordingly. The algorithm to build the graph structure for the representation of $|A|$ is shown as Algorithm 2.

Algorithm 2. buildGraph($A(i, j)$, $Ancestor$)

```
  m ← number of Columns of A
  n ← number of Rows of A
  D ← set of descendants of Ancestor
Ensure: m > 0 ∧ n = m
  function BUILDGRAPH(A, Ancestor)
     if m > 1 then
        prune = 1
        for i = 1 to m do
           for all Columns j that A(i, j) ≠ 0 do
              if BUILDGRAPH(C_ij, Ai, j) ≠ 0 then
                 D = D ∪ Ai, j
                 prune = 0
     else
        if A(i, j) ≠ 0 then
           prune = 0
  return prune
```

With the graph already built, the expression for the determinant is then formulated as in Algorithm 3.

Algorithm 3. Symbolic Determinant from a Graph with node re-use.

---

1. Read the graph in a modified DFS fashion:
   a. Keep track of which columns have been visited
   b. Skip nodes from columns already visited
2. The symbolic expression is the product of a node times the summation of its visited descendants

---

After executing those algorithms, the resulting graph makes re-use of shared nodes from a tree graph. For example, the graph shown in Fig. 5.5 has several repeated nodes, and then they become shared as shown in the graph sketched by Fig. 5.12.

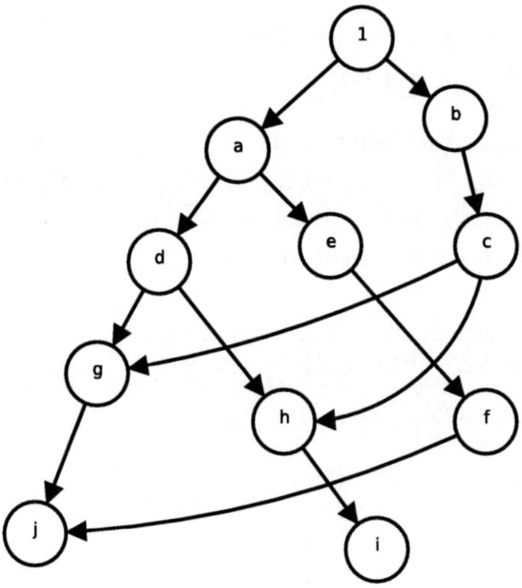

Fig. 5.12 Graph equivalent from Fig. 5.5 with node re-use

5.5.5 Symbolic Derivative in Sensitivity Analysis

In order to obtain the symbolic derivative with respect to a given symbol, we introduce Algorithm 4, which express the determinant as a polynomial stored as an array of sum of products with one entry for each power of W with non zero coefficient.

Algorithm 4. Graph to Polynomial with respect to a desired symbol, e.g. W

1. Expand the nodes of the graph containing symbol W.
2. Read the graph in DFS fashion and preserve only those routes from root to bottom with at least one occurrence of symbol W.
3. The number of occurrences of the symbol is the power of W for a given route.
4. Each route is then expressed as a sum of products replacing the symbol for a one.
5. The summation of all appended routes for each power of W forms the coefficients.

The derivative $\frac{\partial |\boldsymbol{A}|}{\partial W}$ is straightforward as symbol W in the expression $|\boldsymbol{A}|$ has already been factorized. The graph expansion can be performed in the original structure generated when obtaining the determinant thus reducing the memory footprint. Expansions for more than one symbol are possible.

5.5.6 Symbolic Multiparameter Derivative

It is possible to get a multiparameter derivative of the form $\frac{\partial^2 f}{\partial W_1 \partial W_2}$ for symbols W_1 and W_2. For this purpose we make use of the resulting symbolic expression obtained previously. Such expression is in the form of a sum of products, which is further separated in n coefficients (powers of W_1 with coefficient 0 are omitted) of W_1. Each coefficient is expressed as a graph, thus obtaining n graphs. The derivative procedure is repeated for all n graphs now for the symbol W_2. This is useful for some optimization problems where the Hessian matrix needs to be computed [1].

5.5.7 Experiments

For the experimental verification the symbolic transfer function for four CMOS and one BJT (small signal model with CCCS) circuits was computed and time metrics were taken for matrix equations formulation, denominator, numerator and transfer function. The test cases are the differential pair and common source shown in Fig. 5.13, the three stages uncompensated OTA shown in Fig. 5.3 [22], the recycling folded cascode OTA [3], and the 741 OPAMP [29], respectively, as shown in Table 5.3.

The test circuit for the UA741 is a Small-Signal model with R, C and VCCS elements. This same circuit was tested with SCAD3 [29], which does not support reading MOSFETs from a netlist. The running and evaluation time reported by SCAD3 is of 2.97s while our proposed graph-based tool takes 2.7043s, a little bit difference, but sensitivity analysis is straightforward.

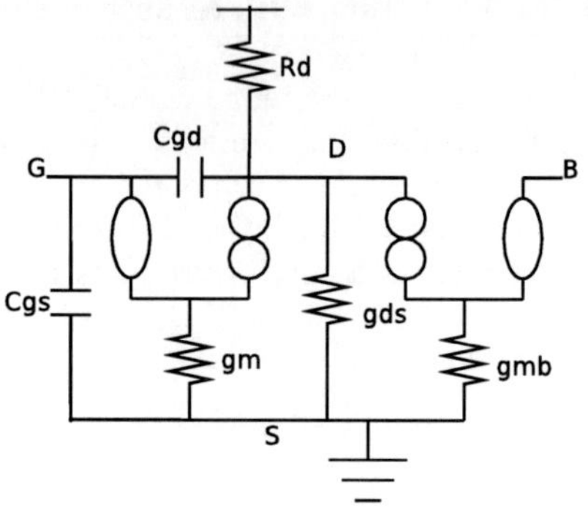

(a) Common Source.

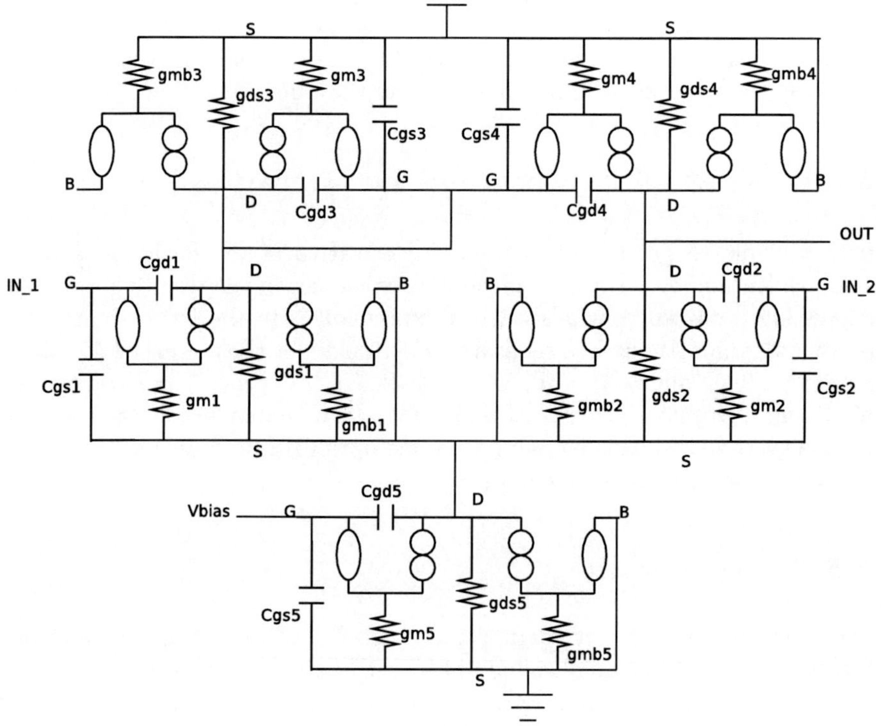

(b) Differential Pair.

Fig. 5.13 Nullor equivalents for (a) common source and (b) differential pair circuits

Numerical evaluation of the resulting symbolic expression is performed in order to provide a comparison among the well known and mature technology (HSpice^{TM}), a DDD-based symbolic tool [30], and our graph-based symbolic tool. The circuit under test is the Differential Pair. The AC analysis output is shown in Fig. 5.14.

Table 5.3 Symbolic and Numerical Evaluation of D(s), N(s) and H(s)

Circuit Features			Computer Time (seg)			
Circuit	Elements	Nodes	Equations	D(s)	N(s)	H(s)
Differential Pair	35	26	1.1235	0.122	0.1464	1.4895
RFC OTA [3]	106	56	1.6603	0.201	0.1869	2.2633
LV Amp	33	18	2.35	0.058	0.0464	2.4544
Common Source	8	6	0.8581	0.041	0.0205	0.9811
741	112	77	0.5123	1.37	0.822	2.7043

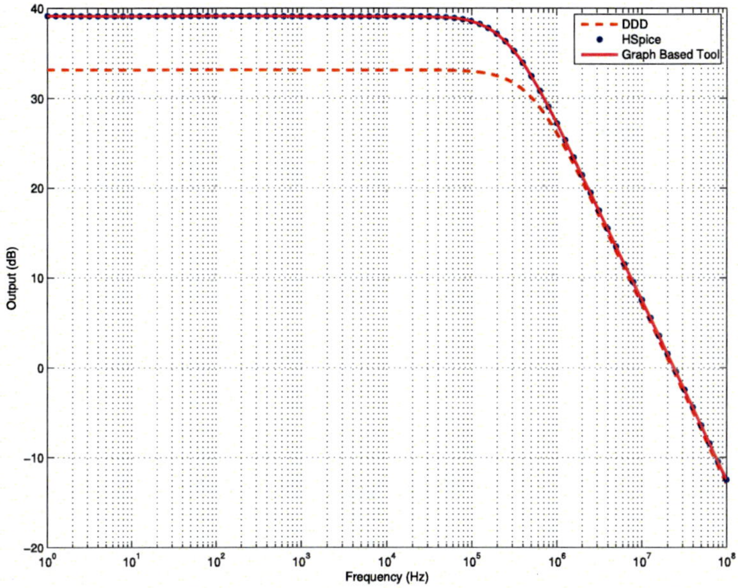

Fig. 5.14 H(s) comparison among DDD [30] (dashed), HSpice^{TM} (dots) and our Graph Based Symbolic Tool (solid) for the differential pair topology

Table 5.4 Sensitivity with respect to gm_1 and R_D

Symbol W	Symbolic Sens(H(s), W)	Value
gm_1	$\frac{gm_1\, r1}{gm_1\, r1 - cgd_1\, r1\, s}$	1.0001
R_D	$\frac{r1 \left(\frac{gm_1 - cgd_1\, s}{gds_1\, r1 + s\,(cgd_1\, r1 + cl\, r1) + 1} - \frac{(gm_1\, r1 - cgd_1\, r1\, s)\,(gds_1 + s\,(cgd_1 + cl))}{(gds_1\, r1 + s\,(cgd_1\, r1 + cl\, r1) + 1)^2} \right) (gds_1\, r1 + s\,(cgd_1\, r1 + cl\, r1) + 1)}{gm_1\, r1 - cgd_1\, r1\, s}$	0.804

Numerical evaluations of the resulting expressions for the RFC OTA [3] are shown in Fig. 5.15. We show the symbolic sensitivity expressions for a Common Source topology. The sensitivity analysis is performed with respect to gm_1 and R_D, and the numerical evaluation is performed as shown in Table 5.4.

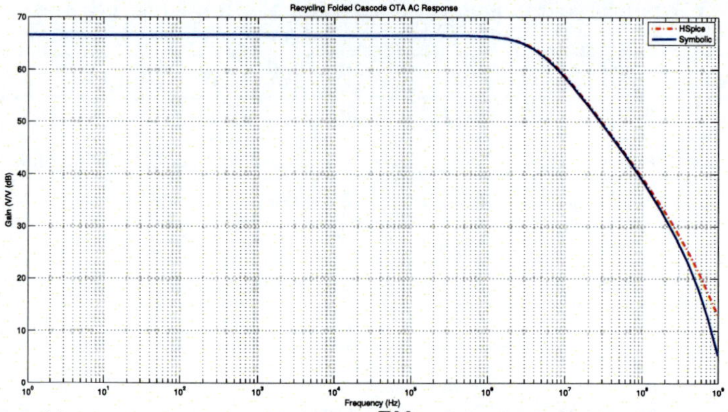

Fig. 5.15 H(s) comparison between HSpice^{TM} (dashed) and Graph Based Symbolic Tool (solid) for the RFC OTA [3]

5.6 Conclusion

A graph-based symbolic technique for the solution of a system of equations for analog ICs, formulated by applying symbolic NA, has been described. The proposed approach is based on the use of nullor equivalents to formulate the nodal admittance matrix, from which exact analytical expressions are derived. The technique has been extended to perform symbolic sensitivity analysis with respect to each symbolic-element and also with respect to multiparameters. Several examples were summarized in the last section and time computations were listed for the formulation and computation of the numerator, denominator and the transfer function of a circuit under test.

Finally, our proposed graph-based symbolic technique was compared with the already known DDD method and with HSPICE^{TM}, for the CMOS differential pair topology. From those results, it is appreciated the good agreement of the graph-based symbolic technique with respect to HSPICE^{TM}.

It is worthy mentioning that as for the DDD method, the graph-based symbolic technique is compact, unique and the complexity to obtain the symbolic expression depends on the size of the graph. This approach highlighted the simplicity for computing AC sensitivities of analog ICs with respect to one or many parameters, but just before computing the symbolic transfer function, as already shown in [20].

Acknowledgement. This work has been supported by UC-MEXUS and CONACyT under the projects CN-11-575 and 131839-Y.

References

1. Agnew, D.: Efficient use of the Hessian matrix for circuit optimization. IEEE Transactions on Circuits and Systems 25(8), 600–608 (1978), doi:10.1109/TCS.1978.1084520

2. Asenova, I., Balik, F.: Multiparameter symbolic sensitivity analysis of active circuits by using nullor model and modified coates flow graph. In: IEEE ELEKTRO, pp. 401–406. IEEE (2012)
3. Assaad, R., Silva-Martinez, J.: The Recycling Folded Cascode: A General Enhancement of the Folded Cascode Amplifier. IEEE Journal of Solid-State Circuits 44(9), 2535–2542 (2009)
4. Balik, F.: Effective optimization of integrated circuits with embedded passive modules by applying semisymbolic lcs ac analysis method. Microelectronics International 28(3), 36–42 (2011)
5. Barmada, S., Musolino, A., Rizzo, R., et al.: Multi-resolution based sensitivity analysis of complex non-linear circuits. IET Circuits Devices & Systems 6(3), 176–186 (2012)
6. Carlin, H.: Singular Network Elements. IEEE Transactions on Circuit Theory 11(1), 67–72 (1964), doi:10.1109/TCT.1964.1082264
7. Chen, J., Shi, G., Tai, A., Lee, F.: A Size Sensitivity Method for Interactive MOS Circuit Sizing. In: IEEE NEWCAS, Montreal, pp. 169–172 (2012)
8. De Camillis, L., Ferranti, F., Antonini, G., et al.: Parameterized Partial Element Equivalent Circuit Method for Sensitivity Analysis of Multiport Systems. IEEE Transactions on Components Packaging and Manufacturing Technology 2(2), 248–255 (2012)
9. Duarte-Villasenor, M., Tlelo-Cuautle, E., de la Fraga, L.G.: Binary genetic encoding for the synthesis of mixed-mode circuit topologies. Circuits, Systems, and Signal Processing 31(3), 849–863 (2012), doi:10.1007/s00034-011-9353-2
10. Fakhfakh, M., Tlelo-Cuautle, E., Fernández, F.: Design of Analog Circuits through Symbolic Analysis. Bentham Science Publishers Ltd. (2012)
11. Gielen, G.G.E., Walscharts, H.C.C., Sansen, W.M.C.: ISAAC: a symbolic simulator for analog integrated circuits. IEEE Journal of Solid-State Circuits 24(6), 1587–1597 (1989), doi:10.1109/4.44994
12. Guerra-Gómez, I., Tlelo-Cuautle, E., Duarte-Villaseñor, M., Sánchez-López, C.: Analysis, Design and Optimization of Active Devices. In: Integrated Circuits for Analog Signal Processing. Springer (2012)
13. Hasegawa, K., Aoki, M., Yamawaki, T., et al.: Modeling transistor variation using alpha-power formula and its application to sensitivity analysis on harmonic distortion in differential amplifier. Analog Integrated Circuits and Signal Processing 72(3), 605–613 (2012)
14. Huang, W.C., Wang, H.Y., Cheng, P.S., et al.: Nullor Equivalents of Active Devices for Symbolic Circuit Analysis. Circuits Systems and Signal Processing 31(3), 865–875 (2012)
15. Izydorczyk, J., Chojcan, J.: High-order sensitivity invariants. International Journal of Circuit Theory and Applications 39(12), 1285–1304 (2011)
16. Kao, C.C.: BDD-based synthesis for mixed CMOS/PTL logic. International Journal of Circuit Theory and Applications 39(9), 923–932 (2011)
17. Kolka, Z., Vlk, M., Horak, M.: Topology reduction for approximate symbolic analysis. Radioengineering 20(1), 252–257 (2011)
18. Li, X., Xu, H., Shi, G., Tai, A.: Hierarchical symbolic sensitivity computation with applications to large amplifier circuit design. In: 2011 IEEE International Symposium on Circuits and Systems (ISCAS), pp. 2733–2736 (2011)
19. Liu, X.X., Palma-Rodriguez, A.A., Rodriguez-Chavez, S., Tan, S.X.-D., Tlelo-Cuautle, E., Cai, Y.: Performance bound and yield analysis for analog circuits under process variations. In: 18th Asia and South Pacific Design Automation Conference (ASP-DAC), pp. 1–5. ACM IEEE (2013)

20. Palma-Rodriguez, A.A., Tlelo-Cuautle, E., Rodriguez-Chavez, S., Tan, S.X.: DDD-based symbolic sensitivity analysis of active filters. In: 2012 8th International Caribbean Conference on Devices, Circuits and Systems (ICCDCS), pp. 1–4 (2012), doi:10.1109/ICCDCS.2012.6188913
21. PierzchaLa, M., Fakhfakh, M.: Transformation of LC-filters to active RC-circuits via the two-graph method. Microelectronics Journal (2011)
22. Rodriguez-Chavez, S., Tlelo-Cuautle, E., Palma-Rodriguez, A.A., Tan, S.D.X.D.: Symbolic DDD-based tool for the computation of noise in CMOS analog circuits. In: 2012 8th International Caribbean Conference on Devices, Circuits and Systems (ICCDCS), pp. 1–4. IEEE (2012), doi:10.1109/ICCDCS.2012.6188912, http://ieeexplore.ieee.org/lpdocs/epic03/wrapper.htm?arnumber=6188912
23. Sánchez-López, C., Fernández, F.V., Tlelo-Cuautle, E.: Generalized admittance matrix models of OTRAs and COAs. Microelectronics Journal 41(8), 502–505 (2010)
24. Sanchez-Lopez, C., Fernandez, F.V., Tlelo-Cuautle, E., Tan, S.X., Sánchez-López, C., Fernández, F.V.: Pathological element-based active device models and their application to symbolic analysis. IEEE Transactions on Circuits and Systems I: Regular Papers 58(6), 1382–1395 (2011), doi:10.1109/TCSI.2010.2097696
25. Sánchez-López, C., Martínez-Romero, E., Tlelo-Cuautle, E.: Symbolic Analysis of OTRAs-Based Circuits. Journal of Applied Research and Technology 9(1), 69–80 (2011)
26. Shi, G.: Computational Complexity Analysis of Determinant Decision Diagram. IEEE Transactions on Circuits and Systems II-Express Briefs 57(10), 828–832 (2010)
27. Shi, G.: A survey on binary decision diagram approaches to symbolic analysis of analog integrated circuits. Analog Integrated Circuits and Signal Processing, 1–13 (2011)
28. Soliman, A.M.: Pathological representation of the two-output CCII and ICCII family and application. International Journal of Circuit Theory and Applications 39(6), 589–606 (2011)
29. Tan, X.D.: Symbolic analysis of large analog circuits with determinant decision diagrams. Ph.D. thesis, University of Iowa (1999)
30. Tan, S.X.D.: Symbolic Analysis by Determinant Decision Diagrams and Applications. In: Design of Analog Circuits through SYMBOLIC ANALYSIS. Bentham Science Publishers Ltd. (2012)
31. Tlelo-Cuautle, E.: Analog Circuits: Applications, Design and Performance. Electrical Engineering Developments. Nova Science Pub. Incorporated (2012), http://books.google.com.mx/books?id=H4S1uAAACAAJ
32. Tlelo-Cuautle, E., Sanchez-Lopez, C., Martinez-Romero, E., et al.: Symbolic analysis of analog circuits containing voltage mirrors and current mirrors. Analog Integrated Circuits and Signal Processing 65(1), 89–95 (2010)
33. Tlelo-Cuautle, E., Sánchez-López, C., Moro-Frías, D.: Symbolic analysis of (MO)(I) CCI (II)(III)-based analog circuits. International Journal of Circuit Theory and Applications 38(6), 649–659 (2010)
34. Vazquez-Leal, H., Filobello-Nino, U., Yildirim, A., Hernandez-Martinez, L., Castaneda-Sheissa, R., Sanchez-Orea, J., Molinar-Solis, J., Diaz-Sanchez, A.: Transient and DC approximate expressions for diode circuits. IEICE Electron. Express 9(6), 522–530 (2012)
35. Wierzba, G.M., Srivastava, A., Joshi, V., Noren, K.V., Svoboda, J.A.: Sspice-a symbolic SPICE program for linear active circuits. In: Proceedings of the 32nd Midwest Symposium on Circuits and Systems, vol. 2, pp. 1197–1201 (1989), doi:10.1109/MWSCAS.1989.102070

Chapter 6
A Designer-Assisted Analog Synthesis Flow

Farakh Javid, Stéphanie Youssef, Ramy Iskander, and Marie-Minerve Louërat

Abstract. This chapter presents a *designer-assisted analog synthesis flow* that is fully controlled by the designer and offers an intuitive design approach. The designer knowledge to conceive an analog IP is the key element of the synthesis flow, it is taken into account to automatically generate the analog IP design procedure and the physical view. Thus both consistency and accuracy of the final design are ensured. The presented flow bridges the gap between the two traditional approaches related to analog synthesis, namely the *simulation-based* and the *knowledge-based* approaches. It combines *accuracy* from simulation-based approaches with *speed of computation* from knowledge-based approaches. The proposed analog synthesis flow is composed of an accurate hierarchical sizing and biasing tool and a parameterizable layout generation tool. To demonstrate the effectiveness of the proposed flow, a fully differential transconductor was completely synthesized in 130nm CMOS technology to respect some performance specifications set by the designer. The obtained very low runtime is due to the introduction of design knowledge during both sizing and layout generation.

6.1 Introduction

During the past few decades, state of the art research in analog circuits synthesis led to the emergence of two major schools : the first school pushing towards *Full Design Automation (FDA)* and the second school pushing towards *Full Design Handcrafting (FDH)*. Several academic and commercial tools have been

Farakh Javid · Stéphanie Youssef · Ramy Iskander · Marie-Minerve Louërat
Université Pierre et Marie Curie, LIP6, 4 Place Jussieu, 75252 Paris, France
e-mail: {Farakh.Javid,Stephanie.Youssef,Ramy.Iskander}@lip6.fr
Marie-Minerve.Louerat@lip6.fr

M. Fakhfakh et al. (Eds.): *Analog/RF & Mixed-Signal Circuit Sys. Design*, LNEE 233, pp. 123–148.
DOI: 10.1007/978-3-642-36329-0_6

developed by the FDA school such as OASYS [9], IDAC [5], OPASYN [13], DELIGHT.SPICE [21], ASTRX/OBLX [22], AMGIE [24], MAELSTROM [14], ANACONDA [23], ASF [15], VASE [6], MOBU [26], WiCkEd [29], Neolinear acquired by Cadence and Analog Design Automation Inc. acquired by Synopsys. Except for OASYS, IDAC and AMGIE which are knowledge-based, the tools were mainly simulation-based for its generality. On the other hand, few academic tools have been developed by FDH school such as COMDIAC [25], PAD [28, 27] and [2, 1], trying to provide analog designers with sufficient insight for full trade-offs optimization.

Moreover, to further reduce the design time, automatic layout generation methodologies have been recently proposed [8]. In [30], an analog layout generator is used to generate individual or matched components layout, taking into account performance considerations. The layout parasitics are studied in [31] with a dedicated layout retargeting approach. LAYGEN [4] considers designer constraints to generate the layout using an evolutionary kernel.

Accumulating experience for several years, analog designers build sufficient expertise to analyze real complex circuits. Though for such complex circuits, FDA tools require important preparatory effort, have large execution times and the resulted design is not always optimal. Nowadays, analog designers seek to develop structured design methodologies that provide : *physics-based design*, *capacity to deal with complex circuits*, *bridging the gap between hand analysis and simulation*, *sufficient design insights*, *performance trade-offs exploration* and *analog design assistance*.

To address these principal needs, the nature of analog design knowledge and the challenges facing today's analog design automation will be further studied in subsequent sections.

6.2 Problem Definition

6.2.1 Analog Design Knowledge Classification

Basically knowledge related to analog design may be classified into two types : *tacit knowledge* and *formal knowledge*. On the one hand, tacit knowledge refers to the designer know-how, experience, intuition, unarticulated models and implicit rules of thumb about circuit design. It exists as the intellectual property of the designer. It is gained over a large period of time with learning and experience. It is difficult to express and may be transferred by the willingness of the designers to share their experience.

On the other hand, formal knowledge is embedded in design documents, repositories, design function and structured description, problem solving routines, computer

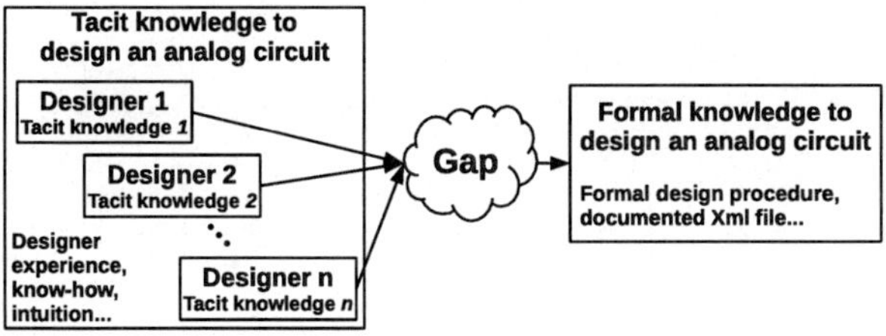

Fig. 6.1 The gap between tacit and formal knowledges in analog design

algorithms, expert knowledge systems, etc. These create the intellectual platform necessary to build and manufacture a design.

It is known that each designer has his own intuition (*i.e.* tacit knowledge) on how to design an analog IP, as illustrated in Fig. 6.1. Thus an interesting issue can be stated as follows : how to bring the designer tacit knowledge to the formal knowledge ? In other words : how to fill the gap shown in Fig. 6.1 ? A possible solution is proposed in the following subsection.

6.2.2 Improved Knowledge-Aware Analog Design Flow

To further analyze the challenges in analog design automation, the tasks involved in both knowledge-based and simulation-based synthesis are presented in Fig. 6.2.

Simulation-based synthesis encapsulates a simulator within an optimization loop. Since the simulator is a verification tool, it starts with a set of sizes and biases (vector 2). Then, it computes small-signal parameters (vector 3) by evaluating transistor models such as BSIM3V3, BSIM4, PSP and EKV. After that, linear and nonlinear performances (vector 4) are evaluated using a set of testbenches. It is to be noticed that model evaluation and performance evaluation are performed by the simulator. The search engine uses sizes and biases (vector 2) as optimization variables. It leaves both model evaluation and performance evaluation to the simulator. This approach is illustrated by the innermost loop of Fig. 6.2. In operating-point-driven formulation [24], the optimizer tunes the circuit design parameters (vector 1) as in the solid-line loop of Fig. 6.2. In this case, complex sizing and biasing procedures are manually written by the designer to ease the mapping from vector 1 to vector 2 [9], [24], [22].

Alternatively, knowledge-based synthesis is very laborious. Using the circuit design parameters (vector 1), the designer codes complex sizing and biasing procedures to compute sizes and biases (vector 2). From vector 2, the designer uses simplified transistor models for model evaluation. Finally, the designer extracts approximate performance equations for performance evaluation.

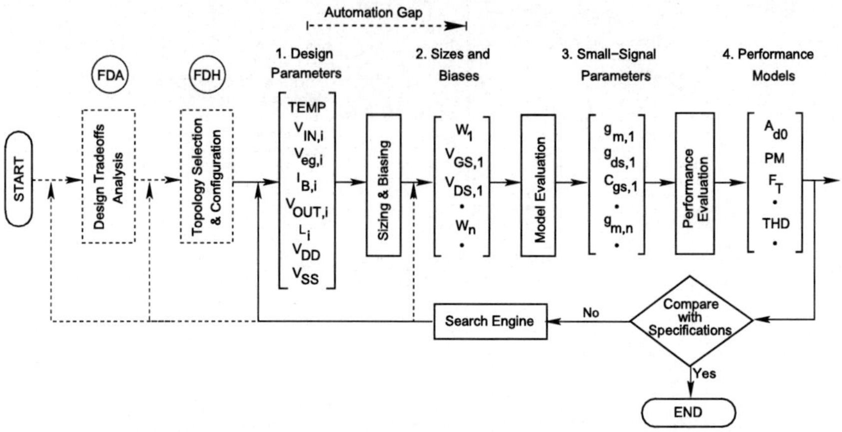

Fig. 6.2 Tasks involved in today's analog design automation efforts

Since the designer tries to transform his own design intuition into a formal sizing procedure, he tends to mix performance equations with sizing and biasing computation. This complicates the synthesis task and affects the clarity and readability of the represented formal knowledge.

Generally, the designer wishes to use more meaningful design parameters (vector 1) to design analog circuits. The mapping to sizes and biases (vector 2) becomes a laborious task that has to be repeated for each newly introduced circuit topology. This step depends mainly on the designer expertise and the complexity of circuit topology. Today this step is not yet formalized, therefore an *automation gap* is identified in the analog design flow, as illustrated in Fig. 6.2. Thus filling this gap helps the designer to express tacit knowledge into a formal knowledge that is structured as suggested by the four vectors in Fig. 6.2. This leads to a more robust knowledge-based analog design flow, that minimizes design errors during the design phase. The use of a formal representation favors the increase of analog design reuse, hence the reduction in design time.

The automation gap is filled by generating design procedures using the *hierarchical sizing and biasing methodology*, which is presented in the next section.

Furthermore, the efforts done by the FDH school in *topology selection and configuration* may be placed on top of the proposed work (FDH in Fig. 6.1). In this case, topology configuration means the selection of appropriate topology parameters among those of vector 1. Tools delivered by FDA school for *design trade-off analysis* would be placed on the very top (FDA in Fig. 6.1). It would be extremely beneficial to explore design trade-offs, while respecting *the designer's choice* of topology and design parameters.

In the proposed work, the concept of *knowledge-aware synthesis* is developed. In this concept, a search engine perform an optimization loop around vector 1 that holds intuitive design parameters. These are mapped to sizes and biases inside the loop using automatically generated sizing and biasing procedure. This procedure computes only feasible dimensions, thus, allowing considerably reduced design space to be examined for the same circuit. The design space is also reduced due to the nature of design parameters in vector 1 that varies in very limited ranges or are set a priori by the designer. Since designer knowledge is involved inside the optimization loop as a sizing and biasing procedure, the loop is qualified as knowledge-aware synthesis. In this optimization loop, it is assumed that performance equations are provided by the designer in order to measure circuit performances in vector 4.

6.2.3 Analog IP Definition

An analog IP is composed of a netlist, a set of design parameters and sizing constraints, and physical constraints for layout generation. These items are specified by the designer. To document an analog IP at the design steps described in sections 6.3 and 6.4, the designer-assisted analog synthesis flow proposes to store the designer knowledge and computed results (the design procedure, the sizes and biases, the layout parameters) in a XML file. Thus, the communication between the different design steps is highly simplified, as for an example the sizing tool can easily perform circuit sizing taking into account layout parameters, without any extractor.

As shown in Fig. 6.3, the sizing tool documents the XML file with the input parameters P_{in} (*<parameters>* section), the list of sizing operators deduced from P_{in} (*<sizing>* section) and computed sizes and biases (*<instance>* section). Among these data, transistor length and width are then used by the layout generation tool to generate the circuit layout. The layout generation tool takes as input the parameters in *<instance>* section plus some additional layout parameters such as the number of dummies, the bulk type, or the current direction. The layout generation tool has a dedicated section in the XML file, to store the designer choices of the layout style, and the floor-planing of the circuit with the matching and symmetrical constraint. Thus, all the information required to design an analog IP can be easily reused, modified and shared among a group of designers, during the whole design process. The next two sections present the hierarchical sizing and biasing methodology and the layout generation methodology.

6.3 The Hierarchical Sizing and Biasing Methodology

In this section, the sizing and biasing tool within the designer-assisted analog design flow is presented. It is based on the *hierarchical sizing and biasing methodology* [10] that exploits the designer tacit knowledge to generate suitable design procedures for analog IPs. The hierarchical sizing and biasing methodology uses a hierarchical

```
<circuit name="GMD">
    <parameters> <!- - P_in Parameters - ->
        <parameters name="GMD.Vdd" value="1.2"/>
        <parameters name="GMD.L" value="1.0e-6"/>
        ...
    </parameters>
    <sizing> <!- - Sizing Operators - ->
        <instance name="m9ap_an" operator="OPW(VG,VS)"/>
        ...
    </sizing>
    <instances> <!- - Devices in GMD - ->
        <instance name="m1ap_an" model="DifferentialPair" mostype="PMOS"
        sourceBulkConnected="True">
            <parameters>
                <parameter name="id" value="40.0e-6"/><!- - parameter obtained from P_in - ->
                <parameter name="w" value="3.4e-6"/><!- - computed parameter - ->
                <parameter name="l" value="3.0e-6"/><!- - parameter obtained from P_in - ->
                <parameter name="nf" value="4"/><!- - input parameter for the layout - ->
                <parameter name="dummies" value="1"/><!- - number of dummies at the end of
                each stack on the layout - ->
                <parameter name="sourcefirst" value="true"/><!- - current direction in the layout - ->
                ...
            </parameters >
        </instance>
    </instances>
    <layout> <!- - Layout Section - ->
    <!- - matching constraint of each devices - ->
        <instance name="m1ap_an" style="Mirror"/ >
        <instance name="m2p_n" style="Mirror"/ >
        <instance name="m9ap_an" style="Mirror"/ >
        <hbtree > <!- - HB*- Tree description - ->
            <group name="gb" align="vertical"><!- - group of symmetrical constraint - ->
                <block name="m1ap_an">
                <block name="m2p_n" position="top">
                <block name="m9ap_an" position="top"/>
                </block>
                </block>
            </group>
        </hbtree>
    </layout>
</circuit>
```

Fig. 6.3 The XML file for GMD subcircuit (Fig.6.17) sizing and layout generation

representation of the circuit (section 6.3.1), a library of *sizing and biasing operators* (section 6.3.2) and a *bipartite graph* (section 6.3.3).

6.3.1 Circuit Hierarchy

An electrical circuit is built as a hierarchy of interconnected subcircuits. A leaf subcircuit in the hierarchy is called a *device* : it is a set of transistors that realize an electrical function (like a differential pair, a current mirror, etc...). A *reference transistor* is defined in each device, it is the only transistor to be sized by an operator.

The reference transistor computed sizes and biases are then propagated, through designer constraints, to the other transistors in the device.

6.3.2 *Sizing and Biasing Operators*

Sizing and biasing operators aim at computing the sizes and biases of reference transistors, they are based on the inversion of the transistor compact model. Each operator has a set of input parameters that are set by the designer and computes unknown widths and biases (see Table 6.1, where $V_{EG} = V_{GS} - V_{TH}$). An operator computes either :

$$W = f_W(Temp, I_D, L, V_{GS}, V_{DS}, V_{BS}) \quad (6.1)$$

or :

$$V_{GS} = f_{V_{GS}}(Temp, I_D, W, L, V_{DS}, V_{BS}) \quad (6.2)$$

f_W and $f_{V_{GS}}$ are two inverse functions of the transistor compact model given in equation (6.3) :

$$I_D = f_{MODEL}(Temp, W, L, V_{GS}, V_{DS}, V_{BS}) \quad (6.3)$$

where *MODEL* is a standard transistor model like BSIM3v3 [19], BSIM4 [19], PSP [20] and EKV [7]. f_W and $f_{V_{GS}}$ are monotonic functions, thus equations (6.1) and (6.2) are solved with the Newton-Raphson method. Convergence criteria for the Newton-Raphson method are the same as in commercial simulators.

Table 6.1 Class definition of sizing & biasing operators

Operator	*Definition*
$OPVS(V_{EG},V_B)(Temp,I_D,L,V_{EG},V_D,V_G,V_B) \mapsto (V_S,W,V_{TH})$	
...	...
$OPVG(V_{EG})$	$(Temp,I_D,L,V_{EG},V_D,V_S) \mapsto (V_G,W,V_{TH},V_B)$
...	...
$OPVGD(V_{EG})$	$(Temp,I_D,L,V_{EG},V_S) \mapsto (V_G,V_D,W,V_{TH},V_B)$
...	...
$OPW(V_G,V_S)$	$(Temp,I_D,L,V_D,V_G,V_S) \mapsto (W,V_{TH},V_B)$
...	...
$OPID(V_G,V_S)$	$(Temp,W,L,V_D,V_G,V_S) \mapsto (I_D,V_{TH},V_B)$
...	...

As shown in Fig.6.4, an electrical simulator is encapsulated within the sizing and biasing operators [11]. Thus the sizing operators directly interface with industrial design kits in order to ensure the accuracy of the computed results. At the bottom in Fig.6.4, there is an electrical netlist that specifies the suitable design kit and contains only two transistors : one PMOS and one NMOS. Both transistors refer to a transistor compact model and are entirely sizable and biasable through simulator interactive commands. This two-transistor netlist is loaded by the electrical simulator

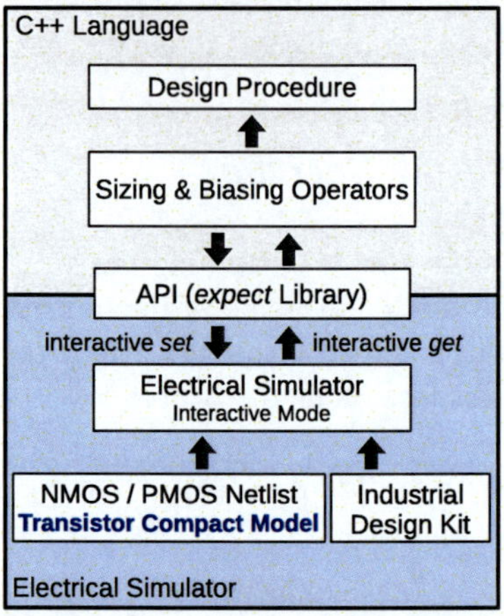

Fig. 6.4 Simulator encapsulation within sizing and biasing operators

launched in interactive mode, to perform sizing and biasing. Three types of interactive commands are evaluated : *set*, *get* and *run*. The *set* command allows to set transistor known parameters (sizes and biases) at the simulator level. The *get* command enables to retrieve currents, voltages and small signal parameters computed by the simulator. After a *set* command, a simulation must be run using the *run* command, in order to compute the DC operating point of the transistor. An application programming interface (API) is developed using *expect* library [17] to automate the *set*, *get* and *run* commands execution. The API is used within the sizing and biasing operators, that are optimized to minimize the number of calls to the simulator, which can reach several hundreds during sizing.

Operator $OPVG(V_{EG})$ implementation is illustrated in Fig. 6.5. First, transistor voltages and width are initialized in lines 5 to 13. *simulator* is the C++ object that encapsulates an electrical simulator. In line 8, the threshold voltage V_{TH} is retrieved from the simulator using *getVth()* function, that invokes the corresponding interactive command for the simulator. Lines 14 to 31 are for the width computation loop, using the Newton-Raphson method. On line 22, *runSimulation()* function performs a single transistor simulation to compute the operating point corresponding to the width computed in line 20. Convergence criteria for the sizing operator are based on $\varepsilon_{reltol} = 10^{-3}$, $\varepsilon_{abstol,v} = 10^{-6}$ and $\varepsilon_{abstol,w} = 10^{-8}$ (lines 26 to 28). Computed width and voltages are returned in line 32.

6.3.3 Bipartite Graph

The hierarchical sizing and biasing methodology handles design parameters dependencies in the form of a bipartite graph [12]. The bipartite graph is the design procedure for an analog IP. It is formally defined in the following subsections.

```
Operator  OPVG($V_{EG}$)
Inputs    $Temp, I_D, L, V_{EG}, V_D, V_S$
Outputs   $V_G, V_B, V_{TH}, W$
Implements
    $V_B = V_S$
    $V_{DS} = V_D - V_S$
    $V_{BS} = V_B - V_S = 0.0$
    $V_{TH}$ = simulator.getVth()
    $V_G \doteq 0.0$
    $W = 10 \cdot W_{min}$
    simulator.setVds($V_{DS}$)
    simulator.setVbs($V_{BS}$)
    iteration_count = 0
    Do
        $W_{prev} = W$
        $V_{G,prev} = V_G$
        $V_G = V_S + V_{EG} + V_{TH}$
        $V_{GS} = V_G - V_S$
        simulator.setVgs($V_{GS}$)
        Solve for $W$ using the Newton-Raphson method
        simulator.setW($W$)
        simulator.runSimulation()
        $V_{TH}$ = simulator.getVth()
        Increment iteration_count
    While (
    ( $| V_G - V_{G,prev} | \geq \varepsilon_{reltol} \cdot max(| V_G |, | V_{G,prev} |) + \varepsilon_{abstol,v}$
        or
    $| W - W_{prev} | \geq \varepsilon_{reltol} \cdot max(| W |, | W_{prev} |) + \varepsilon_{abstol,w}$ )
    and
    iteration_count $\leq$ MAX_ITERATIONS
    )
    Return $[V_G, V_B, V_{TH}, W]$
```

Fig. 6.5 Implementation of operator OPVG(V_{EG}) with simulator calls

6.3.3.1 Bipartite Graph Definition

Definition 1. A *graph* $G(V,E)$ consists of a *vertex set* V (also called *node set*), an *edge set* E, and a *relation* that associates with each edge two vertices.

Definition 2. A *bipartite graph* $G(V_1,V_2,E)$ is a graph whose vertices are divided into two disjoint sets V_1 and V_2 ($V_1 \cap V_2 = \emptyset$) such that each edge $e \in E$ connects a vertex $x \in V_1$ to a vertex $y \in V_2$.

Definition 3. A *directed acyclic graph* (DAG) is a directed graph with no directed *cycles* (closed chain of vertices). In a DAG, all edges are oriented and are called *arcs*.

Definition 4. A bipartite DAG is a bipartite directed graph with no directed cycles.
Definition 5. A bipartite DAG $G = (V_p, V_c, A)$ is defined with the two disjoint sets V_p and V_c. V_p is the set of *parameter nodes*, V_c is the set of *computation nodes*, A is the set of arcs. The bipartite DAG G is the design procedure for an analog IP.

A parameter node is a geometrical or electrical parameter related to a reference transistor, a device or a circuit. The parameter nodes set V_p is split into the three following subsets :

$$V_p = P_{in} \cup P_{out} \cup P_{prop} \tag{6.4}$$

with $P_{in} \cap P_{out} = \emptyset$, $P_{out} \cap P_{prop} = \emptyset$ and $P_{prop} \cap P_{in} = \emptyset$. P_{in} is the set of input parameters for the design procedure. P_{out} is the set of output parameters computed by the design procedure. P_{prop} gathers intermediate parameters propagated between successive computation nodes.

A computation node, that computes one or several parameters, is either a *sizing and biasing operator* (see section 6.3.3.2), a *linear constraint* (see section 6.3.3.3) or a *designer-defined procedure (DDP)*, that allows to express the knowledge that cannot be automatically extracted (see section 6.3.3.4).

The parameters in P_{in}, the linear constraints and the DDPs constitute the designer tacit knowledge. They are set up according to the designer understanding of the circuit behavior, and can be modified to reach the suitable circuit performances.

6.3.3.2 Reference Transistor Sizing and Biasing

To size and bias a reference transistor, a bipartite DAG is associated with it, using a unique sizing and biasing operator. The bipartite graph for the sizing and biasing of the diode-connected transistor using operator *OPVGD(VEG)* (c.f. Table 6.1) is shown in Fig. 6.6(b).

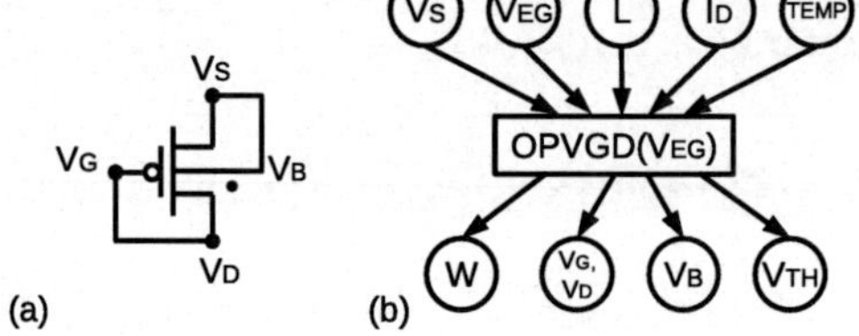

Fig. 6.6 (a) PMOS reference transistor (marked with a dot), (b) its associated bipartite graph

6.3.3.3 Device Sizing and Biasing

From the bipartite DAG of the reference transistor shown in Fig. 6.6, more complex bipartite graphs are built for devices. Fig. 6.7(a) illustrates a current mirror device. The designer can set a current ratio of $1:5$ from transistor M_1 to transistor M_2. Therefore the linear constraint $W_2 = 5 * W_1$ must be respected. The bipartite graph for the current mirror is shown in Fig. 6.7(b). Operator *OPVGD(VEG)* computes the

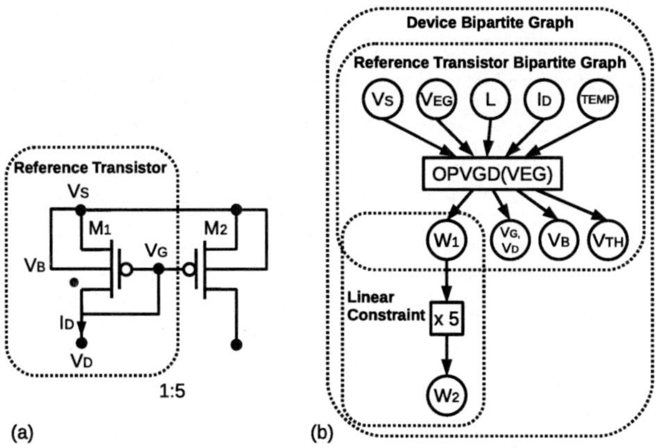

Fig. 6.7 (a) PMOS current mirror device, (b) its associated bipartite graph

width W_1 for the reference transistor as in Fig. 6.6(b), then the designer constraint is added in the graph to compute W_2.

6.3.3.4 Circuit Sizing and Biasing

A bipartite DAG is associated to a circuit. Fig. 6.8 shows the bipartite graph for the GMD circuit within the fully differential transconductor (Fig. 6.17). The designer chooses the P_{in} set of input parameters according to his intent, and propagates their values to the reference transistor of each device using constraints. Next he may declare designer-defined procedures (DDP) to express the current ratios between different transistors. A DDP named *eq1* in Fig. 6.8 is declared as : $I_{D_{M9AP}}(K_1, I_{BIAS}) = K_1 * I_{BIAS}$, with $\{K_1, I_{BIAS}\} \in P_{in}$, where K_1 is a ratio and I_{BIAS} is the biasing current for the circuit. P_{out} mainly contains the widths computed by the sizing operators.

Starting from these data (P_{in} parameters and the designer constraints), that are the designer tacit knowledge, the synthesis routine shown in Fig. 6.9 uses the circuit netlist to automatically generates a formal representation (the bipartite graph) to size and bias an analog IP. Steps (1)-(4) perform a depth-first traversal on the hierarchical instantiation tree of the electrical circuit. The root circuit is described as a parameterized generator. For each child generator, the root generator propagates the known design parameters and asks the synthesis routine to synthesize the child generator. If the child generator is a device, the dependencies of its reference transistor are generated in step (6). Otherwise, if the child generator is a subcircuit, the dependency graphs of its children generators are merged in step (8). Thus, this synthesis routine is a mean to bring the designer tacit knowledge to a formal one.

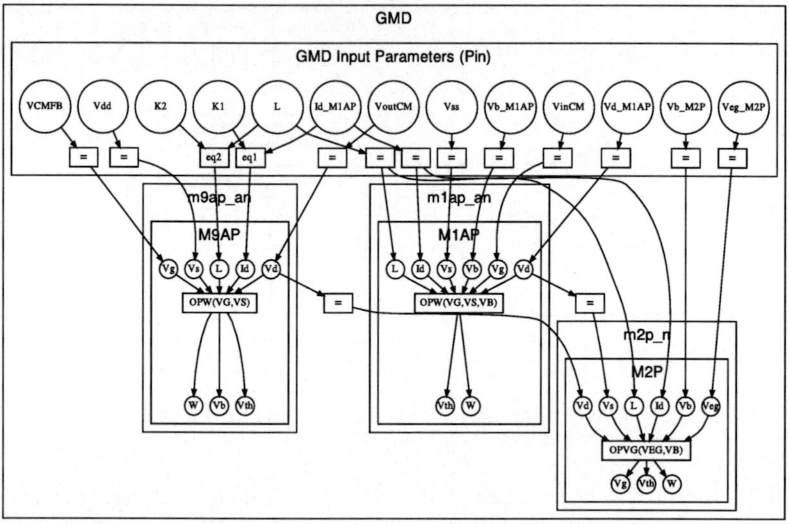

Fig. 6.8 Bipartite graph (*i.e.* design procedure) associated to the GMD subcircuit in the fully differential transconductor (see Fig. 6.17)

```
Procedure synthesize ( generator )
    For Every child Of generator
        Call synthesize(child)
    End For
    If generator is a device
        generate dependencies for the reference transistor
    Else If generator is a subcircuit
        merge dependencies of all children generators
    End If
End Procedure
```

Fig. 6.9 Pseudo-code of the synthesis routine

6.4 The Layout Generation Methodology

This section shows how a *layout description language* helps to generate a parameterized layout for analog basic blocks, and how the layout generation tool provides layout dependent parameters (LDP) that are sent to electrical view to take into consideration their effects during the sizing phase.

6.4.1 *Layout Generation Tool Characterization*

Our layout description language of the smart analog basic blocks is based on Python language. This choice was motivated by the fact that Python is an easy to learn, easy for debugging, object-oriented, portable and interpreted language. This allows the designer to write concise and simple code to describe complex layouts using generic design rules parameters. Actually, each technology process uses a dedicated file to associate a value to each generic layout design rules parameters.

The principle of having a library of smart analog basic blocks is to have the following characteristics :

- A layout description language based on python code that eases and simplifies the layout description code of the analog basic building blocks.
- Our layout generation tool supports different technologies with the new nanometric layout dependent parameters.
- Our layout generation tool supports the layout dependent parasitic parameters (LDPP) of different models BSIM3, BSIM4, etc ...
- Having for each analog basic building, different layout styles, their layout dependent parameters can be compared directly, without the need for an extractor, once the generation of the layout for the same technology or even for different technologies.
- A parameterized and reusable analog basic building block, since the layout description python code is not hard coded thanks to the generic design rules parameters.
- An easy exportation of the layout to a commercial tool.
- A tight link between physical and electrical views.

6.4.2 *Smart Analog Basic Blocks*

This section mainly focuses on the layout generation of the smart analog basic blocks "*devices*" features, and shows the tight link between the electrical and the physical views thanks to the device features.

6.4.2.1 Device Features

A device has an atomic analog function realized by a small set of transistors. The motivation to build a device is the following : the analog electrical behavior of the set of transistors requires a dedicated layout with strong geometrical and robustness constraints. Therefore the layout of the transistor's set has to be designed as a whole. A typical library of analog devices contains : a folded transistor, a differential pair, a current mirror, a cross coupled pair, a level shifter , etc ... Each device may have different layout styles. Each device may contains one or more stacks depending on the chosen style. The goal of the device is to provide an electrical realization along with a physical realization (layout) of an atomic behavior annotated with all the layout dependent parameters. The device is a smart object, since it has two main features : on the one hand a set of methods to study the electrical behavior and on the other hand the set of methods to generate a layout.

6.4.2.2 Stack Object

Folding technique is commonly used in analog circuits. Since this structure is essential, a 'Stack' object has been defined in our layout generation tool. To create the layout of a complete stack, the designer of parameterized analog devices (folded transistors, differential pair and current mirror, etc ...) simply calls `createStack()` method with well specified input parameters.

The input parameters of the `createStack()` method are :

- **Type:** the type of the transistor NMOS or PMOS.
- **W:** the finger width of the stack.
- **L:** the length of each finger.
- **NFs:** the number of stack's fingers (including dummies).
- **NDummies:** the number of dummies at each stack ends.

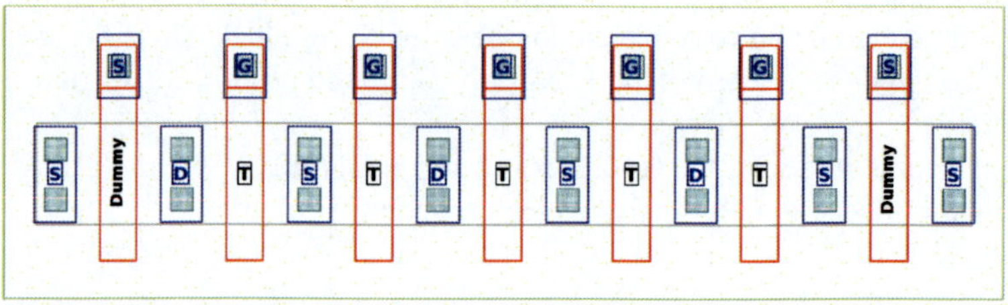

Fig. 6.10 Layout stack example W=2.0μm, L=0.2μm, NFs=7, Type=NMOS and NDummies=1

Fig. 6.10 presents an example of a generated stack layout. The routing is not shown for clarity. The labels "T" on the fingers represent the transistor to which the fingers belong. Once a stack object has been created, it can be queried for useful layout distances as shown in Fig. 6.11.

The distances provided by the stack are :

- **DMCI:** distance from the middle diffusion contact till the isolation edge.
- **DMCG:** distance from the middle diffusion contact till the gate edge.
- **DGG:** distance between two successive gates.
  This is equal to $2 \times DMCG$.
- **DGI:** distance from the edge of the end gate to the isolation edge. This is equal to $DMCI + DMCG$.

Each distance has a method to query it in the stack object. They are used to compute the layout dependent parasitic parameters.

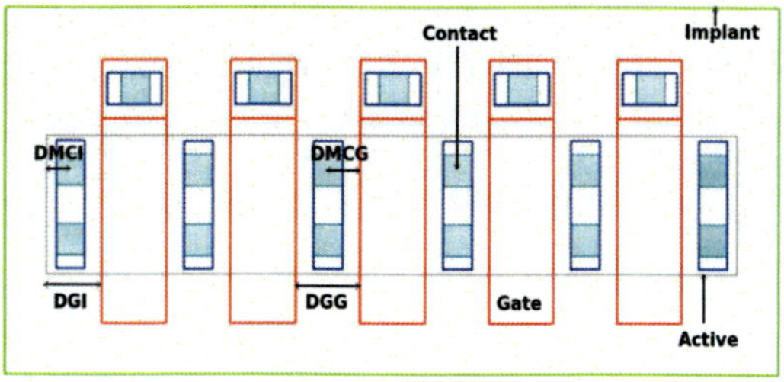

Fig. 6.11 Useful distances provided by the Stack object

6.4.2.3 Device Routing Methodology

As it was mentioned, a layout description language is available. This section presents two parametric generic routing methodologies : *inter-stack* routing methodology and *intra-stack* routing methodology. These methodologies provide the designer a procedural way to perform a symmetrical routing and therefore a symmetrical layout. The main advantage is to handle a large set of device topology based on a generic Python method driven with few input parameters. A device can have one stack as in the interdigitated and mirror styles. To route the nets of one stack the designer has to call *intra-stack* routing methodology by precising : the name of the net to be routed, how it will be routed (line, u-shape or serpentine) and the metal layers of the routed segments. A device also may have many stacks as in the 2D common centroid and the M2 module. To route the nets between two stacks the designer has to call *inter-Stack* routing methodology by precising : the two segments to be routed, the coordinates of the routed segment(s) and the metal layers of the routed segments.

6.4.2.4 Layout Dependent Parameter Computation Methods

A dedicated Python API has been developed to describe the device layout. In addition to the method describing the layout, three special methods have been developed to compute the layout dependent parameters of the MOS transistor model. The first one computes the area and perimeter of the source and drain zones, the second one computes the stress effect parameters introduced in BSIM4 to model nanometric DSM effects and the third one computes the capacitance of the routing wires. The layout Python API offers the possibility to describe technology independent layout generators since it is based on a set of generic layout design rules.

6.4.2.5 Layout Generation Tool Environment

Fig. 6.12 shows how the layout generation environment can help as a device designer debugger environment. The device designer chooses the technology file by clicking on Edit → ConfigureTech. Then the input parameters of device are chosen.

The layout of the device can be seen instantaneously since the layout code description calls the technology file to determine the values of the design rules parameters. All calculated layout dependent parameters can be seen through a message console. In the script editor windows, there is a part of the layout description code to show the layout description languages (here an example for an intra-stack routing is illustrated).

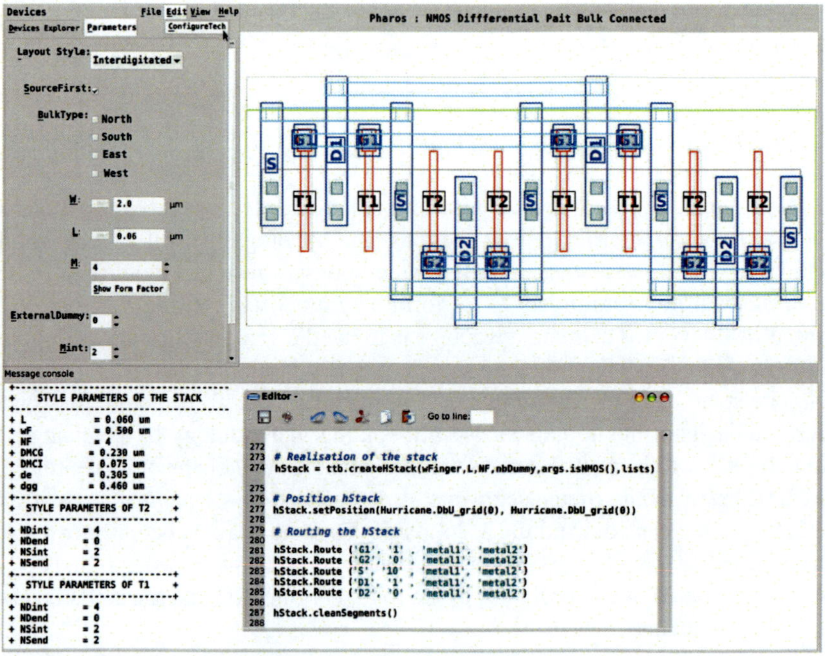

Fig. 6.12 Our layout generation tool environment

6.4.3 Differential Pair Example

For example for a differential Pair Fig.6.13 in 130 nm, W = 3 um, L = 0.15 um, NF = 4, Type = NMOS and NBdummies = 0, there are four different styles.

Interdigitated style Fig.6.14a : a certain number of sub transistor fingers (here it is equal to 2) that belongs to the same transistor "T1" is alternated with the same number of sub-transistor fingers that belongs to the other transistor "T2".

Mirror style Fig.6.14b : the alternated transistor fingers of "T1" and "T2" are put around a symmetrical axis. Since the interdigitated and the mirror styles have only one stack, they are only routed in *intra-Stack routing* methodology. The nets "G1" and "G2" are routed into a line shape, the nets "D1" and "D2" are routed into a u-shape, and the common net "S" is routed into a serpentine.

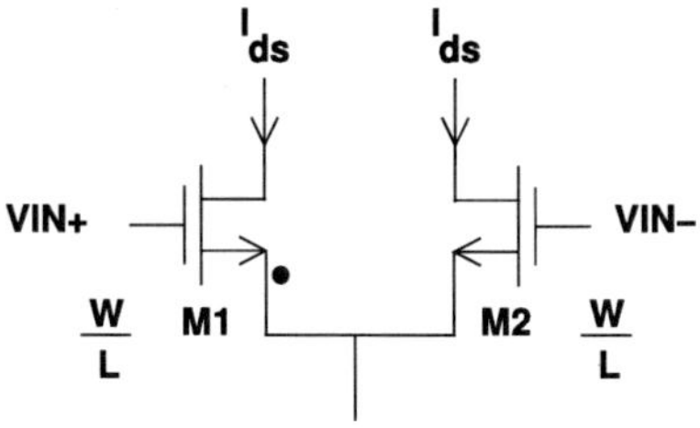

Fig. 6.13 Differential Pair

M2 module style [16] Fig.6.14c : each two transistor fingers that belong to the same transistor are put in a separated stack, then the stack is alternated as wished by the designer. The idea is to easy characterize each stack and easily alternate them. Here the stacks are alternated in a horizontal way.

2D common centroid style Fig.6.14d : the alternated transistor fingers of "T1" and "T2" are put around a common center point. Since the M2 module and the 2D-CommonCentroid styles have multiple stacks, they are routed in both *intra-Stack routing* methodology and *inter-Stack routing* methodology. In the M2 module the net "D1" and "D2" are routed only in an inter-stack routing methodology since they have only one diffusion region in each stack that belongs to their transistor, but for the net "S", it is routed in intra-stack and inter stack routing methodology in a U-shape for both methodologies. For the 2D-common centroid, all the nets are routed in both routing methodologies. The inter stack routing here is performed around and between the stacks.

Table 6.2 compares area, aspect ratio and matching features of the differential pair in a 130 nm technology, with $W = 3.0\mu m$, $L = 0.15\mu m$, $NF = 4$, $Type = NMOS$ and $NB_{dummies} = 0$ respectively for the four layout styles (interdigitated, mirror, horizontal M2 Module and 2D common-centroid). Note that the routing area is larger than the active area in all the cases. This table provides the circuit designer with a clear vision of the advantage and drawbacks of various solutions to draw analog device layouts.

6.4.4 *HB*-TREE*

This section illustrates how this device library can be used to generate the layout of an whole circuit with matching and symmetrical constraint using the algorithm of the HB*-Tree (Hierarchical Binary Tree) described in [18].

6.4.4.1 B*Tree Representation

A B*-tree (Binary Tree) is an algorithm commonly used to represent a *compacted placement* which every module cannot move left and bottom anymore. Every node

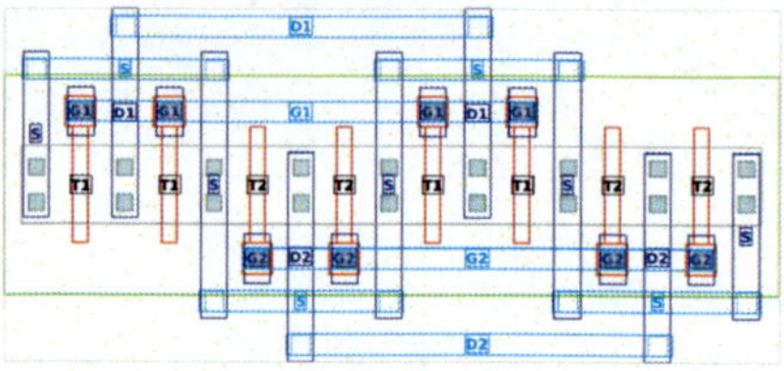

(a) Interdigitated style.

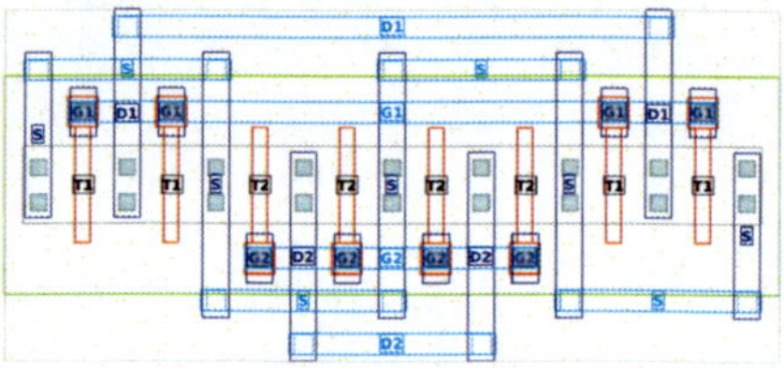

(b) Mirror style.

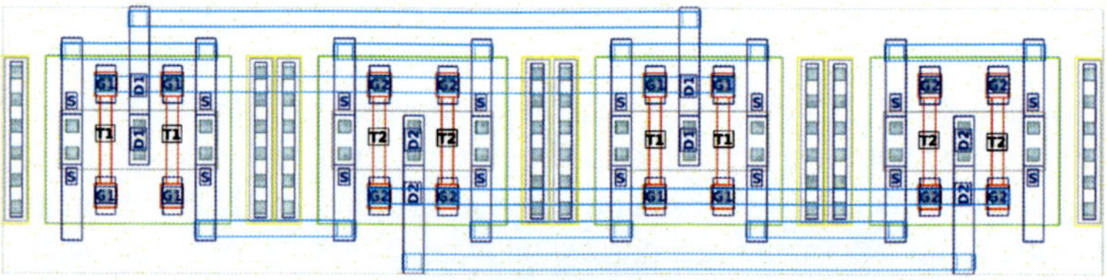

(c) ModuleM2 style.

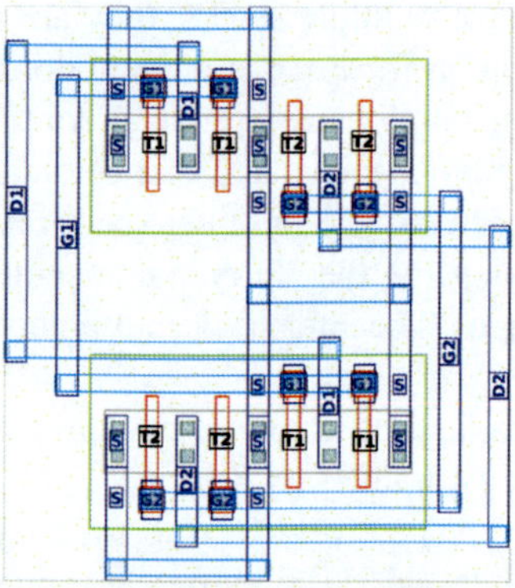

(d) 2D - Common Centroid style

Fig. 6.14 Different Layout Styles of a differential pair

of a B*-tree corresponds to a module of a compacted placement having the following characterization :

- The root of a B*-tree corresponds to the module on the bottom-left corner, in the Fig.6.15a the root is the node "A".
- The left child of a node "A", corresponding to a module "B", represents the lowest, adjacent module on the right side of the module "A" Fig.6.15a.
- The right child of a node "A", corresponding to a module "B", represents the first module above "A" with the same horizontal coordinate Fig.6.15b.

Table 6.2 Area and Matching Features as Function of Layout Styles

Areas in μm 2.				
	Interdigitated	**Mirror**	**Horizontal M2 Module**	**2D Common Centroid**
Total area Active area height/width	26.026 5.505 0.43	26.026 5.505 0.43	47.704 9.555 0.235	43.84 16.578 1.1
horizontal gradient compensation	–	+	+	+
vertical gradient compensation	–	–	–	+
Equivalent routing capacitance	+	–	+	+

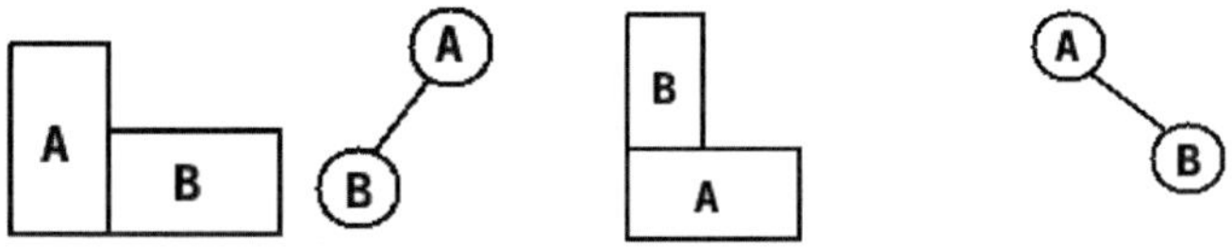

(a) Left child node. (b) Right child node.

Fig. 6.15 The placement of the child node with respective to there parent

6.4.4.2 HB*-Tree Representation

The idea of the implementation of the HB*-tree (Hierarchical Binary Tree) is to give the designer the possibility to make some constraint on a group of devices (modules) to be placed in certain constraints : matching, isolation, symmetry, etc ... The node in the HB*-Tree can be :

- A device node : the node corresponds to one device only.
- A hierarchical node : the node corresponds to a group of a device with certain constraints.

Fig.6.16 represents how an HB*-tree, contained by a hierarchical node "G", with a compact representation, is transformed to a symmetrical placement with respect to a vertical axis.

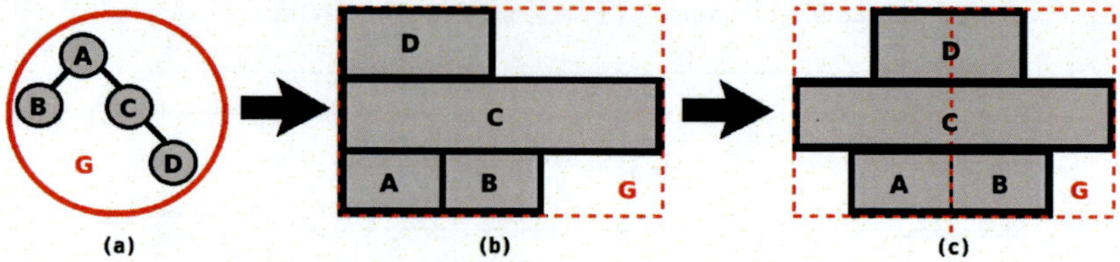

Fig. 6.16 The placement of a group with a symmetrical constraint

6.5 Case Study

In this section, the sizing tool and the layout generation tool of the designer-assisted analog synthesis flow are used to design the fully differential transconductor [3] represented in Fig. 6.17. Each colored rectangle that contains one or two transistors represents a device. Every subcircuit is described in a XML file, the transconductor XML file is defined by calling subcircuit XML files to ensure hierarchical knowledge reuse.

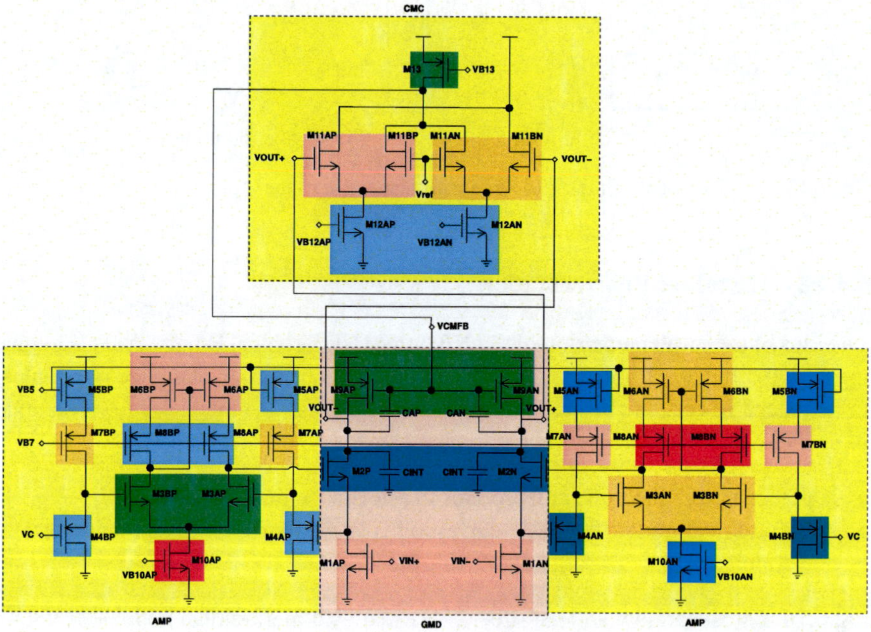

Fig. 6.17 Schematic of the fully differential transconductor. It is composed of four subcircuits : two instances of AMP (AMP1 and AMP2), GMD and CMC.

6.5.1 Knowledge-Aware Transconductor Synthesis

Using the synthesis routine in Fig.6.9, the bipartite graphs for GMD (Fig. 6.8), CMC, and AMP subcircuits are generated. These three bipartite graphs are then merged to get the bipartite graph associated to the fully differential transconductor, that is evaluated from top (P_{in} parameters) to bottom to provide transistor widths and unknown voltages (P_{out} parameters) for the fully differential transconductor. P_{in} mainly consist of the transistor current, length and overdrive voltage, that are part of the designer tacit knowledge.

To validate the concept of knowledge-aware synthesis, the transconductor is synthesized to meet the performance specifications given in Table 6.3. The optimization loop is set as a solid line loop in Fig. 6.2. It comprises vectors 1 to vector 4. The search engine implements a combined Genetic/Nelder-Mead Simplex search. The role of the genetic algorithm is to make a global exploration of the design space. Then the Nelder-Mead Simplex locally refines the global candidate solutions. The automation gap shown in Fig. 6.2 is filled with the bipartite graph of the transconductor shown in Fig. 6.17. The synthesis is performed using 130 nm CMOS technology with VDD=1.2 V. The synthesis results are shown in Table 6.3. The execution runtime is 43 seconds and the number of cost function calls is 100. This very low runtime is due to the fact that parameters in vector 1 vary in a very narrow ranges, hence, reducing the feasible design space of the transconductor. Simulation results are in good agreement with the synthesis results.

The next step consists in generating the layout for the transconductor.

Table 6.3 Synthesis Results of the Transconductor

Circuit Performances	Target	Synthesis
Effective GM@15mV (μA/V)	< 15	7.084
Effective GM@400mv (μA/V)	> 150	163.39
Output Noise (nV/$\sqrt{Hz}$)	< 300	273.5
Third-order Intercept Point (dBVp)	[7,11]	9.18
Transistors in Saturation (except M1AP and M1AN)	= ALL	ALL
Runtime (secs)	-	43
Number of cost function calls	-	100

6.5.2 Transconductor Layout Generation

As Fig.6.17 shows, the circuit designer specified the devices to be matched like for example : the two transistors of the differential pair M11AP and M11BP, the two cascoded transistors M2P and M2N, etc, by precising the layout style of each device, in this example the designer has chosen the mirror style for all the devices. The designer also may select the groups of the devices with certain constraint, in the example the designer has chosen to divide the circuit in group AMP1, AMP2, GMD and CMC and he has precised to place these groups in a symmetrical constraint. Also he may precise to place all the whole circuit in a symmetrical placement.

HB*-Tree Generation
Fig.6.18 presents the HB*Tree generated for the circuit, showing all the sub-circuits symmetrical groups. As the whole circuit is in a global symmetrical constraint so the circuit is represented as one hierarchical node, node "1". In node "1" there is another HB*-Tree that represents the placement of the groups of the circuit AMP1, AMP2, GMD and CMC. In each group there is another HB*-Tree that represents the sub-groups of each group. Finally, each sub-group contains another HB*-Tree that represents the matched devices to be placed in a symmetrical constraint.

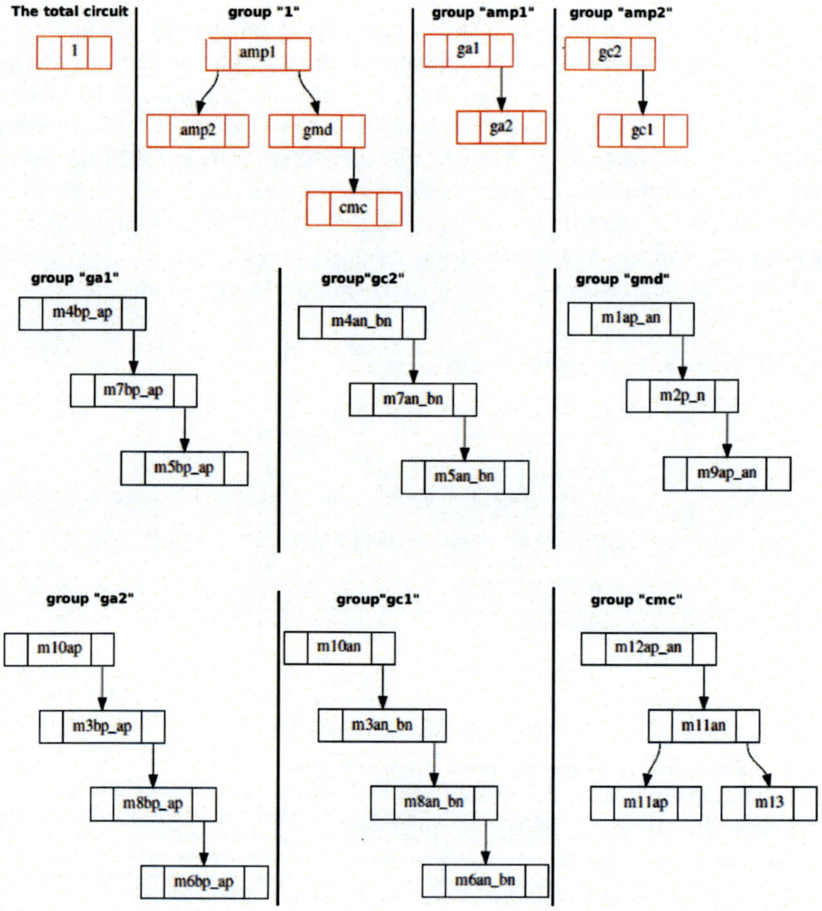

Fig. 6.18 HB _ Tree of the transconductor

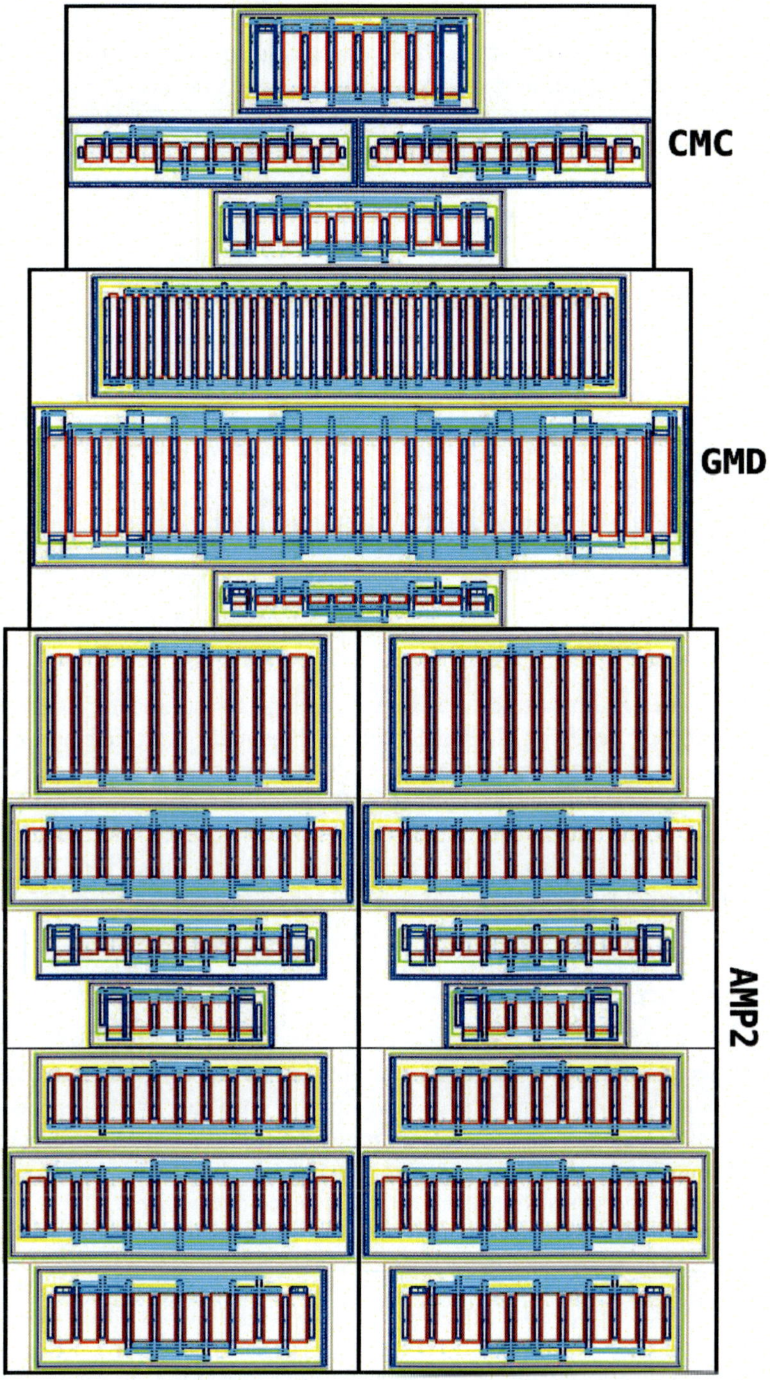

Fig. 6.19 Layout of the transconductor

Layout Generation
Fig.6.19 shows the layout of the whole transconductor circuit with its symmetrical groups. After the generation of the layout, we can see the netlist with the physical W and L and with the layout dependent parameters. Fig. 6.20 shows an example for the differential pair "M1AP_AN"of the subcircuit "GMD".

```
* Differential Pair Bulk Connected
.subckt DifferentialPairBulkConnected 1 2 3 4 5 param: l_val=3.0e-6 w_val=3.4e-6 nf_val=4 tr_name="name" meter=1.0
* NET 1 = d1
* NET 2 = d2
* NET 3 = g1
* NET 4 = g2
* NET 5 = s
* drain1 drain2 gate1 gate2 source
Xtr1 1 3 5 5 M1AP l=l_val w=w_val nfing=nf_val AS=6.4600e-13 AD= 6.4600e-13 PS=1.5200e-06 PD= 1.5200e-06
Xtr2 2 4 5 5 M1An l=l_val w=w_val nfing=nf_val AS=9.0100e-13 AD=6.4600e-13 PS=3.8200e-06 PD= 1.5200e-06
.ends DifferentialPairBulkConnected
```

Fig. 6.20 Netlist of the device M1AP_AN after the layout generation

6.6 Conclusion

An analog design flow centered around the designer was presented. It is composed of a sizing tool, that allows to transfer the designer sizing intent into a formal representation (*i.e.* the bipartite graph), and a layout generation tool, that offers a highly configurable physical view for an analog IP. The designer knowledge and computed results are stored into a XML file that can be easily modified and shared among a group of designers. Using the designer-assisted analog synthesis flow, a fully differential transconductor was successfully synthesized in a relative short time.

6.7 Future Work

Today, techniques to couple the sizing tool based on bipartite graphs and the layout generation tool are currently being developed to introduce layout dependent parasitics during sizing. This ensures that produced layout satisfies performance constraints during sizing and layout generation. Other techniques to produce automatic and compact floorplans are also under research and development.

References

1. Binkley, D.: Tradeoffs and Optimization in Analog CMOS Design. Wiley (2008)
2. Binkley, D.M., Hopper, C.E., Tucker, S.D., Moss, B.C., Rochelle, J.M., Foty, D.P.: A CAD methodology for optimizing transistor current and sizing in analog CMOS design. IEEE Trans. Comput. Aid. Des. 22(2), 225–237 (2003)

3. Chamla, D., Kaiser, A., Cathelin, A., Belot, D.: A $G_m - C$ low-pass filter for zero-IF mobile applications with very wide tuning range. IEEE J. of Solid-State Circuits 40(7), 1443–1450 (2005)
4. Co, N.L., Vianello, M., Guilherme, J., Horta, N.: LAYGEN-Automatic Layout Generation of Analog ICs from Hierarchical Template Descriptions. In: Conference on Ph.D. Research in Microelectronics and Electronics, pp. 213–216 (2006)
5. Degrauwe, M., Nys, O., Dijkstra, E., Rijmenants, L., Bitz, S., Goffart, B., Vittoz, E., Cserveny, S., Meixenberger, C., van der Stappen, G., Oguey, H.: IDAC: An Interactive Design Tool for Analog CMOS Circuits. IEEE Journal of Solid-State Circuits 22(6), 1106–1116 (1987)
6. Doboli, A., Vemuri, R.: Exploration-based high-level synthesis of linear analogue systems operating at low/medium frequencies. IEEE Trans. Comput. Aid. Des. 22(11), 1556–1568 (2003)
7. Enz, C., Krummenacher, F., Vittoz, E.: An Analytical MOS Transistor Model Valid in All Regions of Operation and Dedicated to Low-Voltage and Low-Current Applications. Analog Integrated Circuits and Signal Processing Journal 8(1), 83–114 (1995)
8. Graeb, H., Balasa, F., Castro-Lopez, R., Chang, Y.W., Fernandez, F., Lin, P.H., Strasser, M.: Analog Layout Synthesis - Recent Advances in Topological Approaches. In: Proceedings on Design, Automation and Test in Europe, pp. 274–279 (2009)
9. Harjani, R., Rutenbar, R., Carley, L.: OASYS: A Framework for Analog Circuit Synthesis. IEEE Transactions on Computer-Aided Design of Integrated Circuits and Systems, 1247–1266 (1989)
10. Iskander, R., Louërat, M.M., Kaiser, A.: Automatic DC Operating Point Computation and Design Plan Generation for Analog IPs. Analog Integrated Circuits and Signal Processing Journal 56, 93–105 (2008)
11. Javid, F., Iskander, R., Louërat, M.M.: Simulation-Based Hierarchical Sizing and Biasing of Analog Firm IPs. In: IEEE International Behavioral Modeling and Simulation Conference, pp. 43–48 (2009)
12. Javid, F., Iskander, R., Louërat, M.M., Dupuis, D.: Analog Circuits Sizing Using Bipartite Graphs. In: IEEE Midwest Symposium on Circuits and Systems (2011)
13. Koh, H.Y., Sequin, C.H., Gray, P.R.: OPASYN: A Compiler for CMOS Operational Amplifiers. IEEE Trans. Computer-Aided Design 9(2), 113–125 (1990)
14. Krasnicki, M., Phelps, R., Rutenbar, R.A., Carley, L.R.: MAELSTROM: Efficient simulation-based synthesis for custom analog cells. In: Design Automation Conference, pp. 945–950 (1999)
15. Krasnicki, M.J., Phelps, R., Hellums, J.R., McClung, M., Rutenbar, R.A., Carley, L.R.: ASF: A practical simulation-based methodology for the synthesis of custom analog circuits. In: Proc. IEEE Int. Conf. on Computer-Aided-Design, pp. 350–357 (2001)
16. Lewyn, L., Williams, N.: Is a New Paradigm for Nanoscale Analog CMOS Design Needed? Proceedings of the IEEE 99(1), 3–6 (2011)
17. Libes, D.: EXPECT (2010)
18. Lin, P.H., Chang, Y.W., Lin, S.C.: Analog placement based on symmetry-island formulation. Trans. Comp. -Aided Des. Integ. Cir. Sys. 28(6), 791–804 (2009)
19. Liu, W.: MOSFET Models for SPICE Simulation Including BSIM3v3 and BSIM4. Wiley-Interscience (2001)
20. NXP: MOS Model PSP level 103 (2011), http://www.nxp.com/models/mos_models/psp/
21. Nye, W., Riley, D., Sangiovanni-Vincentelli, A., Tits, A.L.: DELIGHT.SPICE: An Optimization-Based System for Design of Integrated Circuits. IEEE Trans. Computer-Aided Design 7(4), 501–518 (1988)

22. Ochotta, E.S., Rutenbar, R.A., Carley, L.R.: Synthesis of high-performance analog circuits in ASTRX/OBLX. IEEE Trans. Comput. Aid. Des. 15(3), 273–294 (1996)
23. Phelps, R., Krasnicki, M., Rutenbar, R.A., Carley, L.R., Hellums, J.R.: ANACONDA: Robust synthesis of analog circuits via stochastic pattern search. IEEE Trans. Comput. Aid. Des., 567–570 (2000)
24. Van der Plas, G., Debyser, G., Leyn, F., Lampaert, K., Vandenbussche, J., Gielen, G., Sansen, W., Veselinovic, P., Leenarts, D.: AMGIE-A Synthesis Environment for CMOS Analog Integrated Circuits. IEEE Transactions on Computer-Aided Design of Integrated Circuits and Systems 20(9), 1037–1058 (2001)
25. Porte, J.: COMDIAC: Compilateur de Dispositifs Actifs (1997)
26. Rutenbar, R.A., Gielen, G.: Hierarchical modeling, optimization and synthesis for system-level analog and RF designs. Proceedings of the IEEE 95(3), 1556–1568 (2007)
27. Stefanovic, D., Kayal, M.: Structured Analog CMOS Design. Kluwer (2009)
28. Stefanovic, D., Kayal, M., Pastre, M.: PAD: A new interactive knowledge-based analog design approach. Analog Integr. Circ. and Sig. Processing Journal, 291–299 (2005)
29. Stehr, G., Pronath, M., Schenkel, F., Graeb, H., Antreich, K.: Initial sizing of analog integrated circuits by centering within topology-given implicit specification. In: Proc. IEEE Int. Conf. on Computer-Aided-Design, pp. 241–246 (2003)
30. Yilmaz, Y., Dundar, G.: Analog Layout Generator for CMOS Circuits. IEEE Trans. Computer-Aided Design of Integr. Circ. ans Syst. 28(1), 32–45 (2009)
31. Zhang, L., Liu, Z.: A Performance-Constrained Template-Based Layout Retargeting Algorithm for Analog Integrated Circuits. In: ASP-DAC, pp. 293–298 (2010)

Chapter 7
Analog Circuit Design Based on Robust POFs Using an Enhanced MOEA with SVM Models

Nuno Lourenço, Ricardo Martins, Manuel Barros, and Nuno Horta

Abstract. In this chapter, a multi-objective design methodology for automatic analog integrated circuits (IC) synthesis, which enhances the robustness of the solution by varying technological and environmental parameters, is presented. The automatic analog IC sizing tool GENOM-POF was implemented and used to demonstrate the methodology, and to verify the effect of corner cases on the Pareto optimal front (POF). To enhance the efficiency of the tool, a supervised learning strategy, which is based on Support Vector Machines (SVM), is used to create feasibility models that efficiently prune the design search space during the optimization process, thus, reducing the overall number of required evaluations. The GPOF-SVM optimization kernel consists of a modified version of the multi-objective evolutionary algorithm (MOEA), NSGA-II, and uses HSPICE® as the evaluation engine. The usage of standard inputs and outputs eases the integration with other design automation tools, either at system level or at physical level, which is the case of LAYGEN, an in-house layout generation tool. Finally, the approach was validated using benchmark examples, which consist of circuits tested with similar tools, particularly, the former GENOM tool and other tools from literature.

7.1 Introduction

In the last decades, Very Large Scale Integration (VLSI) technologies have been widely improved, allowing the proliferation of consumer electronics and enabling the growth of IC market from $10 billion in 1980 to over than $300 billion in 2013

Nuno Lourenço · Ricardo Martins · Nuno Horta
Instituto de Telecomunicações, Lisboa, Portugal, and Instituto Superior Técnico, Universidade Técnica de Lisboa, Lisboa, Portugal
e-mail: `nuno.horta@lx.it.pt`

Manuel Barros
Instituto de Telecomunicações, Lisboa, Portugal, and Instituto Politécnico de Tomar, Tomar, Portugal

M. Fakhfakh et al. (Eds.): *Analog/RF & Mixed-Signal Circuit Sys. Design*, LNEE 233, pp. 149–167.
DOI: 10.1007/978-3-642-36329-0_7

(according to IC Insights Inc) [1]. IC designers are building systems that are increasingly more complex and integrated. In the System on Chip (SoC) age is common to find devices where the whole system is integrated in a single chip [2].

The need of new functionalities, longer battery times, smaller (thinner) devices, more power efficiency, less production and integration costs and less design cost, makes the design of electronic systems a truly challenging task. The complexity of electronic systems design and the strict time-to-market impose the use of Computer Aided Design (CAD) tools to support the design process. In digital IC design, mature Electronic Design Automation (EDA) tools and design methodologies are available helping the designers to keep up with the new capabilities offered by the technology. Currently almost all low-level phases of the process are automated. The level of automation is far from the push-button stage, but is keeping up reasonably well with the complexity supported by the technology. On the other hand, analog IC design automation tools strive to keep up with the new challenges created by technological evolution [3,4]. Due to the lack of automation, designers keep exploring the solution space almost manually. This method causes long design times, and allied to the non-reusable nature of analog IC, makes analog IC design a cumbersome task.

In this work a multi-objective design methodology and tool for automatic analog IC synthesis, GPOF-SVM, is presented. GPOF-SVM stems from GENOM [5-7] and GENOM-POF [8], the first is a former single objective optimizer enhanced by an SVM feasibility model and the second is a multi-objective circuit optimizer. This chapter is organized as follows: in section 7.2 an overview of related work in analog IC design automation at circuit/system-level sizing is presented; section 7.3 explains the architecture of GPOF-SVM; section 7.4 presents case studies; and finally, in section 7.5 some conclusions are drawn and future work proposed.

7.2 Related Work

Historically, the tools for automated circuit sizing are classified as knowledge-based or optimization based. This classification, illustrated in Fig. 7.1, is based on the fundamental techniques used to address the problem.

Early strategies, like IDAC [9] and BLADES [10], tried to systematize the design by using a design plan derived from expert knowledge. In these methods, a pre-designed plan is built with design equations and a design strategy that produce component sizes that meet the performance requirements. The knowledge-based approach was applied with moderate success to automatic analog IC sizing. The main advantage of this approach is the short execution time. On the other hand, deriving the design plan is hard and time-consuming, the design plan requires constant maintenance in order to keep it up to date with technological evolution, and the results are not optimal, suitable only as a first-cut-design.

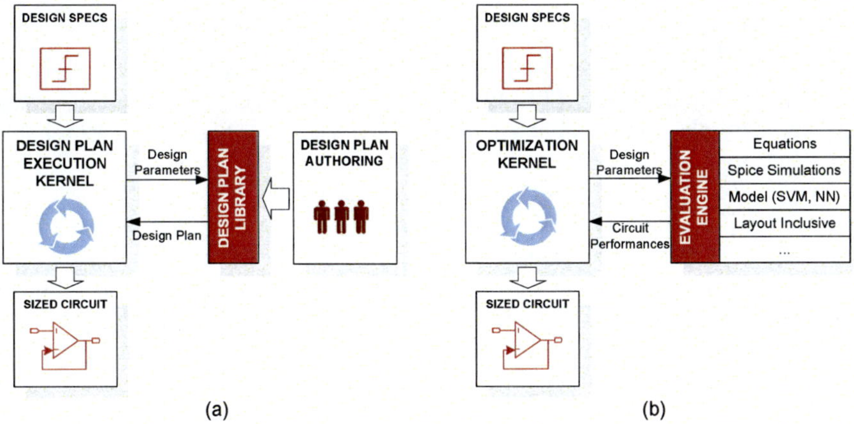

Fig. 7.1 Automatic specification translation approaches: (a) knowledge-based and (b) optimization-based

Aiming for optimality, the next generations of sizing tools apply optimization techniques to analog IC sizing. Based on the evaluation techniques employed, the optimization-based sizing tools can be further classified into three main sub-classes: equation-based, electrical-simulation-based, and numerical-model-based.

- *Equation:* These methods use analytic design equations to evaluate the circuit performance. The strong point of equation-based methods like GPCAD [11], Kuo-Hsuan et. *al.* [12] among others is the short evaluation time, making them, like the knowledge-based approaches, extremely suited to derive first-cut designs. The main drawbacks are: not all design characteristics can be easily mapped by analytic equations and the approximations introduced in the equations yield low accuracy designs. To reduce the long time dispended in model development, automatic techniques were proposed (Gielen et al. in [13] pro-vide a good overview on symbolic analysis applied to analog ICs).
- *Electrical Simulation:* These sizing techniques use a circuit simulator to evaluate the circuit's performance. The strong points of this approach are generality and easy-and-accurate model, however, typified by long execution time. To cope with this limitation Kuo-Hsuan et. *al.*[12] used equations to derive an approximate initial solution, Cheng et *al.* [14] solving the bias of the transistors first, the transistor sizes are then derived from the bias point using electric simulation. In MAELSTROM and ANACONDA [15] the evaluation is done using a parallel mechanism that shares the evaluation load among multiple computers.
- *Numerical Model:* The numerical-model-based tools like Alpaydin et. *al.* [16] and Barros et. *al.* [6] use macro models, e.g., neural-networks or support vector machines, to speed up the evaluation of the circuit's performance, reducing the high execution times caused by the exclusive use of electrical simulation inside the optimization loop, especially at system-level. A different approach is the usage of POF, where a suitable solution is selected from the pre-generated set

of optimal solutions, these models are then used hierarchically for system level sizing [17,18].

In MINLP [19], DARWIN [20], SEAS [21] and MOJITO [22,23] device sizing and topology selection are done simultaneously. These methods are more reliable than other topology selection techniques, as they treat the problem in a unified manner. The computation time, however, is extremely high. Koza [24], Lohn [25], Sripramong [26], Shoou-Jin [27] and more recently Hongying [28] presented a design methodology that creates new topologies. This approach is typified by high computation time, which limits the number of components in the circuit. Another issue with bottom-up generation is that designers are suspicious of the generated structures as they may differ "too much" from well-known trusted analog circuits [29]. Fig. 7.2 shows the panorama of analog circuit synthesis contributions.

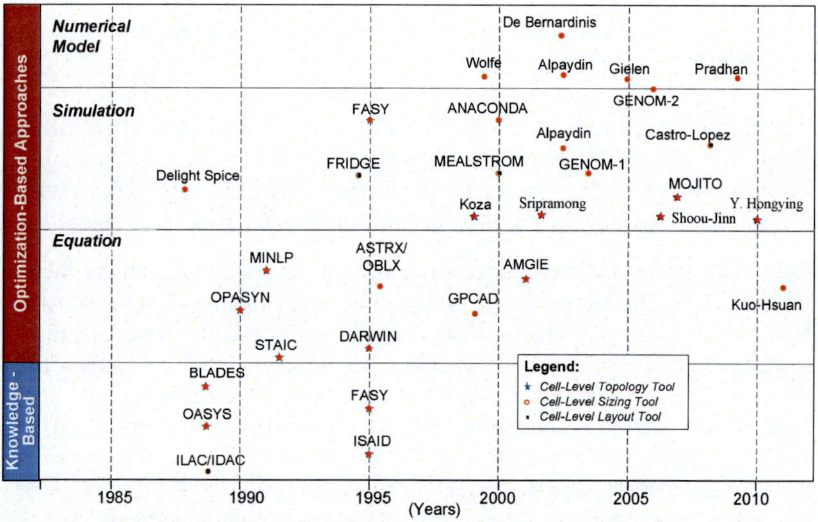

Fig. 7.2 Overview of analog design automation tools

7.3 GPOF-SVM Architecture

GPOF-SVM addresses the problem of automatic specification translation at circuit level, also known as circuit sizing, where from the set of specifications, the designer finds out the sizes for the components (widths and lengths of the transistors, resistors, capacitors, etc.). To verify if the design is robust, i.e., the vast majority of the fabricated circuits will work according to specifications, corner analysis is employed. Corner analysis is among the most common techniques for analog IC design centering, and consists in a worst-case approach where the circuit is simulated over multiple combinations of process parameters variations (power supply, temperature, etc.). In Fig. 7.3 the 27 corners cases obtained by considering 3 values for power supply, operating temperature and library models are shown.

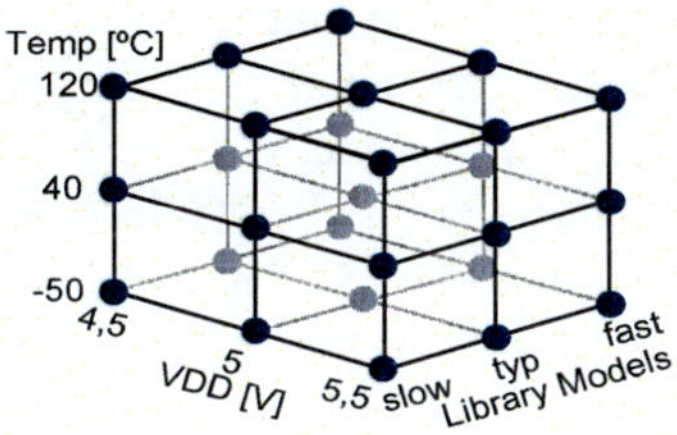

Fig. 7.3 Corner cases example

GPOF-SVM, whose architecture is shown in Fig. 7.4, is based on the elitist multi-objective evolutionary optimization kernel NSGA-II [30], and uses the industrial grade simulator HSPICE® [31] to evaluate the performance of the design. GPOF-SVM targets the design of robust circuits, by allowing the consideration of corner cases during optimization. In addition, an SVM [32], which models the functional feasibility of the circuit, is used to speed up the convergence to feasible areas on the design space.

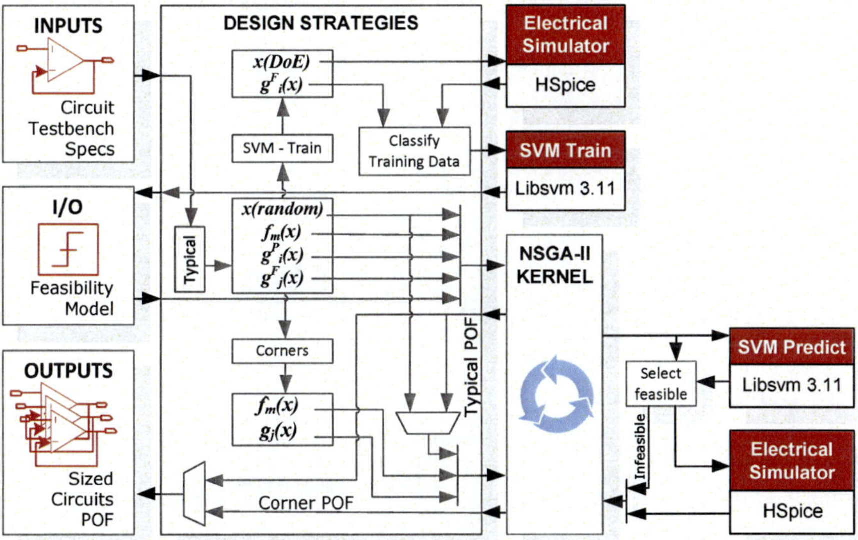

Fig. 7.4 GPOF-SVM architecture

In order to use GPOF-SVM, the designer inputs the circuit netlist and testbench, defines the optimization variables, design constraints and objectives, and the corners cases. Then, GPOF-SVM, models the circuit as an optimization problem, defined by the tuple *{x,F,G}*, where *x* is the vector of design variables, *F* is the vector objectives and *G* is the vector of inequality constraints, suitable to be optimized by the NSGA-II kernel. The functional constraints, which are a sub-set of G, are used to define the functional feasibility regions used to train the feasibility model, where the training data is obtained using fractional Design of Experiments (DOE) to generate a set of circuits that are simulated to evaluate how well they met the functional constraints. The tools' output is a family of Pareto optimal circuits that fulfill all the constraints and represent the feasible tradeoffs between the different optimization objectives. The next subsections provide the details of the architecture using a simple circuit to illustrate the descriptions.

7.3.1 Inputs and Outputs

The inputs from the designer are the circuit and testbench in the form of HSPICE® netlists. The netlist must have the optimization variables as parameters, and must include means to measure the circuit's performance; the corner's parameter variations are also included in the netlist. Fig. 7.5 shows a simple differential amplifier with the testbench schematic and parts of the corresponding netlist.

Circuit.cir

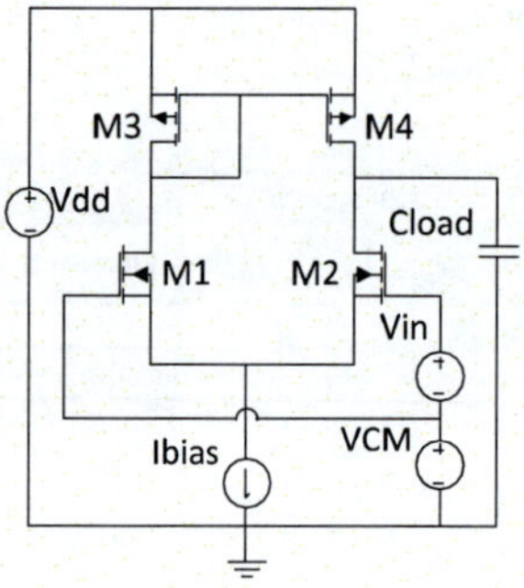


```
** Testbench
* define the DC voltage sources and bias
Vdd 6 0 5V
VCM 2 0 2.5V
...

* AC analysis
.ac dec 20 1 1000Meg SWEEP DATA = PIPEdata
...

* Circuit
M1 5 3 1 0 nmos L=_l1 W=_w1
M2 4 7 1 0 nmos L=_l1 W=_w1
M3 5 5 6 6 pmos L=_l2 W=_w2
M4 4 5 6 6 pmos L=_l2 W=_w2

* Performance Measurements
.MEASURE ac 'gain_dc' max vdb(4) from=0.1k to=10k
.MEASURE ac 'gbw' trig at=0.1k targ vdb(4) val=0 fall=1
.MEASURE ac 'ph' find vp(4) at=gbw
.MEASURE 'fase' PARAM = 'ph'
.MEASURE 'rmspower' rms power

* Overdrives
.measure AC vov_m1   = param('VGS(m1)-VTH(m1)')
...

* Margins
.measure AC delta_m1   = param('VDS(m1)-VDSAT(m1)')
...
```

(a) (b)

Fig. 7.5 Example circuit: (a) schematic; (b) partial view of netlist

In addition, the designer defines ranges for the optimization variables, design constraints, and optimization objectives. Tables 7.1 and 7.2 illustrate these definitions for the circuit in Fig. 7.5. The output is a family of sized circuits representing the possible tradeoffs between the objectives being optimized.

Table 7.1 Variable ranges for the example in Fig. 7.5

Var.	**W1**	**W2**	**L1**	**L2**	**Ib**
Max.	500.0e-6	500.0e-6	15.0e-6	15.0e-6	400.0e-6
Min.	1.0e-6	1.0e-6	0.35e-6	0.35e-6	30.0e-6

Table 7.2 Objectives and design constraints for the example in Fig.7.5

Constraint	**Measure**	**Target**	**Units**	**Description**
Performance	gbw	≥ 35	MHz	Unit-gain frequency
	pm	$65 \leq \text{pm} \leq 90$	Degree	Phase margin
Functional	vov_m1	$50 \leq \text{vov\_m1} \leq 200$	mV	Vgs –Vt
	vov_m2	$50 \leq \text{vov\_m2} \leq 200$	mV	Vgs –Vt
	vov_m3	$100 \leq \text{vov\_m3} \leq 300$	mV	Vgs –Vt
	vov_m4	$100 \leq \text{vov\_m4} \leq 300$	mV	Vgs –Vt
	delta_m1	≥ 50	mV	Vds – Vdsat
	delta_m2	≥ 50	mV	Vds – Vdsat
	delta_m3	≥ 50	mV	Vds – Vdsat
	delta_m4	≥ 50	mV	Vds – Vdsat
Objective	gain_dc	maximize	dB	Gain DC
	rms_power	minimize	W	RMS power

7.3.2 Optimization Kernel

The optimization engine in GPOF-SVM is a modified NSGA-II to interface with HSPICE® and SVM, used to estimate feasibility and evaluate the individual objective and constraint functions. The NSGA-II was selected over SPEA and other multi-objective evolutionary algorithms because of the good characteristics of the output Pareto [30]. The option of using HSPICE® to evaluate the circuit's performance was due to the accuracy of the results and the availability of models for the devices provided by the foundries. The multi-objective optimization kernel module was designed to solve the problem:

$$\begin{aligned} &\textit{find x that minimize } f_m(x) && m = 1,2,\ldots M \\ &\textit{subject to } g_j(x) \geq 0 && j = 1,2,\ldots J \\ &x_i^L \leq x_i \leq x_i^U && i = 1,2,\ldots N \end{aligned} \tag{7.1}$$

where, x is a vector of N optimization variables, $g_j(x)$ one of the J constraints and $f_m(x)$ one of the M objective functions. Except for minor changes, it was implemented as in [30], using simulated binary crossover and mutation operators [33], tournament selection, and constrained based dominance check.

7.3.3 Design Strategies

GENOM-POF supports three design strategies: Typical, Corners, and Typical plus Corners. The next subsections describe each of the strategies.

7.3.3.1 Typical (T)

As the name states, in this strategy the circuit is evaluated using only typical conditions, this strategy is faster, and despite the output does not consider the limitations imposed by the corners it is useful for design tradeoffs analysis. First the design problem is described as an optimization problem, and then the NSGA-II optimization kernel can be executed. In order to satisfy the problem formulation in eq. (7.1), the design objectives being minimized are used directly as one of the $f_m(x)$, and the ones being maximized are multiplied by −1. The design constraints are normalized and multiplied by −1, if necessary, according to eq. (7.2).

$$g_i(x) = \begin{cases} \dfrac{p_i - P_i}{|P_i|} & \text{when the design constraint is } p_i \geq P_i \\ \dfrac{P_i - p_i}{|P_i|} & \text{when the design constraint is } p_i \leq P_i \end{cases} \tag{7.2}$$

where, p_j is the measured circuit characteristic, and P_j is the correspondent acceptable limit. Table 7.3 illustrates the objective and constraint functions for the circuit in Fig. 7.5 using the design specifications in Table 7.2.

Table 7.3 $f_m(x)$ and $g_j(x)$ for the example from Fig. 7.5

Performance Constraints	$g_0(x) = \dfrac{gbw}{35\times10^6} - 1$	$g_1(x) = \dfrac{pm}{65} - 1$	$g_2(x) = 1 - \dfrac{pm}{90}$
Functional Constraints	$g_3(x) = \dfrac{vov\_m1}{50\times10^{-3}} - 1$	…	$g_{15}(x) = \dfrac{delta\_m4}{50\times10^{-3}} - 1$
Objectives	$f_0(x) = -gain\_dc$	$f_1(x) = rmspower$	-

7.3.3.2 Corners (C)

In the Corners strategy, the design is optimized from the beginning using all the corners, i.e., for each evaluation the circuit is simulated once for each corner case, this makes it the slower strategy, but the output circuits are feasible in all tested corner conditions. To handle the multiple corners, the objective and constraint functions are modified using eq. (7.3).

$$\hat{f}_m(x) = \max_{c=1,2,..C}\left(f_m^c(x)\right)$$
$$\hat{g}_j(x) = \sum_{c=1}^{C} c_j^c(x) \text{ with } c_j^c(x) = \begin{cases} 0 \text{ if } g_j^c(x) \geq 0 \\ g_j^c(x) \text{ if } g_j^c(x) < 0 \end{cases} \tag{7.3}$$

where, C is the number of corners, and $f_m^c(x)$ and $g_j^c(x)$ are respectively the objective $f_m(x)$ and the constraint $g_j(x)$, as defined for the typical case evaluated in corner case *c*. In this worst case approach, each objective, which is being minimized, is evaluated using the maximum value obtained from the simulation of circuit in all the corner cases, and each constraint is evaluated as the sum of the normalized violation in all the corner cases where it is violated.

7.3.3.3 Typical Plus Corners (TC)

In this strategy, typical optimization is done until it stops evolving or the maximum number of generation is reached. Then, the typical POF is used as starting point for corner optimization. This strategy is a tradeoff between the execution time and robustness of the solution, and the reduction of the genetic information (localization of the search) imposed by the use of the typical POF as starting point for the corner optimization.

7.3.4 Functional Feasibility Model

In order to improve the convergence of the optimization kernel and reduce the time consumed in the evaluation using HSPICE® a functional feasibility model, which is reusable for different objectives and performance constraints, is used to avoid the simulation of infeasible solutions.

The derivation of the functional and performance constraints depends of the circuit in question, and is up to the designer to define which constraints are functional constraints and which are performance constraints. For example, for the circuit in Fig. 7.5 the performance constraints impose limits to the DC gain and phase margin, while the functional constraints impose limits to the overdrives voltages and saturation of the devices.

This separation has to do with the intrinsic behavior of the circuit, the functional constraints relates to the topology of the circuit and represent design strategies used to ensure proper behavior, whereas the performance constraints relate to some performance metric usually defined from the design specifications. Another way to view this separation is that to ensure the linearity of an amplifier it is recommended to have the transistors in saturation (functional requirement), and for a given design the DC gain must be greater than 30 dB (performance requirement), whereas in another design the unity frequency must be larger 40 MHz (performance requirement).

An important property of the functional constraints is that, for a given process and topology, they must be valid for a wide range of designs, otherwise the functional feasibility model is not reusable. The next sections describe how the functional feasibility model is derived and how it is used to enhance GPOF-SVM performance.

7.3.4.1 Building the Feasibility Model

The feasibility model, which follows the approach taken by GENOM [7], now for the multi-objective case, uses a SVM classifier [32] to estimate the compliance with functional constraints. To train the classifier, a training set is obtained using a fractional DOE strategy, then those points are simulated and their functional feasibility evaluated. The overall functional feasibility is computed as shown in eq. (7.4).

$$ffeas(x) = \frac{1}{J^F}\sum_{j=1}^{J^F} c_j^F(x), \text{ with } c_j^F(x) = \begin{cases} 0 \text{ if } g_j^F(x) \geq 0 \\ g_j^F(x) \text{ if } g_j^F(x) < 0 \end{cases} \tag{7.4}$$

where J^F is the number of functional constraints and $g_j^F(x)$ the functional constraint j.

The sampled points are then sorted into 3 classes, feasible, quasi-feasible, and infeasible, based on the value of *ffeas(x)*. The limits are $ffeas(x)=0$, $ffeas(x)\geq -T$, and $ffeas(x)<-T$ respectively.

As noticed on GENOM [7], the data sets are highly unbalanced with very few feasible points, and unbalanced data creates difficulties to the classifier. The main reason is that most classifiers, like SVM, tend to optimize the overall accuracy without considering the weight of relative distribution of each class and they are designed to generalize from sample data to avoid the noise, and in this case they would treat the feasible points as noise and would ignore then. To overcome this issue, strongly infeasible points, i.e., points where $ffeas(x)<-T_2$ with $T_2<T$, are discarded and not used to train the model.

To select the values T and T_2, first a fractional sampling [34] is performed (for small problems the full combinatory sampling can be used). At this point is likely to have few (or none) feasible points in the data set. Then, set the values T and T_2 in such way that the number of feasible plus quasi-feasible samples is approximately the same as the number of not-discarded-infeasible samples, and at least 5% of the total available samples. This last condition is to ensure that there are a reasonable number of points in each class. The SVM classifier is then trained using the grid search technique suggested in [35].

7.3.4.2 Evaluation with the Feasibility Model

The integration of the feasibility model in the evaluation is done in the trivial way. First the model is used to classify the individuals being evaluated as feasible, quasi-feasible and infeasible, and then unfeasible ones are discarded, and not put for electrical simulation. Due to the nature of the analog IC design space, some precautions must be taken when using the feasibility model: first the model can consider infeasible areas of feasibility that were not sampled, second in the beginning of the optimization, is likely to have only infeasible points, and without electrical evaluation there is no way to tell which ones are better and to guide the evolution of the population properly.

To accommodate these two factors, the functional feasibility pruning is enabled at each generation with 50% change. Moreover, when full feasibility is attained, i.e., both performance and functional constraints are met; it was noticed that the evolution of the algorithm do not generate a significant amount of infeasible points, and therefore the classification despite being much faster that electrical simulation is not pruning and represents an extra cost, therefore after full feasibility is attained the feasibility model is no longer used.

7.4 Case Study

The GPOF-SVM tool is here demonstrated for typical and corner cases by designing a single ended folded cascode amplifier and a fully differential telescopic amplifier. The folded cascode circuit is described in section 7.4.1, and the telescopic amplifier in section 7.4.2.

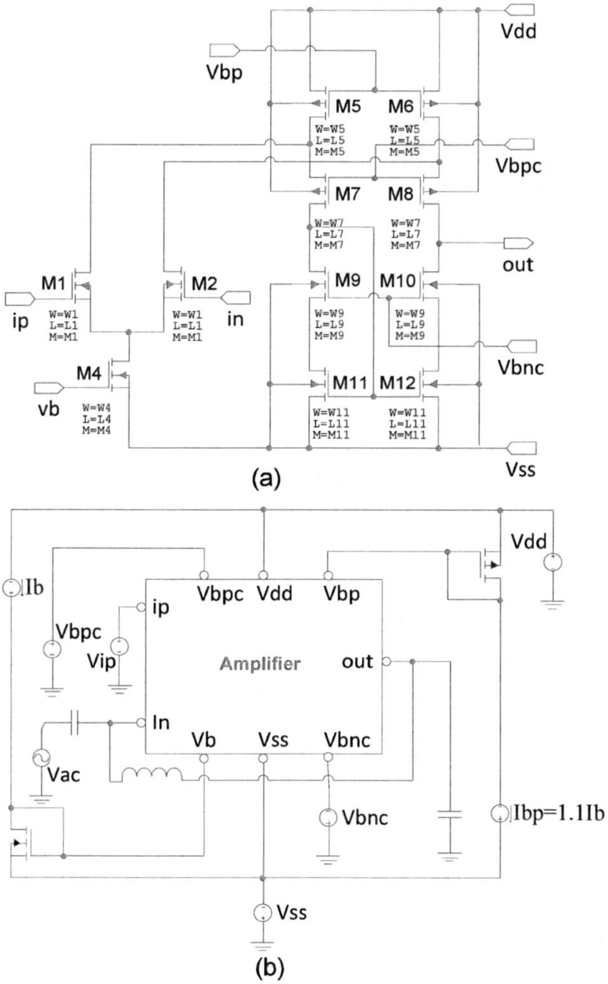

Fig. 7.6 Single-ended folded cascode amplifier (a) schematic (b) testbench

7.4.1 *Single Ended Folded Cascade Amplifier*

The circuit schematic is shown in Fig. 7.6, and the ranges, objectives and constraints are listed in Tables 7.4 and 7.5. The problem has 15 optimization variables, 2 objectives and 19 constraints. In addition, 9 corner cases were defined using the combination of technology models (typical, fast and slow) and temperature values (-40°C, 50°C, 120°C). All the presented results are for *UMC 0,13μm* technology and include only feasible solutions.

Table 7.4 Variable ranges

Var. [1]	**l1, l4, l5, l7, l9, l11**	**w1, w4, w5, w7, w9, w11**	**Ib [μA]**	**Vbcn [V]**	**Vbcp [V]**
Max.	0.80 μ*m*	400.0 μ*m*	500	0.0	0.4
Min.	0.12 μ*m*	0.24 μ*m*	30	-0.4	0.0

[1] The variables l1 and w1 are dimensions, in [μm], of M1 and M2; l4 and w4 of M4; l5 and w5 of M5 and M6; l7 and w7 of M7 and M8; l9 and w9 of M9 and M10; l11 and w11 of M11 and M12.

Table 7.5 Objectives and design constraints

Constraints	Measure	Target	Units	Description
Performance	gbw	≥ 24	MHz	Unit-gain frequency
	a0	≥ 40	dB	DC Gain
	sr	≥ 10	V/μs	Slew Rate
	pm	55 ≤ pm ≤ 90	Degree	Phase margin
Functional	ov[1]	≥ 30	mV	Vgs –Vt
	d[1]	≥ 1.2	V/V	(Vds – Vdsat)/Vdsat
	osp	≥ 0.3	V	
	osn	≤ -0.3	V	
Objectives	area[2]	minimize	μm	Area
	a0	maximize	dB	DC Gain

[1] The constraint applies to: M1, M4, M5, M7, M9 and M11.
[2] The area is the sum of all the devices gate area (WxL) excluding bias devices.

7.4.1.1 Synthesis

Figure 7.7 shows the Pareto fronts and execution time that were obtained by running the 3 strategies T, C and TC until the max generation limit. The algorithm parameters were: a population of 32 elements, crossover and mutation rate of 90% and 10%, respectively, and 400 generations for T and C, and 200/200 for TC (200 for the first step and 200 for the second step). The functional feasibility model was not used because finding a feasible solution in this example is easy (less than 5 generations), and after finding a feasible solution the optimizer tends to generate only feasible solutions rendering the feasibility model useless.

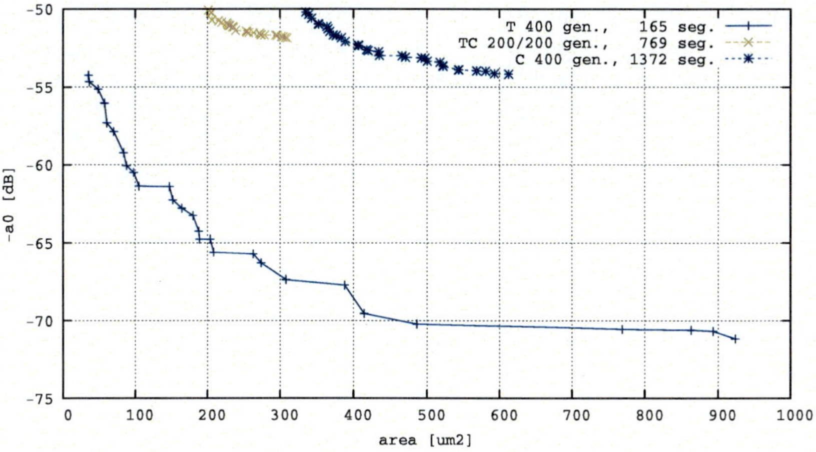

Fig. 7.7 POF obtained using the 3 design strategies T, TC and C

The POF obtained using T was found faster (in 165 seconds), and dominates the others (because it has fewer constraints). TC strategy was faster than C and in a region of the POF provided circuits with smaller area for the same gain; however it does not dominate the one obtained with C completely. By starting the corner

Table 7.6 Summary of the Corner and Typical plus Corner run illustrated in Fig. 7.7

Strategy	Corner (C)			Typical plus Corner (TC)		
	Smaller	Middle	Larger	Smaller	Middle	Larger
Time [s]	1372			769		
Area [μm²]	335.62	471.75	613.05	201.07	254.72	310.11
Gain [dB]	50.27	53.05	54.16	50.08	51.47	51.84
L1 [μm]	0.44	0.57	0.61	0.42	0.42	0.41
W1 [μm]	27.59	52.01	52.04	16.45	16.44	16.44
L4 [μm]	0.42	0.58	0.63	0.62	0.60	0.72
W4 [μm]	19.88	31.53	54.07	22.36	35.31	50.96
L5 [μm]	0.24	0.27	0.32	0.28	0.28	0.26
W5 [μm]	225.74	187.58	344.72	100.83	104.66	156.58
L7 [μm]	0.33	0.41	0.42	0.40	0.51	0.57
W7 [μm]	188.92	265.16	259.84	126.70	145.28	145.35
L9 [μm]	0.59	0.59	0.60	0.36	0.36	0.36
W9 [μm]	55.30	59.76	60.12	15.59	15.59	15.59
L11 [μm]	0.18	0.17	0.18	0.16	0.16	0.16
W11 [μm]	13.26	14.51	12.93	5.34	5.34	5.34
Ib [μA]	313.51	321.97	291.61	156.65	148.86	154.59
Vbcn [V]	-0.106	-0.1062	-0.1059	-0.0789	-0.0789	-0.0789
Vbcp [V]	0.0862	0.0870	0.0870	0.0890	0.0912	0.0950

optimization from the already optimized typical POF, it is easier to fulfill the additional constraints imposed by the corners and leads to better and faster results, however there is some biasing that narrow the search range. Table 7.6 summarizes the output of C and TC strategies.

In Fig. 7.8 the layout, which was generated using LAYGEN [36,37], is shown, for the extreme points from Table 7.6, i.e. the point with larger gain C larger, and the point with smaller area TC smaller.

7.4.2 Fully Differential Telescopic Amplifier

The fully differential telescopic amplifier circuit including bias schematic is shown in Fig. 7.9, and the variable ranges, objectives and constraints are listed in Table 7.7 and Table 7.8. The problem has 16 optimization variables, 3 objectives, 6 performance constraints and 32 functional constraints. In addition, the same 9 corner cases were defined using the combination of technology models (typical, fast and slow) and temperature values (-40ºC, 50ºC, 120ºC). All the presented results are for *UMC 0,18μm* technology and include only feasible solutions.

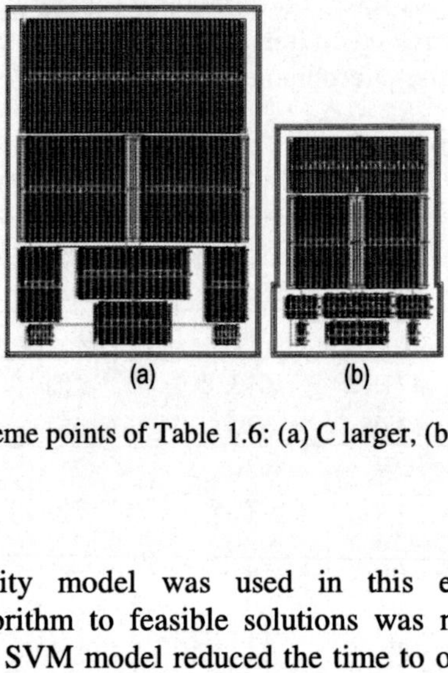

Fig. 7.8 Layout for the extreme points of Table 1.6: (a) C larger, (b) TC smaller

7.4.2.1 Synthesis

The functional feasibility model was used in this example because the convergence of the algorithm to feasible solutions was not immediate. In the single objective case the SVM model reduced the time to obtain the first feasible solution in 15-20%, in the multi-objective case, after 100 runs, the drop was around 10%. The multi-objective optimization explores the solution space more efficiently, reducing the effects of pruning infeasible solutions.

Two 2-D projections of the 3 objective POF, obtained using the TC strategy with parameters: population of 80 elements, crossover and mutation rate of 90% and 10% respectively and 300/200 generations, are shown in Fig. 7.10, and illustrate the effect of corners cases in decreasing the performance achieved by the circuit, Table 7.9 summarizes the synthesis results.

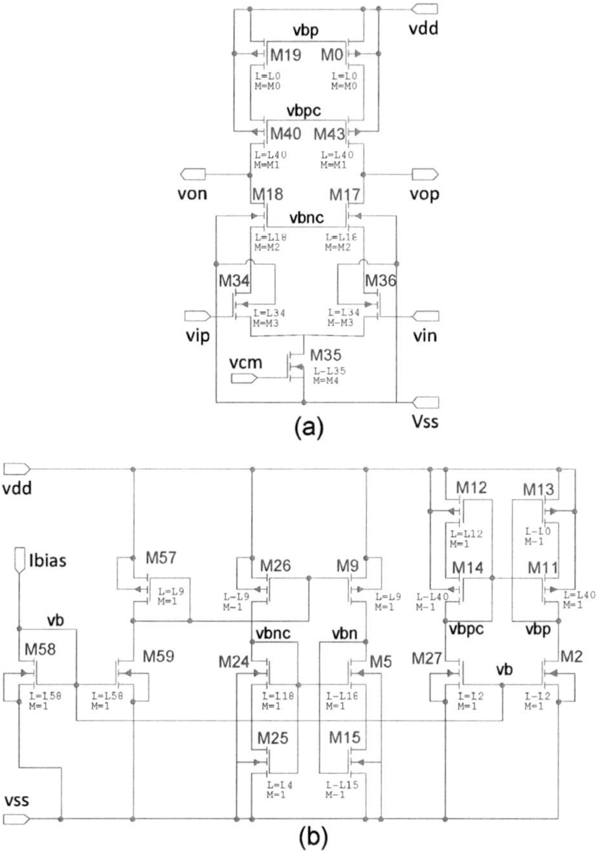

Fig. 7.9 Fully differential telescopic amplifier schematic: (a) amplifier, (b) bias

Table 7.7 Variable ranges

Var.	**L0, L2, L9, L12, L15, L18, L24, L34, L35, L40, L58**	**M0, M1, M2, M3, M4**
Max.	10.0 μ*m*	110
Min.	0.18 μ*m*	1

By having more than 2 objectives the 2-D projections are not monotonic like in the previous example. This is due to the fact that there are solutions that seem to be dominated but have better performance in the objectives not present in the 2-D projection. In Fig. 7.10, the projections of Gain *vs.* Power and GBW *vs.* Power are overlapped, each point in the Pareto front is represented by the 2 points in the graphic that have the same value of Power, one from each of the projections.

Table 7.8 Objectives and design constraints

Constraints	Measure	Target	Units	Description
Performance	gain_dc	≥ 75	dB	DC Gain
	gbw	≥ 100	MHz	Unity-gain frequency
	fase	60 ≤ fase ≤ 90	Degree°	Phase margin
	power	≤ 10	mW	Power
	iavdd	≤ 10	mA	Vdd current
Functional	vov[1]	≥100	mV	Vgs –Vt
	vov_m18, vov_m17	≥45	mV	Vgs –Vt
	vov_m34, vov_m36	≥50	mV	Vgs –Vt
	vov[2]	≤200	mV	Vgs –Vt
	vov[3]	≤300	mV	Vgs –Vt
	d[4]	50 ≤ d ≤ 200	mV	Vds – Vdsat
	d[5]	≥ 50	mV	Vds – Vdsat
Objectives	Power	Minimize	W	Power
	gbw	Maximize	Hz	Unity-gain frequency
	gain_dc	Maximize	dB	DC Gain

[1] The constraint applies to: M19, M0, M40, M43 and M35.

[2] The constraint applies to: M19, M0, M18 and M17.

[3] The constraint applies to: M40, M43, M34, M36 and M35.

[4] The constraint applies to: M40, M43, M17, M18 and M35.

[5] The constraint applies to: M19, M0, M34, and M36.

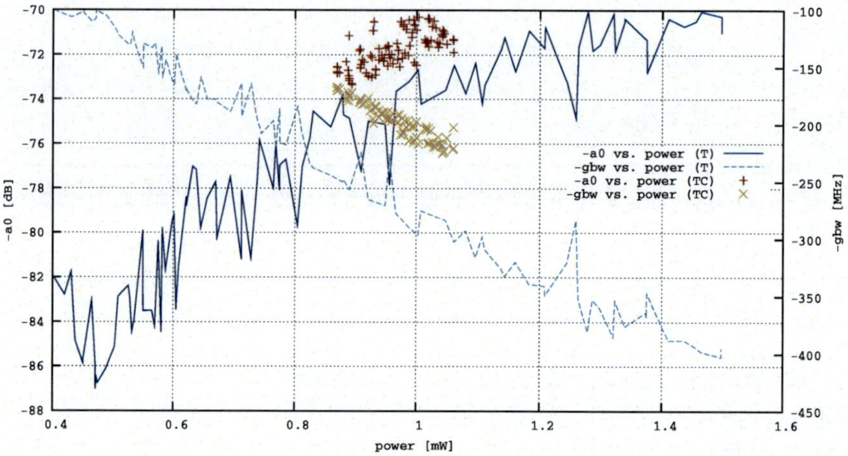

Fig. 7.10 2-D Projections of the 3-D POF obtained using T and TC

Table 7.9 Summary of the synthesis results

Strategy	Typical step (637 [s])			Corner step (5363 [s])		
	Less Power	Larger GBW	Larger Gain	Less Power	Larger GBW	Larger Gain
Power[mW]	0.402	1.498	0.470	0.869	1.04	0.894
GBW [MHz]	100.8	394.8	100.8	170.1	224.6	172.0
Gain[dB]	81.98	71.02	87.02	72.53	70.87	73.32

7.5 Conclusion

The proposed methodology and tool, GPOF-SVM, were used to successfully design well known analog circuits, taking into account robustness consideration by the inclusion of corner cases. Moreover, the multi-objective nature of the IC design synthesis makes it well suited for automatic design using multi-objective optimization strategies. In this approach, the output is not one solution, but a set of completely designed non-dominated solutions, all meeting the specifications. It is up to the designer to select the tradeoff between the concurrent objectives that is more interesting for the target project. The usefulness of GPOF-SVM to designers was shown using different design strategies. First, using the Typical (T) design strategy, the designer explores several design tradeoffs in a matter of minutes, which is useful for system level design. Then, using the Corners (C) or TC strategies the designer can obtain a family of optimum robust circuits that comply with the specification in all corner cases considered. Additionally, in order to enhance the efficiency of the NSGA-II based optimization kernel, a supervised learning strategy, which is based on a SVM approach, is used to create functional feasibility models. These models allow the efficient pruning of the design search space during the optimization process with absolute gains ranging from 10 to 20% in terms of the overall number of required evaluations and larger gains in terms of time consumption once electrical simulation, particularly for large circuits, is clearly more time expensive than the SVM model evaluation. Finally, the layout generation is demonstrated by linking the GPOF-SVM output with the entry of the in-house tool LAYGEN-II.

Acknowledgments. This work was supported by the Portuguese Science and Technology Foundation (Grant FCT-DFRH-SFRH/BD/72698/2010) and by the Instituto de Telecomunicações (Research project AIDA - IT/LA/1112/2011).

References

[1] McClean, B.: IC Market to Top $300 Billion for First Time in 2013, `http://www.icinsights.com/` (accessed December 13, 2011)
[2] Mori, K., Yamada, H., Takizawa, S.: System on Chip Age. In: The International Symposium on VLSI Technology, Systems, and Applications, Taipei, Taiwan (1993)

[3] ITRS: International Technology Roadmap for Semiconductors - 2010 Update. ITRS, http://www.itrs.net (accessed December 10, 2011)
[4] Gielen, G.G.E., Rutenbar, R.A.: Computer-aided design of analog and mixed-signal integrated circuits. Proc. IEEE 88(12), 1825–1852 (2000)
[5] Barros, M., Guilherme, J., Horta, N.: Analog circuits optimization based on evolutionary computation techniques. Integration, The VLSI Journal 43(1), 136–155 (2010)
[6] Barros, M.F.M., Guilherme, J.M.C., Horta, N.C.G.: Analog Circuits and Systems Optimization based on Evolutionary Computation Techniques. SCI, vol. 294. Springer, Heidelberg (2010)
[7] Barros, M., Guilherme, J., Horta, N.: GA-SVM feasibility model and optimization kernel applied to analog IC design automation. In: ACM Great Lakes Symposium on VLSI, Stresa-Lago Maggiore, Italy (2007)
[8] Lourenco, N., Horta, N.: GENOM-POF: Multi-Objective Evolutionary Synthesis of Analog ICs with Corners Validation. In: Genetic and Evolutionary Computation Conference, Philadelphia, USA (2012)
[9] Degrauwe, M.G.R., Nys, O., Dijkstra, E., Rijmenants, J., Bitz, S., Goffart, B., Vittoz, E.A., Cserveny, S., Meixenberger, C., Stappen, G., Oguey, H.J.: IDAC: an interactive design tool for analog CMOS circuits. IEEE Journal of Solid State Circuits 22(6), 1106–1116 (1987)
[10] El-Turky, F., Perry, E.E.: BLADES: an artificial intelligence approach to analog circuit design. IEEE Transactions on Computer Aided Design 8(6), 680–692 (1989)
[11] Hershenson, M.D.M., Boyd, S.P., Lee, T.H.: GPCAD: a tool for CMOS op-amp synthesis. In: IEEE/ACM International Conference on Computer-Aided Design, San Jose, CA, USA (1998)
[12] Kuo-Hsuan, M., Po-Cheng, P., Hung-Ming, C.: Integrated hierarchical synthesis of analog/RF circuits with accurate performance mapping. In: International Symposium on Quality Electronic Design, Santa Clara, California, USA (2011)
[13] Gielen, G., Wambacq, P., Sansen, W.M.: Symbolic analysis methods and applications for analog circuits: a tutorial overview. Proc. of IEEE 82(2), 287–304 (1994)
[14] Cheng-Wu, L., Pin-Dai, S., Ya-Ting, S., Soon-Jyh, C.: A bias-driven approach for automated design of operational amplifiers. In: International Symposium on VLSI Design, Automation and Test, Hsinchu, Taiwan (2009)
[15] Phelps, R., Krasnicki, M., Rutenbar, R.A., Carley, L.R., Hellums, J.R.: Anaconda: simulation-based synthesis of analog circuits via stochastic pattern search. IEEE Transactions on Computer Aided Design 19(6), 703–717 (2000)
[16] Alpaydin, G., Balkir, S., Dundar, G.: An evolutionary approach to automatic synthesis of high-performance analog integrated circuits. IEEE Transactions on Evolutionary Computation 7(3), 240–252 (2003)
[17] Deniz, E., Dundar, G.: Hierarchical performance estimation of analog blocks using Pareto Fronts. In: Conference on Ph.D. Research in Microelectronics and Electronics, Berlin, Germany (2010)
[18] Roca, E., Castro-Lopez, R., Fernandez, F.V.: Hierarchical synthesis based on pareto-optimal fronts. In: European Conference on Circuit Theory and Design, Antalya, Turkey (2009)
[19] Maulik, P.C., Carley, L.R., Rutenbar, R.A.: Integer programming based topology selection of cell-level analog circuits. IEEE Transactions on Computer Aided Design 14(4), 401–412 (1995)
[20] Kruiskamp, W., Leenaerts, D.: DARWIN: CMOS opamp synthesis by means of a genetic algorithm. In: ACM/IEEE Design Automation Conference, San Francisco, California, USA (1995)

[21] Ning, Z.Q., Mouthaan, T., Wallinga, H.: SEAS: a simulated evolution approach for analog circuit synthesis. In: IEEE Custom Integrated Circuits Conference, San Diego, California, USA (1991)
[22] McConaghy, T., Palmers, P., Steyaert, M., Gielen, G.G.E.: Trustworthy Genetic Programming-Based Synthesis of Analog Circuit Topologies Using Hierarchical Domain-Specific Building Blocks. IEEE Transactions on Evolutionary Computation 15(4), 557–570 (2011)
[23] Palmers, P., McConnaghy, T., Steyaert, M., Gielen, G.: Massively multi-topology sizing of analog integrated circuits. In: Design, Automation & Test in Europe, Nice, France (2009)
[24] Koza, J.R., Bennett, F.H., Andre, D., Keane, M.A., Dunlap, F.: Automated synthesis of analog electrical circuits by means of genetic programming. IEEE Transactions on Evolutionary Computation 1(2), 109–128 (1997)
[25] Lohn, J.D., Colombano, S.P.: A circuit representation technique for automated circuit design. IEEE Transactions on Evolutionary Computation 3(3), 205–219 (1999)
[26] Sripramong, T., Toumazou, C.: The invention of CMOS amplifiers using genetic programming and current-flow analysis. IEEE Transactions on Computer Aided Design 21(11), 1237–1252 (2002)
[27] Shoou-Jinn, C., Hao-Sheng, H., Yan-Kuin, S.: Automated passive filter synthesis using a novel tree representation and genetic programming. IEEE Transactions on Evolutionary Computation 10(1), 93–100 (2006)
[28] Hongying, Y., Jingsong, H.: Evolutionary design of operational amplifier using variable-length differential evolution algorithm. In: International Conference on Computer Application and System Modeling, Taiyuan, China (2010)
[29] Martens, E., Gielen, G.: Classification of analog synthesis tools based on their architecture selection mechanisms. Integration, the VLSI Journal 41(2), 238–252 (2008)
[30] Deb, K., Pratap, A., Agarwal, S., Meyarivan, T.: A fast and elitist multiobjective genetic algorithm: NSGA-II. IEEE Transactions on Evolutionary Computation 6(2), 182–197 (2002)
[31] Synopsis, Hspice, `http://www.synopsys.com` (accessed December 14, 2011)
[32] Chang, C.C., Lin, C.J.: LIBSVM: A library for support vector machines. ACM Transactions on Intelligent System Technology 2(3), 1–27 (2011)
[33] Deb, K., Agrawal, R.B.: Simulated binary crossover for continuous search space. Complex Systems 9(2), 115–148 (1995)
[34] Montgomery, D.C.: Design and analysis of experiments, 6th edn. John Wiley & Sons, Hoboken (2005)
[35] Hsu, C.W., Chang, C.C., Lin, C.J.: A practical guide to support vector classification. Technical Report, Department of Computer Science, National Taiwan University (2003)
[36] Lourenço, N., Vianello, M., Guilherme, J., Horta, N.: LAYGEN - Automatic Layout Generation of Analog ICs from Hierarchical Template Descriptions. In: Conference on Ph.D. Research in Microelectronics and Electronics, Otranto (Lecce), Italy (2006)
[37] Martins, R., Lourenco, N., Horta, N.: LAYGEN II: Automatic Analog ICs Layout Generator based on a Template Approach. In: Genetic and Evolutionary Computation Conference, Philadelphia, USA (2012)

Part II
Techniques

Chapter 8
Applications of Symbolic Analysis in the Design of Analog Circuits

Francesco Grasso, Antonio Luchetta, and Maria Cristina Piccirilli

Abstract. The use of symbolic techniques in the realization of efficient automated tools for designing analog circuits is described in this chapter. In particular, three phases of the design cycle of an integrated circuit are considered: the simulation phase, the design centering phase, and the fault diagnosis phase. A biquadratic RC active filter was chosen in order to show the potency of several symbolic programs developed by the authors for particular use in these three phases of analog circuit design.

8.1 Introduction

As is the case for the vast majority of engineering products, the realization of an integrated circuit follows a typical development cycle. Starting from an initial set of specifications, a preliminary design is derived, aimed at fulfilling these requirements. This initial design needs to be verified, and this is done at different levels. The first level is simulation, where mathematical models of the device (of increasing complexity) are developed and verified by means of dedicated software tools. The results of this simulation are then used in order to refine the initial set of specifications and to update the design accordingly. Then a new simulation campaign is carried out, with a loop that is repeated until satisfactory results are obtained.

The next step is the realization of a physical prototype that undergoes a series of electrical tests, always aimed at the verification of the requirements. Again, the results of these tests are used to refine the specifications and the design until sufficient confidence in the design is gained, allowing the initiation of the actual production phase. The goal of this iterative process is not only to ensure that all the design parameters fall within the specified tolerance intervals, but also that this is done with sufficient margins. In fact, the larger the margins on the design, the

Francesco Grasso · Antonio Luchetta · Maria Cristina Piccirilli
Department of Electronics and Telecommunications,
University of Firenze
Florence, Italy
e-mail: {francesco.grasso,antonio.luchetta}@unifi.it,
mariacristina.piccirilli@unifi.it

M. Fakhfakh et al. (Eds.): *Analog/RF & Mixed-Signal Circuit Sys. Design*, LNEE 233, pp. 171–194.
DOI: 10.1007/978-3-642-36329-0_8

higher the probability that every single device works correctly; in other words, the higher the final product yield. It is evident that the closer the parameter value to the central point of its tolerance interval, the higher its performance margins, and consequently, the higher the probability that the device fulfills the specifications. This is the reason why the process of design optimization is also termed design centering. It is evident as well that a good design centering carried out in the simulation phase allows to minimize the number of physical prototypes and the number of tests that these prototypes will undergo, with significant cost savings. However, the simulation itself, even if more economical than the physical testing, is an activity that has its costs, typically proportional to the duration of the activity itself. Tools minimizing the simulation activity are therefore also extremely important in the circuit development economy.

When the prototype is tested, it can also happen that some components do not work properly, typically when subjected to excessive stress; in extreme cases, they may even be permanently damaged. Such failures need to be absolutely removed by a proper redesign of the circuit. The problem is how to identify the failed component and the failure cause, since the only available information is the presence of anomalous behavior in the circuit outputs. Identifying the components from a simple check of the test results is in most cases very difficult. The availability of tools able to automatically carry out this kind of identification would therefore be a significant advantage in the design development process.

At present the integrated circuit market is characterized by an increasing level of integration complexity due, for example, to the simultaneous integration of both analog and digital circuits, shorter life cycles, and faster time-to-market requests. Today these needs are hardly achievable in the analog world, characterized by long design cycles. This is due to its inherently more complex nature, to the larger number of elements to be accounted for and traded, and to the lack of supporting CAD tools. In order to keep the pace with the digital world, it was found necessary to drastically cut the design time and cost of analog circuits. The result has been an increasing demand for tools able to assist, improve, automate, and, in general, make the analog design more efficient. Tools of this kind are well established in the digital world; on the contrary, much less advanced is the situation in the case of analog circuits, where the available tools (such as the classical numerical simulators), mostly dedicated to pure analysis, need to be complemented by others, specifically dedicated to design optimization. A way of constructing efficient tools is to exploit symbolic techniques, which allow to gain insight and experience in the behavior and tradeoffs of analog circuits. These techniques, in combination with other ones, such as frequency response analysis and neural networks, can become an important part of the automation of essential design tasks.

The goal of this chapter is to describe some symbolic programs dedicated to the above mentioned design optimization tasks, whose good efficiency is the direct outcome of their symbolic nature. In particular, the simulation phase is discussed in Section 1, the design centering phase in Section 2, and the fault diagnosis phase in Section 3. A biquadratic RC active filter was chosen in order to show the potency of several symbolic programs developed by the authors for particular use in these three phases of analog circuit design.

8.2 Simulation Phase

The first step of analog design is the generation of the theoretical circuit starting from the set of specifications. Subsequently there is the step of design validation. This is an important phase, aimed at ensuring that the pen-and-paper circuit obtained as a result of the theoretical design process is correct. The main tool in this phase is circuit simulation. In fact, the strict connection between theoretical design and its validation relies on both designer knowledge and extensive simulation.

The generic term "simulation" indicates the analysis of the theoretical circuit with software tools. Three different kinds of analysis can be considered: qualitative analysis, numerical analysis, and symbolic analysis. Qualitative analysis defines conceptual relationships between output and input, which assist the designers to choose the proper directions during the design procedure. This kind of analysis is typically performed in a theoretical way, without the aid of software tools. Numerical analysis consists in evaluating the behavior of the considered circuits by using numerical values of the component parameters to obtain numerical values as results of the analysis procedure. This kind of analysis is typically performed with software tools, such as SPICE-like programs. Symbolic analysis provides closed-form symbolic expressions for the characteristics of a circuit. This kind of analysis can be performed by hand, but also and much more efficiently with software tools. In fact a symbolic simulator yields, as its output, analytic expressions in a much shorter time and for more complex circuits with respect to what is achievable by manual procedures.

Symbolic analysis together with numerical analysis can achieve very efficient software tools for the automation of the design validation phase. In fact symbolic simulation [1]-[9] is an important complement to numerical simulation for several reasons. First of all numerical simulation returns as output a series of numbers in tabulated or plotted form, which, even if accurately simulating the circuit behavior, are specific for a particular set of parameter values. Then, if the value of the parameter changes, it is necessary to perform the simulation again. With numerical simulation, given a set of parameters, the behavior of the circuit can be verified in a very short time, but it is much more difficult to understand which circuit elements determine the observed performance and no solution is suggested when the behavior does not meet the specifications. Furthermore a great number of simulations is required in order to verify the performance and check the influence of parameter changes. By contrast, symbolic simulation outputs symbolic expressions valid for any parameter value, thus allowing to obtain real insight into the behavior of the circuit. Furthermore symbolic simulation can give as output not only expressions with all (fully symbolic analysis) or some (semi-symbolic analysis) component parameters expressed as symbols, but also simplified symbolic expressions, where only significant terms appear in the final expressions. In this last case the availability of expressions containing only the dominant contributions allows better understanding of the circuit behavior and helps to choose the most appropriate tradeoffs for reaching the desired final results. In conclusion, numerical simulators serve to verify the performance of already sized circuits, while symbolic simulators are useful to predict the behavior of unsized circuits through the use of relationships such as, for example, transfer functions or poles and zeros, by taking into account design specifications. In this sense numerical and symbolic simulators are complementary and, together, can contribute to create efficient automatic tools for designing analog circuits.

Several symbolic simulators have been developed (see, for example, [10]-[16]) with different capabilities and different analysis tasks. In this section the symbolic program SAPWIN (Symbolic Analysis Program for WINdows), developed by the authors, is presented [17]-[21]. It has been developed by taking into account the following considerations. At present few technical PC programs are available to perform symbolic analysis of analog circuits. Furthermore, even if several of the most famous general purpose commercial or open-source mathematical software packages include symbolic computation capabilities, they are usually neither easy to learn nor agile to use. Moreover they are not optimized for the symbolic calculation of large systems and they do not have a standalone working capability. In the following, SAPWIN is briefly described and its capabilities are shown through its application to a biquadratic RC active filter.

SAPWIN is an integrated package (currently at version 3.0) of schematic capture, symbolic analysis, and graphic post-processing for linear analog circuits. It is completely coded in the C++ programming language and its simulation engine is based on a two-graph method [22]. The program provides several tools to create the scheme of a linear analog circuit, to perform its symbolic analysis, and to show the results in graphic form. In the schematic capture option, the main screen is a white sheet where the user can draw a circuit by using typical Windows tools to copy, cut, paste, move and edit a component or a part of the circuit. All the passive components, controlled sources, and many linear models of active devices (operational amplifiers and small-signal equivalent models of BJT and MOSFET transistors) are available. The program can produce symbolic network functions where each component can appear with its symbolic name or with a numerical value. The graphical post processor is able to show the network function and to plot gain, phase, delay, pole and zero position, time domain step, and impulse response (see [23] for the algorithms of poles/zeros calculation and inversion of the network function).

The symbolic expressions generated by SAPWIN are also saved, in a particular format, in a binary file. Several applications have been developed using SAPWIN as symbolic simulator engine, such as symbolic sensitivity analysis, transient analysis of power electronic circuits, testability evaluation, circuit fault diagnosis. All the programs presented in this chapter are based on the use of SAPWIN.

In order to show the capabilities of SAPWIN in the phase of circuit simulation, let us consider the biquadratic RC active filter named KHN, where the band-pass (BP) output has been chosen. The schematic entry is shown in Fig. 8.1. As soon as the simulation ends, an output window is opened and the symbolic expression is presented. On the right hand side of the output window is the list of relevant parameters and their current values. On the top right the end time for the time response and the frequency interval for the frequency response can be inserted (see Fig. 8.2). The resulting expression can be viewed in both symbolic and numerical form (in this last case the values are those assigned in the schematic editor and now appearing in the parameter edit boxes). The simulation result is given in the Laplace domain, but it is also possible to evaluate it as a phasor at a given frequency (or angular frequency).

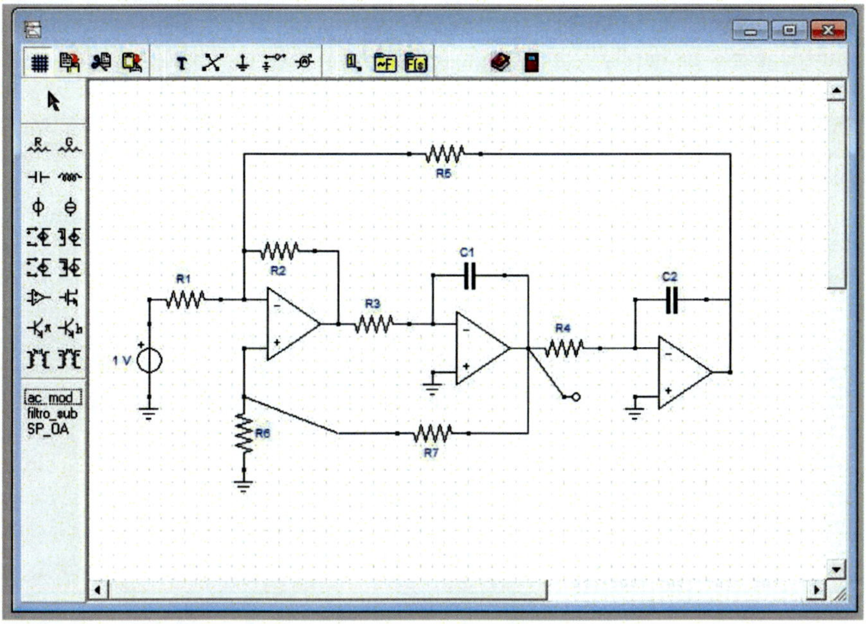

Fig. 8.1 KHN schematic entry

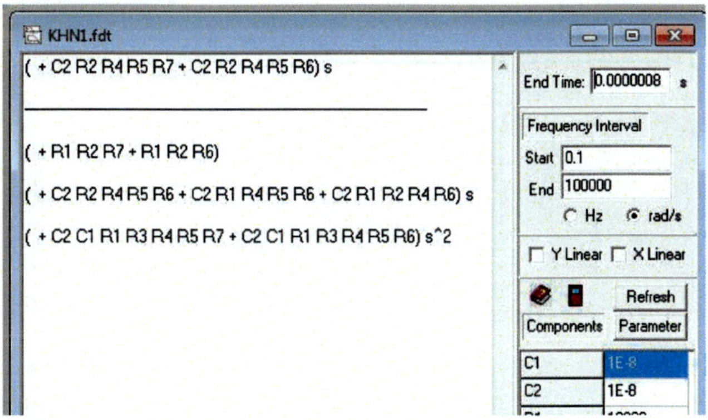

Fig. 8.2 Output window for KHN circuit

The phasor of the response is shown with both the real/imaginary and amplitude/phase representations at the frequency selected in the right side window. For this kind of output, in the list of the component values it is possible to specify complex values for the symbolic independent sources, in the form of amplitude and phase (in degrees).

It should be observed that the symbolic analysis of a circuit, also for medium sized networks, does result in a very large output expression that could be unmanageable or hardly readable. In order to improve the manageability of the generated functions, approximation methods can be used. SAPWIN is able to approximate the output functions, choosing a given error over a range of frequencies and evaluating it after the generation by means of an efficient approximation method.

KHN gain and the corresponding component values (the operational amplifiers are considered ideal) are shown in Fig. 8.3. If the obtained results are unsatisfactory, it is sufficient to assign a new set of values to the component parameters without simulating the circuit again. This allows to obtain the results in a very short time. The gain of KHN for another set of component values is shown in Fig. 8.3, while the corresponding poles and zeros in the complex plane are shown in Fig. 8.4. When a diagram is visualized in the window, it is also possible to add an X-Y cursor to evaluate the numerical values on the diagram. A window, containing the diagrams of the gain and the impulsive response of KHN circuit parameterized with respect to the component C1, is shown in Fig. 8.5. In the figure also a cursor is present on the impulsive response diagram.

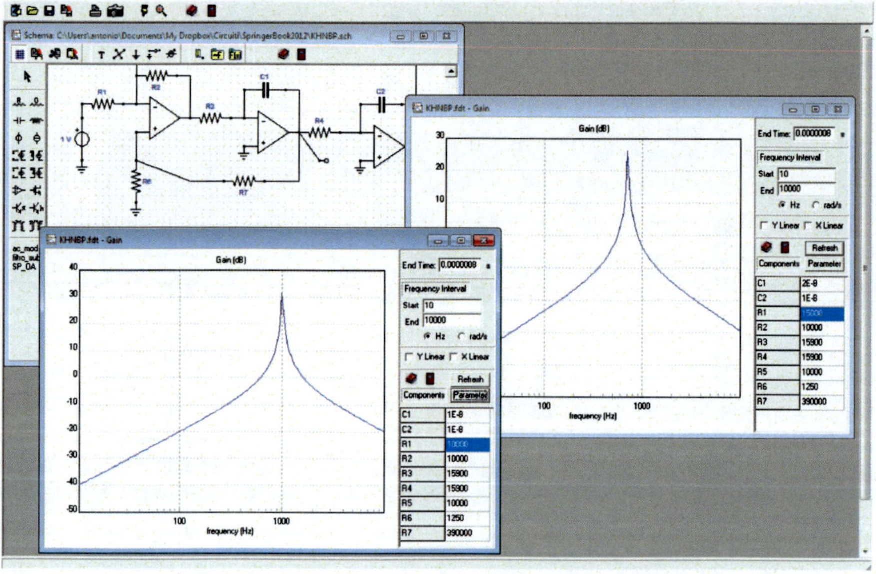

Fig. 8.3 Two KHN gain curves relevant to different sets of component values

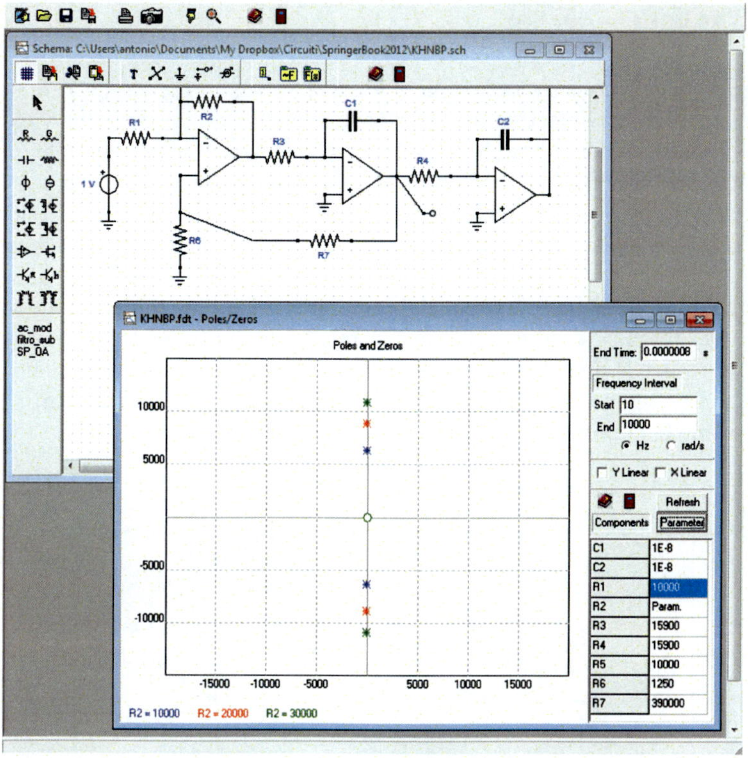

Fig. 8.4 Poles and zeros diagram relevant to the component value set of Fig. 8.3

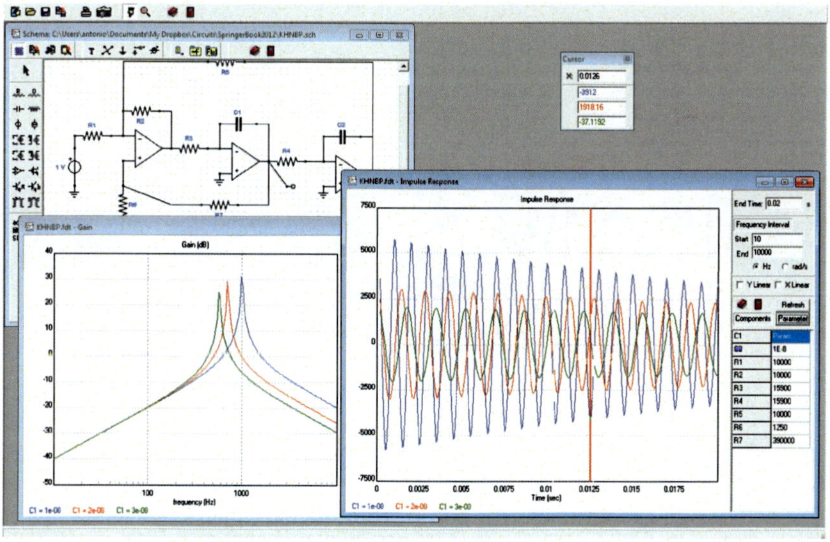

Fig. 8.5 Window containing the diagrams of the gain and impulsive response of KHN circuit parameterized with respect to the component C1

8.3 Design Centering

Design centering consists in assigning a set of values to the circuit parameters to maximize tolerance-intervals for these parameters or to maximize the yield for an assumed statistical distribution [24]-[28]. The interpretation of yield maximization for a circuit with two parameters is reported in Fig. 8.6, where $p_{x,nom}$ are the nominal values as from the design and $p_{x,opt}$ are the parameter values that maximize the yield. The region inside the bold curve indicates the acceptability region, the region inside the ellipse includes all values that are within manufacturing tolerance. Consequently, the grey zone, i.e., the intersection of the above two regions, describes the achievable yield. In order to have an optimal yield it is necessary to select nominal values in the innermost zone of the acceptability region. Design centering is precisely the search for this optimal point, while the optimal point itself is called the design center.

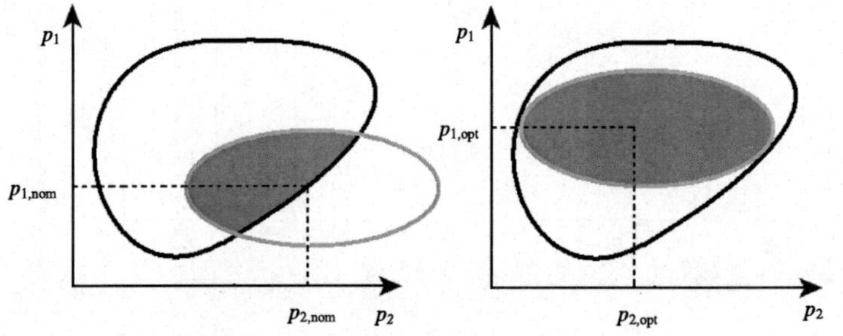

Fig. 8.6 Graphical interpretation of yield maximization

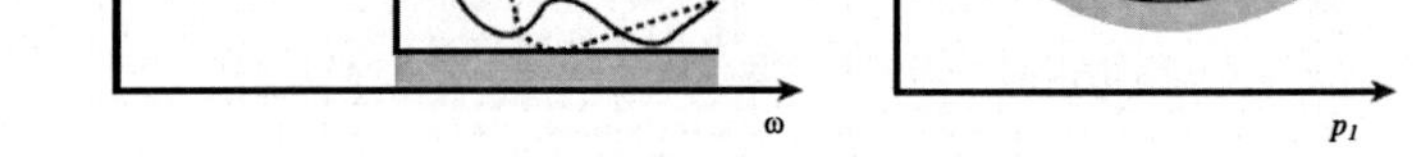

Fig. 8.7 Illustration of the correlation between constrains and acceptability region

A good design centering carried out in the simulation phase allows to minimize the number of physical prototypes and the number of tests that these prototypes will undergo, with significant cost savings. However, the simulation itself, even if more economic that the physical testing, is an activity that has its costs, typically proportional to the duration of the activity itself. Tools allowing to minimize the simulation activity are therefore also extremely important in the circuit development economy. Symbolic analysis can satisfy this requirement not only for the specific problem of design centering, but also for the problem of acceptability region representation. In the following, after a theoretical introduction, two symbolic programs, developed by the authors, are considered. The first one, named SAR (Symbolic Acceptability Region), is able to represent the acceptability region of analog linear circuits, the second one, named DESCEN (DESign CENtering), is dedicated to the design centering problem.

8.3.1 Theoretical Foundations

The acceptability region is defined as a region in the space of circuit parameters for which all the constraints derived by the design specifications are fulfilled. For the design centering and other statistical design tasks the knowledge of acceptability region is required. The correlation between constraints and acceptability region is shown in Fig. 8.7, where **e** is the vector of the parameter values and the acceptability region cross-section is referred to the parameters p_1 and p_2.

The problem of design centering has been traditionally faced in two main ways, the geometrical approach and the statistical approach, even if several methods exist that hybridize these approaches [29], [30]. In the geometrical approach, the acceptability region is approximated by a known geometrical body, such as a polytope or an ellipsoid, then the center of this body, even if approximated, is taken to be the design center. This method has several disadvantages, such as: some limitations associated with the types of geometric bodies used to approximate the acceptability region, the design center dependent on the exact probability distributions of the variables, the hypothesis of convex acceptability region [31]. In the statistical approach, the overall performance of the solution space is estimated by simulating the circuit behavior for a sample of feasible points. The larger the sample, the more accurate the estimation will be. This method has the disadvantage of being computationally expensive when the targeted yield is high [32], but it would give a perfect accuracy if every point in the space was sampled. In any case, the design centering problem starts from the approximation of the acceptability region, otherwise it is necessary to verify if a point **p** in the parameter space is feasible or not, and this requires several circuit simulations. Since symbolic analysis techniques can give noteworthy advantages with respect to numerical techniques in all the applications that require the repetition of a high number of simulations performed on the same circuit topology, they can be advantageously applied also in design centering problems for both the statistical and the geometrical approaches. In the following the symbolic approach to acceptability region representation [33], [34] and to design centering [35], [36] is summarized.

Consider the problem of designing a linear, time-invariant circuit with k design parameters and h constraints. The circuit is represented by a network function

$F(s,\mathbf{p})$ in the s domain, which, generally, is a transfer function. For the sake of simplicity, consider only constraints referred to the amplitude response of the circuit as follows:

$$|F(j\omega_i,\mathbf{p})| \le K_i \text{ for } i = 1\cdots h \tag{8.1}$$

where $\mathbf{p} = [p_1, p_2, \ldots p_k]$ is a point in the parameter space. A generic $F(j\omega,\mathbf{p})$ can be expressed as:

$$F(j\omega,\mathbf{p}) = \frac{a_n(\mathbf{p})(j\omega)^n + \cdots + a_1(\mathbf{p})(j\omega) + a_0(\mathbf{p})}{b_m(\mathbf{p})(j\omega)^m + \cdots + b_1(\mathbf{p})(j\omega) + b_0(\mathbf{p})} \tag{8.2}$$

where the coefficients of both the numerator and the denominator can be considered as a sum of products (SOP) of the circuit parameters, and the degree of this expression, with respect to a single parameter, is always equal to one. Then it is possible to write the expression (8.2) in bilinear form with respect to the q-th parameter as:

$$F(j\omega, p_q) = \frac{A(j\omega) + p_q B(j\omega)}{C(j\omega) + p_q D(j\omega)} \tag{8.3}$$

where the polynomials A, B, C, and D depend only on the frequency, because the numerical value for all the other parameters has been fixed. Equation (8.3) can be expressed as a function of the real and imaginary parts of A, B, C, and D as:

$$F(j\omega, p_q) = \frac{A_r(j\omega) + jA_i(j\omega) + p_q B_r(j\omega) + jp_q B_i(j\omega)}{C_r(j\omega) + jC_i(j\omega) + p_q D_r(j\omega) + jp_q D_i(j\omega)} \tag{8.4}$$

where the subscripts r and i of the polynomials A, B, C, and D indicate the real and imaginary parts of the corresponding polynomials. At this point it is possible to solve the h inequalities in equation (8.1) with respect to the q-th parameter, when the other parameters have a fixed numerical value. Using expression (8.4), the generic j-th inequality in equation (8.1) can be expressed in the following way:

$$|A_r + jA_i + p_q B_r + jp_q B_i| \le K_j |C_r + jC_i + p_q D_r + jp_q D_i| \tag{8.5}$$

That is:

$$\sqrt{(A_r + p_q B_r)^2 + (A_i + p_q B_i)^2} \le K_j \sqrt{(C_r + p_q D_r)^2 + (C_i + p_q D_i)^2} \tag{8.6}$$

where $A_r = A_r(j\omega_j)$, $A_i = A_i(j\omega_j)$, $B_r = B_r(j\omega_j)$, $B_i = B_i(j\omega_j)$, $C_r = C_r(j\omega_j)$, $C_i = C_i(j\omega_j)$, $D_r = D_r(j\omega_j)$, and $D_i = D_i(j\omega_j)$. Taking the square form of the expression (8.6), after easy calculations, the following inequality can be obtained:

$$a_j p_q^2 + b_j p_q + c_j \le 0 \tag{8.7}$$

where:

$$\begin{aligned} a_j &= B_r^2 + B_i^2 - K_j^2 D_r^2 - K_j^2 D_i^2 \\ b_j &= 2\left(A_r B_r + A_i B_i - K_j^2 C_r D_r - K_j^2 C_i D_i\right) \\ c_j &= A_r^2 + A_i^2 - K_j^2 C_r^2 - K_j^2 C_i^2 \end{aligned} \tag{8.8}$$

In case of constrains defined as $\left|F\left(j\omega_j,\mathbf{p}\right)\right| \geq K_j$, it is again easy to verify that $a_j^* p_q^2 + b_j^* p_q + c_j^* \leq 0$, where $a_j^*=-a_j$, $b_j^*=-b_j$, $c_j^*=-c_j$. For each constraint, a range of values for the parameter p_q can be derived by the expression (8.7), once the values of A_r, A_i, B_r, B_i, C_r, C_i, D_r, and D_i have been evaluated for each frequency of interest. This operation can be easily performed as follows, if the symbolic form of the polynomials *A*, *B*, *C*, and *D* is available. Once the unknown parameter p_q has been chosen and the frequency ω_j and all the other parameter values have been fixed, the numerator and the denominator of the expression (8.4) are evaluated for p_q=1 and for p_q=0, bringing to the following equations:

$$\begin{aligned} N_1 &= N\left(j\omega_j, p_q\right)\Big|_{p_q=1} = A_r + jA_i + B_r + jB_i; \\ D_1 &= D\left(j\omega_j, p_q\right)\Big|_{p_q=1} = C_r + jC_i + D_r + jD_i; \\ N_2 &= N\left(j\omega_j, p_q\right)\Big|_{p_q=0} = A_r + jA_i; \\ D_2 &= D\left(j\omega_j, p_q\right)\Big|_{p_q=0} = C_r + jC_i; \end{aligned} \tag{8.9}$$

Now the numerical values of A_r, A_i, B_r, B_i, C_r, C_i, D_r, and D_i are the following:

$$\begin{aligned} A_r &= \mathrm{Re}\{N_2\}; A_i = \mathrm{Im}\{N_2\}; \\ B_r &= \mathrm{Re}\{N_1 - N_2\}; B_i = \mathrm{Im}\{N_1 - N_2\}; \\ C_r &= \mathrm{Re}\{D_2\}; C_i = \mathrm{Im}\{D_2\}; \\ D_r &= \mathrm{Re}\{D_1 - D_2\}; D_i = \mathrm{Im}\{D_1 - D_2\}; \end{aligned} \tag{8.10}$$

This procedure, when applied to each parameter p_q and to each frequency ω_j, allows to determine the range of values satisfying the constraints for each circuit parameter at a time.

Referring to the acceptability region problem, starting from the previous theoretical considerations, the acceptability region can be represented through the determination of two-dimensional (2D) sections relevant to couples of parameters. The algorithm for determining a 2D section is explained in detail in [34]. In this algorithm a couple of parameters, p_x and p_y, is chosen and the shape of the acceptability region is investigated as a 2D cross-section relevant to these two parameters. Only p_x and p_y vary, while all the other parameters are fixed. A pseudo 3D representation of the acceptability region can be simply obtained by the same procedure, through the superposition of the 2D sections relevant to prefixed variations of a third parameter. Of course the procedure is applicable to the case of more than three variable parameters, but, obviously, it is not possible to plot the obtained region.

Referring to the design centering problem, the approach can be summarized as follows [35], [36]. Starting from an initial point $\mathbf{p_0}$ belonging to the acceptability region, the transfer function is evaluated in the frequencies of the constraints by considering only one parameter in symbolic form and assigning to all the others the corresponding value in $\mathbf{p_0}$. By means of the above described algorithm, all the constraints are solved with respect to the symbolic parameter and a range of values satisfying them is determined for it. This range is explored in order to locate the center of the acceptability region with respect to the considered parameter. This exploration is performed by sampling the range and, for each sample, determining the range of values satisfying the constraints for all the other parameters considered one at a time. Also in this phase the above described algorithm is used. By adding the hypervolumes relevant to each sample and considering the half of the whole hypervolume, the numerical value of the symbolic parameter is determined. The initial point $\mathbf{p_0}$ is updated by replacing the initial value of the parameter in $\mathbf{p_0}$ with the freshly-determined numerical value. This procedure is repeated for each parameter. When all the parameter values have been updated, a new value $\mathbf{p_1}$ is obtained and the procedure is repeated again. The iterations end when the difference between the vector $\mathbf{p}$ obtained in the i-th iteration and the one obtained in the (i+1)-th iteration is lower than a fixed value ε. The vector $\mathbf{p}$ obtained in the last iteration is output as the design center.

It is important to note that, by exploiting symbolic simulation techniques, the circuit is simulated only once for determining the symbolic transfer function. During the procedures, only evaluations of the symbolic transfer function are required. Finally, it is worth pointing out that here only the constraints referred to the amplitude response have been considered, because this kind of constraint is the most commonly used. Of course, other kinds of constraints can be considered, such as the phase response or the delay, and the extension to this kind of responses is straightforward.

8.3.2 SAR and DESCEN

The procedure described in the previous section has been implemented in the programs SAR [33], [34] and DESCEN [35], [36], whose input is the symbolic transfer function from SAPWIN.

Referring to the SAR program, once it has loaded the symbolic form of the transfer function from the file generated by SAPWIN, the circuit specifications can be introduced in the constraint window and the acceptability region relevant to two chosen circuit parameters can be obtained. For KHN circuit, whose constraints and nominal values are reported in Table 8.1 and in Table 8.2 respectively, a 2D cross-section of the acceptability region, relevant to the parameters C2 and R1, is shown in Fig. 8.8(a) (all the other parameters are at their nominal value). By selecting the parameter R7, the 3D plot is obtained (Fig. 8.8(b)). The 2D and 3D plots of the acceptability region referred to the same parameters, but with all the other parameters fixed at their design center value, are shown in Fig. 8.8(c) and (d) respectively. The program allows also to represent the contribution of a single constraint to the acceptability region.

Table 8.1 Gain constraints for KHN circuit

Frequency (Hz)	900	1000	1100
Lower limit (dB) >	10	40	12
Upper limit (dB) <	20	43	15

Table 8.2 Nominal values and design center of KHN circuit

Components	C1 nF	C2 nF	R1 kΩ	R2 kΩ	R3 kΩ	R4 kΩ	R5 kΩ	R6 kΩ	R7 kΩ
Nominal	10	10	10	10	15.9	15.9	10	1.25	390
Design center	13.05	10	10	11.29	13.79	15.88	10.00	1.26	509.19

DESCEN allows to use both the above described geometrical and statistical approaches for searching the design center. Furthermore it allows to perform a yield estimation by using Monte Carlo analysis. The statistical approach consists of the implementation of the center of gravity method [24], using the symbolic approach in Monte Carlo analysis. The verification of the constraints for each set of parameter values is performed by replacing numerical values in the symbolic transfer function without simulating the circuit for each set of parameter values.

Referring to KHN circuit and using the constraints reported in Table 8.1, we obtain the design center reported in Table 8.2, where the nominal component values, i.e. the starting point $\mathbf{p}_0$ of the above described procedure, are also reported (with 10% tolerance – E12). The evaluation of the yield using the Monte Carlo analysis gives the following results: nominal values 40.1%, design center values 90.6%.

8.4 Fault Diagnosis

In the prototype characterization phase, it can happen that some components do not work properly, typically when subjected to excessive stress; in extreme cases, they may even be permanently damaged. Such failures need to be absolutely removed by a proper redesign of the circuit. The availability of tools able to automatically carry out the faulty component identification would be a significant advantage in the design development process. Also in this case symbolic analysis can play a very useful role not only in the fault diagnosis phase, but also in the test point selection.

For analog circuits, multiple parametric faults are as significant as (or even more significant) large single parameter variations or catastrophic faults. Parametric fault diagnosis is based on a set of equations which are nonlinear in the unknown component parameter values. These equations are related to a set of measurements carried out on specific points of the circuit, called test points. The test point selection is a non-trivial operation, because not all the possible test points can be easily accessed. For example, it is usually very difficult to measure currents without breaking connections. In other words, the test point selection must take into account practical measurement problems strictly tied with the used technology and with the application field of the circuit. So, in order to perform a good test point selection, it is necessary to have a quantitative index for comparing the different possible choices. The testability measure concept meets this requirement.

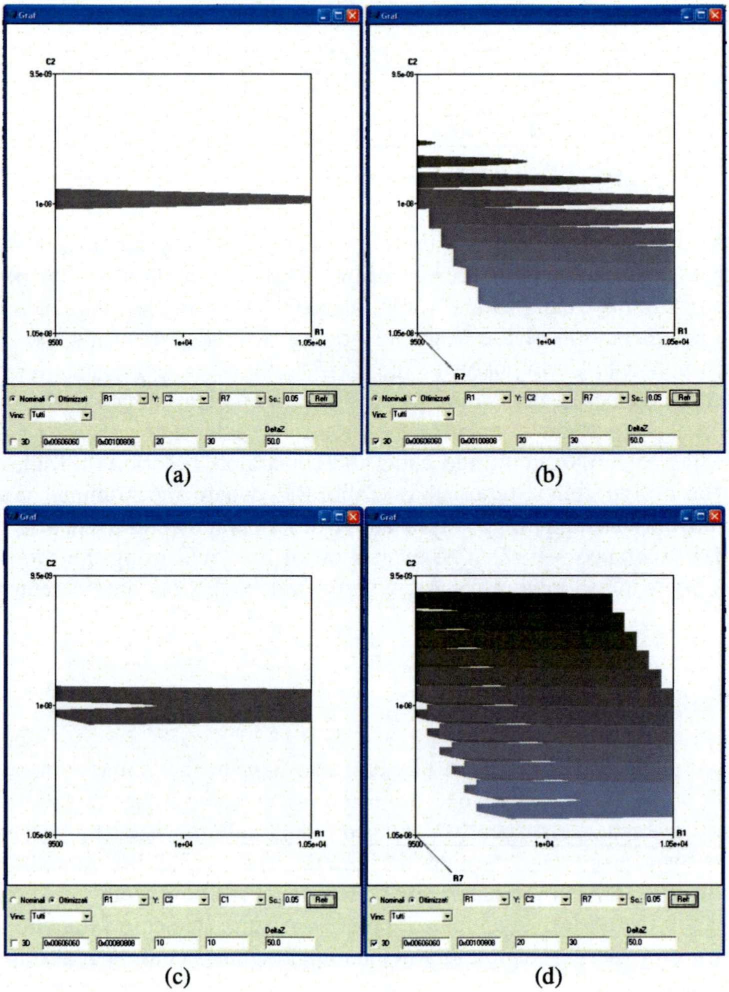

Fig. 8.8 Screenshots of SAR program: (a) 2D (nominal values), (b) 3D (nominal values), (c) 2D (design center values), (d) 3D (design center values)

Consequently, fault diagnosis can be split into two different steps: testability analysis and fault location. Testability analysis consists of testability evaluation and ambiguity group determination; by contrast, fault location involves determining the faulty components. In the following, these two steps are considered for analog, linear, time-invariant circuits, highlighting the advantages of symbolic analysis. After a theoretical introduction, two symbolic programs, developed by the authors, are considered. The first one, named TAGA (Testability and Ambiguity Group Analysis), is able to determine testability and canonical ambiguity groups, the second one, named AMUD (Automatic MUltifrequency Diagnosis), is able to perform the fault diagnosis by selecting the best measurement frequencies. Both the programs are included in the SAPWIN package.

8.4.1 Testability and Ambiguity Groups

One of the most used definitions of testability is associated with the solvability degree of the nonlinear fault equations used in parametric fault diagnosis techniques and it indicates the ambiguity resulting from an attempt to solve such equations in a neighborhood of almost any failure [37]-[40]. Therefore, this testability measure allows to know a priori if a unique solution of the fault diagnosis problem exists. Furthermore, if this solution does not exist, it gives a quantitative measure of how far we are from it, i.e. how many components cannot be diagnosed with the given test point set.

When testability is low, an important concept is that of ambiguity group. An ambiguity group is, essentially, a group of components where, in the case of a fault, it is not possible to uniquely identify the faulty component. A canonical ambiguity group is a "minimal" ambiguity group, i.e. a group that does not contain, within it, ambiguity groups of lower order. The canonical ambiguity groups give information about the solvability of the fault diagnosis problem with respect to each component, in case of bounded number of faults (*k*-fault hypothesis) [41].

Summarizing, once a set of test points has been selected, independently of the method effectively used in the fault location phase, the testability measure gives a theoretical and rigorous upper limit to the degree of solvability of fault diagnosis problem at global level, while the ambiguity group determination gives the solvability degree at component level. If these important concepts are not properly taken into account, the quality of the obtained results is severely limited [42]. So, testability analysis is essential to both the designer, who must know which test points have to be accessible, and the test engineer, who must know how many and which parameters can be uniquely isolated by the planned tests.

Referring to parametric fault diagnosis, the testability measure T is given by the maximum number of linearly independent columns of the Jacobian matrix associated with the fault diagnosis equations. By considering the circuit network functions as fault diagnosis equations, it has been demonstrated [43] that the testability T is equal to the rank of a matrix $\mathbf{B}_C$, independent of the complex frequency s, whose entries are constituted by the derivatives of the coefficients of the fault diagnosis equations with respect to the potentially faulty circuit parameters. The availability of network functions in symbolic form strongly reduces the computational effort in the determination of the $\mathbf{B}_C$ matrix entries, because they can be simply led back to derivatives of sums of products. Furthermore the testability and ambiguity group determination can be performed by assigning arbitrary values to the components, because testability does not depend on component values [38].

Let us consider the following K fault diagnosis equations:

$$h_l(\mathbf{p},s)=\frac{N_l(\mathbf{p},s)}{D(\mathbf{p},s)}=\frac{\sum_{i=0}^{n_l}\frac{a_i^{(l)}(\mathbf{p})}{b_m(\mathbf{p})}\cdot s^i}{s^m+\sum_{j=0}^{m-1}\frac{b_j(\mathbf{p})}{b_m(\mathbf{p})}\cdot s^j}\qquad l=1,....K \tag{8.11}$$

where $\mathbf{p}=[p_1,p_2,...,p_R]^t$ is the vector of the potentially faulty parameters. The Jacobian matrix $\mathbf{B_C}$ of this system can be considered as the testability matrix:

$$\mathbf{B_C}=\begin{bmatrix} \frac{\partial \frac{a_0^{(1)}}{b_m}}{\partial p_1} & \frac{\partial \frac{a_0^{(1)}}{b_m}}{\partial p_2} & \cdots & \frac{\partial \frac{a_0^{(1)}}{b_m}}{\partial p_R} \\ \vdots & \vdots & \ddots & \vdots \\ \frac{\partial \frac{b_{m-1}}{b_m}}{\partial p_1} & \frac{\partial \frac{b_{m-1}}{b_m}}{\partial p_2} & \cdots & \frac{\partial \frac{b_{m-1}}{b_m}}{\partial p_R} \end{bmatrix} \tag{8.12}$$

Independently of the used fault location method, the testability value $T=\text{rank}(\mathbf{B_C})$ gives information on the solvability degree of the problem, as explained below:

- if $T=R$, where R is the number of unknown elements, the parameter values can be theoretically uniquely determined starting from a set of measurements carried out on the test points;
- if $T<R$, a locally unique solution can be determined only if $R–T$ components are considered not faulty.

Generally T is not maximum and the hypothesis of a bounded number k of faulty elements is made (k-fault hypothesis), where $k \leq T$. Then, the testability gives the solvability degree of the fault diagnosis problem and, consequently, the maximum possible k.

The matrix $\mathbf{B_C}$ gives other information besides the global solvability degree of the fault diagnosis problem. In fact, by observing that each column is relevant to a specific parameter of the circuit and by considering the linearly dependent columns of $\mathbf{B_C}$, other information can be obtained. For example, if a column is linearly dependent with respect to another one, this means that a variation of the corresponding parameter provides a variation on the fault equation coefficients indistinguishable with respect to that produced by the variation of the parameter corresponding to the other column. This means that the two parameters are not testable and they constitute an ambiguity group of the second order. By extending this reasoning to groups of linearly dependent columns of $\mathbf{B_C}$, ambiguity groups of higher order can be found. In the case of low testability and k-fault hypothesis, whatever fault location method is used, it is necessary to be able to select, as potentially faulty parameters, a set of elements that represents at best all the circuit components. To this end, the determination of both the canonical ambiguity groups and surely testable group is of fundamental importance. As reported in [41], a set of k parameters constitutes a canonical ambiguity group of order k if the corresponding k columns of the testability matrix $\mathbf{B}_C$ are linearly dependent and every sub-set of this group of columns is constituted by linearly independent columns. A set of n parameters, whose corresponding columns in the testability matrix $\mathbf{B}_C$ do not belong to any ambiguity group, constitutes a surely testable group of order n.

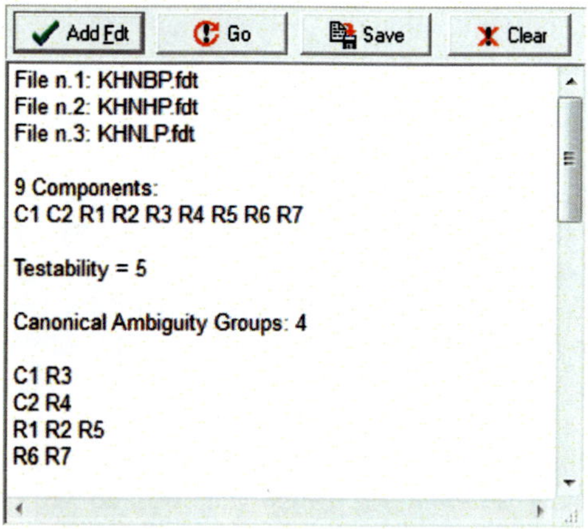

Fig. 8.9 TAGA results for the circuit in Fig. 8.1

In order to better understand the importance of canonical ambiguity groups, let us consider KHN circuit shown in Fig. 8.1, where the outputs of the three operational amplifiers are the test points. The program TAGA determines testability and canonical ambiguity groups. It exploits the SVD (Singular-Value Decomposition), which allows to obtain the effective numerical rank of the Jacobian matrix, and an efficient procedure for canonical ambiguity group determination [44]. The program results are shown in Fig. 8.9. As it can be seen, there are four canonical ambiguity groups without elements in common. Three groups are of the second order and in this case it is impossible to select a set of components giving a unique solution [41]. There are no surely testable elements. As the testability is equal to five, we can take into account at most a 5-fault hypothesis, i.e. a possible solution can be obtained if only five component values are considered as unknowns. The elements to be selected as representative of the circuit components are just one component from each second order canonical ambiguity group and two components from the third order canonical ambiguity group. This group of components is defined as testable. Suppose, for example, a single fault case. Independently of the used fault location method, if the obtained solution gives as potentially faulty element a component belonging to a second order canonical ambiguity group, it is only possible to know that there is a fault in this ambiguity group, but this fault cannot be exactly located, because there is not a unique solution. On the contrary, if a component belonging to the third order ambiguity group is located as faulty element, there is a unique solution, and then the fault can be located with certainty. In fact a fault in a component of this group can be counterbalanced only by simultaneous faults on all the other components of the same group. However, by the hypothesis of single fault, this situation cannot occur.

From the example, it is possible to understand that a testable group is a group of potentially faulty components giving solution to the problem of fault location. It permits to confine the presence of faults to well-defined groups of components [41]. A testable group is easily obtainable through a combinatorial procedure, starting from the canonical ambiguity group knowledge.

8.4.2 Fault Location

During fault diagnosis, a symbolic approach can yield advantages not only in computational terms, but also because it automatically includes the testability analysis in the fault diagnosis procedure.

Symbolic techniques are used in parametric fault diagnosis methods based on frequency domain measurements and on the k-fault hypothesis. The fault location techniques can be substantially split in two groups: techniques based on a bilinear decomposition of the fault equations and techniques based on a Newton-Raphson approach. The techniques based on a bilinear decomposition of the fault equations [45]-[48] are usually suitable for single and double fault cases, because they become excessively complex for a larger number of faults. The Newton-Raphson based techniques are generally suitable for any possible fault hypothesis [49]. Furthermore they can also be useful for the search of the best measurements frequencies. In fact, this choice influences the fault location, because the solution of the fault diagnosis equations is perturbed by measurement errors and component tolerances. The choice of a suitable set of measurement frequencies allows to minimize the effect of these perturbations. The use of symbolic techniques may turn out to be very useful in solving this problem. In [50]-[53] some procedures for selecting the set of frequencies which leads to a good location of parametric faults in analog linear circuits are reported. In [53] the fault diagnosis procedure used in the program AMUD is described in detail and in the following a brief description is reported.

Starting from the testability analysis of the circuit under test, a set of testable components is determined and a frequency dependent sensitivity matrix relevant to these components is built. At this point, by referring to the algebraic indices of this matrix as the condition number and the norm of its inverse, it is possible to define a Test Index (TI) that gives us a quantitative measure of the requirements that the set of measurement frequencies has to satisfy to minimize the effects of measurement errors and component tolerances. The frequency set determination is automatically performed by a genetic algorithm (GA) that minimizes TI. The Newton-Raphson algorithm is used to solve fault diagnosis equations similar to those in equation (8.11). Starting from the nominal component values, the genetic algorithm determines the test frequencies. The fault equations are solved with the

Newton-Raphson algorithm by using measurements carried out on the frequencies determined by GA. Starting from the obtained solution, GA determines new test frequencies. The Newton-Raphson algorithm determines a new solution and, if the relative difference with the value found at the previous step is lower than a fixed value δ, the procedure ends. The fault diagnosis solution corresponds to the last determined parameter values and the test frequency set is the one used in the last application of the Newthon-Raphson algorithm. The whole procedure, apart from testability analysis, is performed by AMUD. The availability of network functions in symbolic form strongly reduces the computational effort in both the testability analysis phase and the determination of the sensitivity matrix.

Let us consider the nominal values (10% tolerance – E12) in Table 8.2 for KHN circuit in Fig. 8.1. Only one test point, corresponding to BP output, has been selected. SAPWIN performs the symbolic analysis and yields the circuit network function in symbolic form. TAGA gives a testability value T = 3 and the testable group constituted by R3, R5, and R6 is chosen. Since the unknowns are three and the network function is one, it is necessary to add other two fault equations for solving the problem. This is made by choosing three frequencies in which the network function and measurements are considered. By considering for the components their numerical value, GA finds the first set of frequencies: f_1=800.71 Hz, f_2=1111.54 Hz, f_3=1296.0 Hz. At this point a double fault can be simulated by replacing the nominal values of R3 and R5 with the values R3=17 kΩ and R5=8.5 kΩ. The measures of the amplitude of V_{BP} with the faulty component values performed at the frequencies f_1, f_2, and f_3 are affected by an error with Gaussian distribution ($\mu = 0.0$, $\sigma = 0.05$). The constant δ has been chosen to be equal to 2%. The program AMUD gives the following values for the testable components: R3'=16.916 kΩ, R5'= 6.721 kΩ and R6'=1236 Ω. The analysis of the results already suggests that R3 and R5 are the faulty components. Then the calculation of the best frequencies with this value of the components is performed again by AMUD. The new set of frequencies is: f_1= 750.0 Hz, f_2=1050.0 Hz, f_3=1620.0 Hz. By repeating the diagnosis procedure with this new set of frequencies, the following values of the testable components are determined: R3"=17.026 kΩ, R5"= 8.543 kΩ, and R6"=1255 Ω. Comparing these values with the previous ones, the following percentage deviations are obtained:

$$\Delta(R3'')\% = (|R3'' - R3'|/ R3') \cdot 100 = 0.65\%$$
$$\Delta(R5'')\% = (|R5'' - R5'|/ R5') \cdot 100 = 27.11\%$$
$$\Delta(R6'')\% = (|R6'' - R6'|/ R6') \cdot 100 = 1.54\%$$

Considering, in particular, the value of Δ(R5"), the diagnosis procedure is not completed. Therefore, it is repeated again by using as starting point the previously obtained component values. The new set of frequencies is now: f_1= 923.0 Hz, f_2=1070.0 Hz, f_3=1384.0 Hz. With this set of frequencies the diagnosis procedure gives the following values of the testable components: R3'"= 16.890 kΩ, R5'"=8.582 kΩ, and R6'"= 1258 Ω. Comparing again the last values with the previous ones, the new percentage deviations are obtained: Δ(R3''')% = 0.80%, Δ(R5''')% = 0.46%, Δ(R6''')% = 0.24%.

Now all the percentage deviations are less than δ, then the procedure is ended. By comparing the obtained values with the nominal ones, the faulty components are R3 and R5. Comparing these values with the actual fault values, we have 0.65% and 0.96% error, respectively.

A detailed overview of the procedures of fault location based on the use of symbolic techniques is reported in [54], where additional information on testability analysis and on the application of symbolic techniques to nonlinear analog circuits are given [55]. Furthermore, a fault diagnosis technique applicable to nonlinear analog circuits is summarized [56].

Finally, it is important to remark that the most important results achieved today in the application of symbolic techniques to fault diagnosis are relevant to testability analysis, where the symbolic approach gives excellent results. For fault location, the symbolic approach is not the only possible one [57]-[59]; for example, good results have been obtained also by using neural networks or genetic algorithms. However, since testability analysis is the initial step in the analog fault diagnosis, necessary for any kind of fault location procedure, the symbolic approach is indeed very useful and can give a noteworthy contribution for reducing the gap between analog and digital fields.

8.5 Conclusion

The use of symbolic techniques in the realization of efficient automatic tools for designing analog circuits has been described in this chapter. In particular three phases of the design cycle of an integrated circuit have been considered: the simulation phase, the design centering phase, and the fault diagnosis phase. A biquadratic RC active filter was chosen in order to show the potency of several symbolic programs developed by the authors in order to optimize efficiency during these three phases. All the programs presented in this chapter work with linear or linearized analog circuits. However the applicability only to linear circuits is a minimal restriction, since, at present, the analog part of modern complex systems is almost all linear, while the nonlinear functions are moved toward the digital part [60]. Furthermore the programs can simulate circuits of small or medium dimensions and can be applied to large dimension circuits by partitioning them into smaller parts. However, symbolic analysis allows to locate the dominant parameters of a circuit. Then, for large circuits, an alternative to partition could be to select only the dominant terms of the circuit under analysis and to consider approximate expression of the network functions.

The software packages mentioned in the chapter are available for downloading at the page http://www.cirlab.unifi.it/Sapwin.

References

[1] Gielen, G., Wambacq, P., Sansen, W.M.: Symbolic analysis methods and applications for analog circuits: a tutorial overview. Proceedings of the IEEE 82(2), 287–304 (1994)

[2] Wambacq, P., Dobrovolny, P., Gielen, G., Sansen, W.: Symbolic analysis of large analog circuits using a sensitivity-driven enumeration of common spanning trees. IEEE Transactions on Circuits and Systems - II: Analog and Digital Signal Processing 45(10), 1342–1350 (1998)

[3] Gielen, G., Wambacq, P., Sansen, W.M.: Symbolic network analysis methods for practical analog integrated circuits: a survey. IEEE Transactions on Circuits and Systems - II: Analog and Digital Signal Processing 45(10), 1331–1341 (1998)

[4] Daems, W., Verhaegen, W., Wambacq, P., Gielen, G., Sansen, W.: Evaluation of error-control strategies for the linear symbolic analysis of analog integrated circuits. IEEE Transactions on Circuits and Systems - I: Fundamental Theory and Applications 46(5), 594–606 (1999)

[5] Verhaegen, W., Gielen, G.: Efficient DDD-based symbolic analysis of linear analog circuits. IEEE Transactions on Circuits and Systems - II: Analog and Digital Signal Processing 49(7), 474–487 (2002)

[6] Vanassche, P., Gielen, G., Sansen, W.: Symbolic modeling of periodically time-varying systems using harmonic transfer matrices. IEEE Transactions on Computer-Aided Design of Integrated Circuits and Systems 21(9), 1011–1023 (2002)

[7] Daems, W., Gielen, G., Sansen, W.: Circuit simplification for symbolic analysis of analog integrated circuits. IEEE Transactions on Computer-Aided Design of Integrated Circuits and Systems 21(4), 395–407 (2002)

[8] Fernandez, F.V., Rodriguez-Vazquez, A., Huertas, J.L., Gielen, G.: Symbolic Analysis Techniques – Applications to Analog Design Automation. IEEE Press, New York (1998)

[9] Lin, P.M.: Symbolic Network Analysis. Elsevier, New York (1991)

[10] Gielen, G., Walscharts, H., Sansen, W.: ISAAC: A symbolic simulator for analog integrated circuits. IEEE Journal Solid-State Circuits 24(6), 1587–1597 (1989)

[11] Swings, K., Sansen, W.: ARIADNE: A constraint-based approach to computer-aided synthesis and modeling of analog integrated circuits. Analog Integrated Circuits and Signal Processing 3(3), 197–215 (1993)

[12] Li, B., Gu, D.: SSCNAP: A program for symbolic analysis of switched capacitor circuits. IEEE Transactions on Computer-Aided Design of Integrated Circuits and Systems 11(3), 334–340 (1992)

[13] Cheng, Y., Fujii, R.: SAUCES: A sensitivity analysis program for analog circuit design. In: Proc. IEEE International Symposium on Circuits and Systems, San Diego, pp. 1175–1178 (May 1992)

[14] Fernandez, F.V., Rodriguez-Vazquez, A., Huertas, J.L.: ASAP: A program for the symbolic analysis of analog integrated circuits. In: Proc. EUROASIC Conference, Paris, France, May 29-June 1, pp. 80–85 (1990)

[15] DeGrauwe, M., Goffart, B., Joss, B., Rijmenants, J., Meixenberger, C., Schwarz, T., Litsios, J., Seda, S., Pierre, M., Jongsma, J., Scarnera, C., Hornstein, T., Deck, P.: The ADAM analog design automation system. In: Proc. IEEE International Symposium on Circuits and Systems, New Orleans, pp. 820–822 (May 1990)

[16] Wierzba, G.M., Srivastava, A., Joshi, V., Noren, K.V., Svoboda, J.A.: SSPICE – A symbolic SPICE program for linear active circuits. In: Proc. 32nd Midwest Symposium on Circuits and Systems, Champaign, IL, August 14-16, pp. 1197–1201 (1989)

[17] Liberatore, A., Luchetta, A., Manetti, S., Piccirilli, M.C.: A new symbolic program package for the interactive design of analog circuits. In: Proc. IEEE International Symposium on Circuits and Systems (ISCAS 1995), Seattle WA, April 30-May 3, pp. 2209–2212 (1995)
[18] Luchetta, A., Manetti, S., Reatti, A.: SAPWIN - A symbolic simulator as a support in electrical engineering education. IEEE Transactions on Education and Enclosed CD-ROM Issue 44(2), 213–213 (2001)
[19] Huelsman, L.P.: SAPWIN, symbolic analysis program for Windows - PC programs for engineers. IEEE Circuits and Devices Magazine 6(2), 4–6 (1996)
[20] Huelsman, L.P.: Symbolic analysis – A tool for teaching undergraduate circuit theory. IEEE Transaction on Education 39(2), 243–250 (1996)
[21] Fedi, G., Giomi, R., Luchetta, A., Manetti, S., Piccirilli, M.C.: SAPWIN 2.0: a symbolic software tool for educational purposes in analysis and synthesis of analog circuits. In: Proc. of International Conference on System Science and Engineering, ICSSE, Los Angeles, CA, May 16-22, pp. 33–36 (1999)
[22] Grimbleby, J.B.: Algorithm for finding the common spanning trees of two graphs. Electronics Letters 17(13), 470–471 (1981)
[23] Vlach, J., Singhal, K.: Computer Methods for Circuit Analysis and Design. Van Nostrand Reinhold, New York (1994)
[24] Spence, R., Soin, R.S.: Tolerance Design of Integrated Circuits. Addison-Wesley, Reading (1988)
[25] Styblinski, M.A.: Statistical design optimization. In: Chen, W.K. (ed.) The Circuits and Filters Handbook, ch. 50. CRC Press, Boca Raton (2003)
[26] Graeb, H.E.: Analog Design Centering and Sizing. Springer (2007)
[27] Abdel-Malek, H.L., Hassan, A.-K.S.O., Heaba, M.H.: A boundary gradient search technique and its applications in design centering. IEEE Transactions on Computer-Aided Design of Integrated Circuits and Systems 18(11), 1654–1661 (1999)
[28] Sapatnekar, S.S.: Design by optimization. In: Chen, W.K. (ed.) The Circuits and Filters Handbook, ch. 49. CRC Press, Boca Raton (2003)
[29] Feldman, P., Director, S.W.: Accurate and efficient evaluation of circuit yield and yield gradients. In: Proc. International Conference on Computer-Aided Design, Santa Clara, CA, November 11-14, pp. 120–123 (1991)
[30] Bernardo, M.C., Buck, R., Liu, L., Nazaret, W.A., Sacks, J., Welch, W.J.: Integrated circuit design optimization using a sequential strategy. IEEE Transactions on Computer-Aided Design of Integrated Circuits and Systems 11(3), 361–372 (1992)
[31] Sapatnekar, S.S., Vaidya, P.M., Kang, S.-M.: Convexity-based algorithms for design centering. IEEE Transactions on Computer-Aided Design of Integrated Circuits and Systems 13(12), 1536–1549 (1994)
[32] Abdel-Malek, H.L., Hassan, A.-K.S.O., Heaba, M.H.: The ellipsoidal technique for design centering and region approximation. IEEE Transactions on Computer-Aided Design of Integrated Circuits and Systems 10(8), 1006–1014 (1991)
[33] Grasso, F., Manetti, S., Piccirilli, M.C.: SAR: a symbolic program for acceptability region representation in analog circuit design. In: Proc. Xth International Workshop on Symbolic and Numerical Methods, Modeling and Applications to Circuit Design (SM2ACD 2008), Erfurt, Germany, October 7-8, pp. 142–148 (2008)
[34] Grasso, F., Manetti, S., Piccirilli, M.C.: A method for acceptability region representation in analogue linear networks. International Journal of Circuit Theory and Applications 37, 1051–1061 (2009)
[35] Grasso, F., Manetti, S., Piccirilli, M.C.: A new geometrical approach to design centering of analog circuits. In: Proc. SPIE, Sevilla, vol. 5837, pp. 937–944 (May 2005)

[36] Grasso, F., Manetti, S., Piccirilli, M.C.: A symbolic approach to design centering of analog circuits. Microelectronics Reliability 47(8), 1288–1295 (2007)
[37] Saeks, R.: A measure of testability and its application to test point selection theory. In: Proc. 20th Midwest Symposium on Circuits and Systems, Lubbock, TX, August 15-17, pp. 576–583 (1977)
[38] Sen, N., Saeks, R.: Fault diagnosis for linear systems via multifrequency measurement. IEEE Transactions on Circuits and Systems 26(7), 457–465 (1979)
[39] Chen, H.M.S., Saeks, R.: A search algorithm for the solution of multifrequency fault diagnosis equations. IEEE Transactions on Circuits and Systems 26(7), 589–594 (1979)
[40] Saeks, R., Sen, N., Chen, H.M.S., Lu, K.S., Sangani, S.: Fault analysis in electronic circuits and systems. Technological Report, Institute for Electronics Science, Texas Technological University, Lubbock (1978)
[41] Fedi, G., Manetti, S., Piccirilli, M.C., Starzyk, J.: Determination of an optimum set of testable components in the fault diagnosis of analog linear circuits. IEEE Transactions on Circuits and Systems - I: Fundamental Theory and Applications 46(7), 779–787 (1999)
[42] Fedi, G., Manetti, S., Piccirilli, M.C.: Comments on Linear circuit fault diagnosis using neuromorphic analyzers. IEEE Transactions on Circuits and Systems - II: Analog and Digital Signal Processing 46(4), 483–485 (1999)
[43] Liberatore, A., Manetti, S., Piccirilli, M.C.: A new efficient method for analog circuit testability measurement. In: Proc. IEEE Instrumentation and Measurement Technology Conference, Hamamatsu, Japan, pp. 193–196 (May 1994)
[44] Manetti, S., Piccirilli, M.C.: A singular-value decomposition approach for ambiguity group determination in analog circuits. IEEE Transactions on Circuits and Systems - I: Fundamental Theory and Applications 50(4), 477–487 (2003)
[45] Fedi, G., Liberatore, A., Luchetta, A., Manetti, S., Piccirilli, M.C.: A symbolic approach to the fault location in analog circuits. In: Proc. IEEE International Symposium on Circuits and Systems (ISCAS 1996), Atlanta, GA, May 14, pp. 810–813 (1996)
[46] Catelani, M., Fedi, G., Giraldi, S., Luchetta, A., Manetti, S., Piccirilli, M.C.: A new symbolic approach to the fault diagnosis of analog circuits. In: Proc. IEEE Instrumentation and Measurements Technical Conference, Brussels, Belgium, June 4-6, pp. 1182–1185 (1996)
[47] Catelani, M., Fedi, G., Giraldi, S., Luchetta, A., Manetti, S., Piccirilli, M.C.: A fully automated measurement system for the fault diagnosis of analog electronic circuits. In: Proc. XIV IMEKO World Congress, Tampere, Finland, June 1-6, pp. 52–57 (1997)
[48] Fedi, G., Luchetta, A., Manetti, S., Piccirilli, M.C.: Multiple fault diagnosis of analog circuits using a new symbolic approach. In: Proc. VI International Workshop on Symbolic Methods and Applications in Circuit Design, Lisbon, Portugal, October 12-13, pp. 139–143 (2000)
[49] Fedi, G., Giomi, R., Luchetta, A., Manetti, S., Piccirilli, M.C.: On the application of symbolic techniques to the multiple fault location in low testability analog circuits. IEEE Transactions on Circuits and Systems - II: Analog and Digital Signal Processing 45(10), 1383–1388 (1998)
[50] Grasso, F., Luchetta, A., Manetti, S., Piccirilli, M.C.: Recent advances in symbolic techniques for analog fault diagnosis. In: Proc. VII International Workshop on Symbolic Methods and Appications in Circuit Design, Sinaia, Romania, 10, October 10-11, pp. 21–24 (2002)

[51] Grasso, F., Luchetta, A., Manetti, S., Piccirilli, M.C.: Symbolic techniques for the selection of test frequencies in analog fault diagnosis. Analog Integrated Circuits and Signal Processing 40(3), 205–213 (2004)

[52] Grasso, F., Manetti, S., Piccirilli, M.C.: An approach to analog fault diagnosis using genetic algorithms. In: Proc. Mediterranean Electrotechnical Conference (MELECON 2004), Dubrovnik, Croatia, May 12-15, pp. 111–114 (2004)

[53] Grasso, F., Luchetta, A., Manetti, S., Piccirilli, M.C.: A method for the automatic selection of test frequencies in analog fault diagnosis. IEEE Transactions on Instrumentation and Measurements 56(6), 2322–2329 (2007)

[54] Manetti, S., Piccirilli, M.C.: Symbolic function approaches for analogue fault diagnosis. In: Sun, Y. (ed.) Test and Diagnosis of Analogue, Mixed-Signal and RF Integrated Circuits, ch. 2. IET-The Institution of Engineering and Technology, London (2008)

[55] Fedi, G., Giomi, R., Manetti, S., Piccirilli, M.C.: A symbolic approach for testability evaluation in fault diagnosis of nonlinear analog circuits. In: Proc. IEEE International Symposium on Circuits and Systems (ISCAS 1998), Monterey, CA, May 31-June 3, pp. 9–12 (1998)

[56] Manetti, S., Piccirilli, M.C.: Symbolic simulators for the fault diagnosis of nonlinear analog circuits. Analog Integrated Circuits and Signal Processing 219, 59–72 (1993)

[57] Luchetta, A., Manetti, S., Piccirilli, M.C.: Critical comparison among some analog fault diagnosis procedures based on symbolic techniques. In: Proc. Design, Automation & Test in Europe (DATE 2002), Paris, France, March 4-8, p. 1105 (2002)

[58] Grasso, F., Luchetta, A., Manetti, S., Piccirilli, M.C.: Symbolic techniques in parametric fault diagnosis of analog circuits. In: Proc. Baltic Electronics Conference (BEC 2002), Tallin, Estonia, October 6-9, pp. 271–274 (2002)

[59] Cannas, B., Fanni, A., Manetti, S., Montisci, A., Piccirilli, M.C.: Neural network-based analog fault diagnosis using testability analysis. Neural Computing & Applications 13(4), 288–298 (2004)

[60] Bushnell, M.L., Agrawal, V.D.: Essentials of Electronic Testing for Digital, Memory and Mixed-Signal VLSI Circuits, ch. 10, p. 310. Kluwer, Norwell (2000)

Chapter 9
Synthesis of Electronically-Controllable Signal Processing/Signal Generation Circuits Using Modern Active Building Blocks

Raj Senani, D.R. Bhaskar, A.K. Singh, and V.K. Singh

Abstract. There is a great need for evolving electronically-controllable analog signal processing and signal generation circuits suitable for integration in both bipolar and CMOS technology such that their parameters of interest can be adjusted electronically by means of external DC bias currents or voltages. Traditionally, such circuits have been developed through the so-called operational transconductance amplifier-capacitor (OTA-C) or G_m–C circuits. In bipolar OTAs, their transconductance is linearly tunable through an external DC bias current whereas in CMOS OTAs/transconductors, electronic tunability usually comes through an external DC bias voltage. With the realization of the translinear current conveyor which is based upon a mixed translinear cell (MTC) and exhibits an input resistance looking into terminal X which is electronically-controllable, there has been a surge of research activities in exploiting this property through the so-called second generation Controlled Current Conveyors (CCCII) in several applications in analog circuit design. The objective of this chapter is to present, in a tutorial-review format, significant developments on electronically-controllable analog circuits using CCCIIs as active elements. Too, analog circuit designs using other building blocks of more recent origin, which employ MTC as an input stage and

Raj Senani
Division of Electronics and Communication Engineering,
Netaji Subhas Institute of Technology, Sector 3, Dwarka, New Delhi 110078, India
e-mail: senani@nsit.ac.in

D.R. Bhaskar
Department of Electronics and Communication Engineering, Faculty of Engineering and Technology, Jamia Millia Islamia, Jamia Nagar, New Delhi 110025, India
e-mail: dbhaskar@jmi.ac.in

A.K. Singh
Department of Electronics and Communication Engineering, ITS Engineering College, Plot # 46, Knowledge Park III, Greater Noida U.P. 201308, India
e-mail: abdheshks@yahoo.com

V.K. Singh
Department of Electronics Engineering, Institute of Engineering and Technology, Sitapur Road, Aliganj, Lucknow 226021, India
e-mail: vksingh@ietlucknow.edu

M. Fakhfakh et al. (Eds.): *Analog/RF & Mixed-Signal Circuit Sys. Design*, LNEE 233, pp. 195–221.
DOI: 10.1007/978-3-642-36329-0_9

therefore, are closely related to CCCIIs, have also been taken up. Finally, directions of further work in this area have been pointed out.

9.1 Introduction

Operational transconductance amplifier-capacitor (OTA-C) circuits have been the popular choices for synthesizing electronically-controllable universal voltage mode and current-mode biquads, sinusoidal oscillators as well relaxation oscillators capable of generating non-sinusoidal signals such as square wave, triangular wave etc. However, since the advent of the Second generation Controlled Current Conveyors (CCCII) which provide electronic control of the X-port input resistance R_x through an external bias current, CCCIIs have attracted lot of attention of several research groups and circuit designers around the world as alternative building blocks to synthesize a variety of electronically-controllable signal processing and signal generation circuits. The electronic-controllability in a CCCII emanates from a specific circuit architecture known as a mixed translinear cell (MTC) which forms the input cell of a typical CCII hardware. This MTC apart from the electronic-control also empowers the CCCIIs with higher slew rate and larger input signal handling capability. The above mentioned features have been the primary reasons for stimulating research on the use of CCCIIs in the synthesis of a variety of analog signal processing and signal generation circuits. The electronic tunability of R_x is also manifested in a number of other analog building blocks of relatively more recent origin such as current-controlled current feedback amplifiers (CC-CFA), current-controlled current differencing buffered amplifiers (CC-CDBAs), current-controlled current differencing transconductance amplifiers (CC-CDTA), current-controlled current conveyors transconductance amplifiers (CCCCTA) and CCCIIs with negative intrinsic resistance etc. which also employ the MTC as a basic unit in their internal architecture. Furthermore, all the considered building blocks are implementable in both bipolar and CMOS IC technologies and hence, the circuits employing them are attractive from the point of view of IC implementation.

This chapter focuses on the synthesis of various electronically-controllable signal processing/signal generation circuits such as simulated impedances, current-controlled grounded and floating resistors, summers, subtractors, instrumentation amplifiers, universal biquad filters, sinusoidal signal generators, precision rectifiers, frequency doubler and multipliers/dividers etc. The coverage includes the basics and hardware implementation of various building blocks mentioned above and includes some elegant representative applications using them. Finally, some comments on the recent directions of the research in this area have been made.

9.2 The Second Generation Current-Controlled Conveyors (CCCII) and Their Applications

A CCCII is a special case of the second generation Current Conveyor[1] in which the finite non-zero input resistance at port X (R_x) is taken into consideration. The

[1] The *Current Conveyors* were introduced as new circuit building blocks by Sedra and Smith in the seventies in their classic papers [31, [32].

basic translinear circuit proposed by Fabre, Saaid, Weist and Boucheron [1], employing the so-called mixed-translinear-cell (MTC) comprising of transistors Q_1-Q_2-Q_3-Q_4, is shown in Fig. 9.1. A straight forward analysis of this circuit shows that it is characterized by

$$i_y = 0 \tag{9.1}$$

$$\frac{V_y - V_x}{V_T} = \sinh^{-1}\left(\frac{-i_x}{2I_B}\right) \cong -\frac{i_x}{2I_B}, \text{ for } i_x << 2I_B \tag{9.2}$$

$$\text{or } V_x \approx V_y + i_x R_x \text{ where } R_x = \frac{V_T}{2I_B} \text{ and } i_z = i_x \tag{9.3}$$

It is, thus, seen that in a CCCII, R_x is electronically-tunable through I_B.

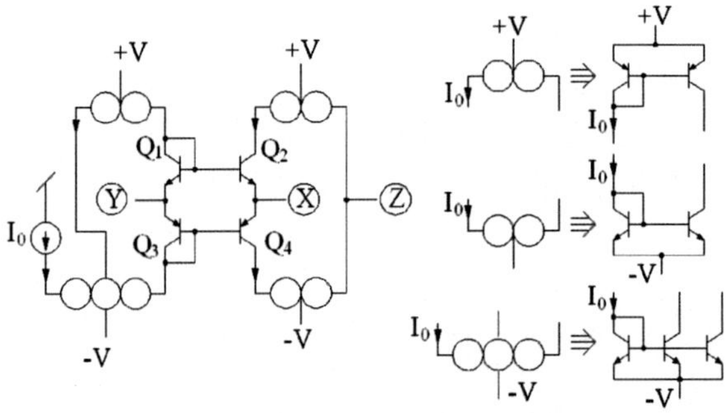

Fig. 9.1 Bipolar circuit architecture of the CCCII+ (adapted from [1] ©1996 IEEE)

Based upon the above equations, a CCCII can be represented by the symbolic notation and equivalent circuit as shown in Fig. 9.2 with V = ± 2.5 volts, this circuit has been reported to have a voltage gain of 0.9984, a current gain of 1.002 and a bandwidth of about 615MHz.

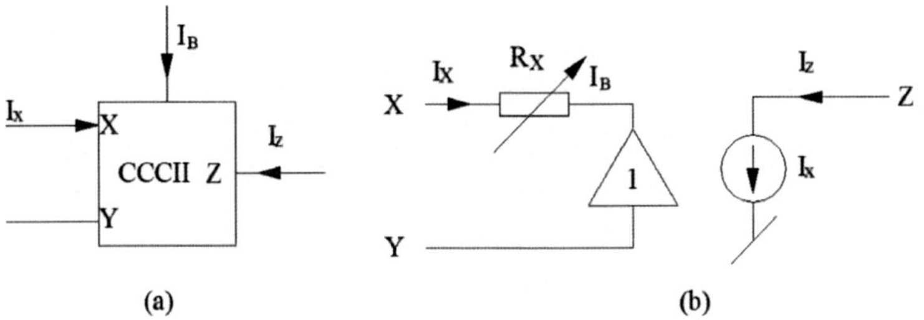

Fig. 9.2 Symbolic notation and simplified equivalent circuit of the CCCII

In the following sections, we demonstrate how a CCCII+ or CCCII- or a combination of both can be used to realize a number of useful electronically-controllable linear as well as non-linear circuits. It may be mentioned that from

Fig. 9.1, a CCCII- can be obtained by breaking the link at the junction of Z-terminal and adding one more pair of current mirrors in a cross-coupled manner.

It is interesting to note that a CCCII can also be realized by a CMOS structure analogous to the circuit of Fig. 9.1 as shown Fig. 9.3 which was proposed by Chaisricharoen, Chipipop and Sirinaovakul [2]. For other interesting CMOS-based CCCII configurations, see [3] and [4].

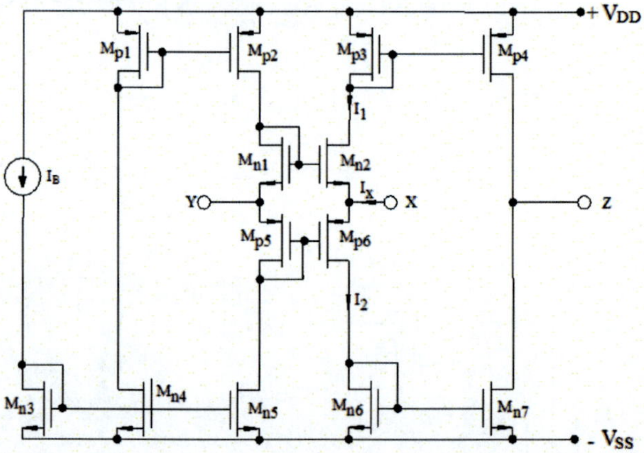

Fig. 9.3 An exemplary CMOS realization of CCCII+ (adapted from [2] © 2010 Elsevier)

The basic translinear cell in the above CMOS implementation is re-drawn in the following and can be seen to be a CMOS analog of the bipolar MTC.

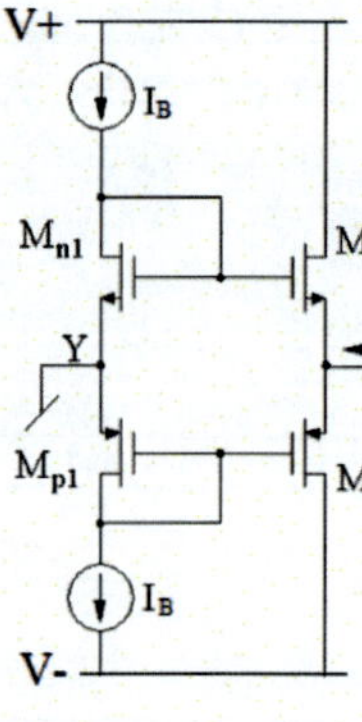

Fig. 9.4 Core of translinear CMOS CCCII (adapted from [2] © 2010 Elsevier)

An analysis of the circuit of Fig. 9.3, for a linear relationship between V_x, V_y and I_x, reveals that with

$$\frac{\mu_p W_p}{L_p} \cong \frac{\mu_n W_n}{L_n} \tag{9.4}$$

the characterizing equations of the circuit are given as

$$V_X \cong V_Y + \frac{i_X}{\sqrt{2I_B C_{0x}}\left(\sqrt{\frac{\mu_p W_p}{L_p}} + \sqrt{\frac{\mu_n W_n}{L_n}}\right)} \tag{9.5}$$

$$R_x \cong \frac{1}{\sqrt{2I_B C_{0x}}\left(\sqrt{\frac{\mu_p W_p}{L_p}} + \sqrt{\frac{\mu_n W_n}{L_n}}\right)} \tag{9.6}$$

$$i_y = 0 \text{ and } i_z = i_x \tag{9.7}$$

From the above, it is seen that in case of CMOS CCCII also, R_x is controllable by DC-bias current I_B. For a more detailed analysis of this CMOS CCCII the reader is referred to [2].

A 2 GHz CCCII in standard 0.8 μm BiCMOS technology was proposed by Seguin and Fabre [5]. A BiCMOS implementation based upon this is shown here in Fig. 9.5.

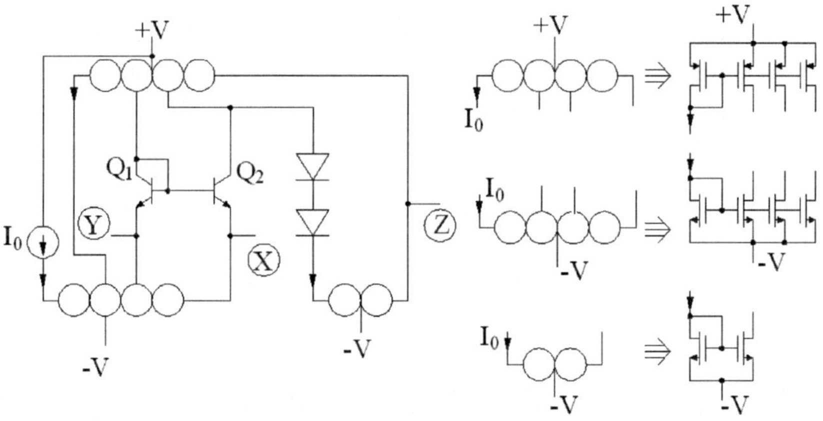

Fig. 9.5 A complete Bi-CMOS CCCII implementation (adapted from [5] © 2001 IEE)

With a bias current of 500μA and a DC supply voltage of = ± 2V the circuit exhibits a –3dB bandwidth of 4.5 GHz for the voltage transfer between terminal Y and X and 2 GHz for the current transfer between terminals X and Z.

We now show a number of interesting applications of CCCIIs in the realization of linear as well as non-linear circuits. In most of the circuits, no external resistors are needed and hence, no component-matching is required. The analog circuits based upon CCCIIs provide the following advantageous features:

- All functions can be realized by using either only CCCIIs or CCCII and capacitors.

- The significant parameters of the realized circuits can be controlled by varying externally supplied DC bias currents.
- CCCII-based circuits can be operated from low power supply voltages (±2.5 V) in contrast to op-amp circuits which generally require ±12V.
- CCCII has much higher slew rate (over 200V/μs) as compared to op-amps (0.5V/μs)
- CCCII-based circuits can be operated at much higher frequencies because of very large bandwidth.
- CCCII-based circuits are highly suitable for IC implementation due to complete absence of resistors and availability of electronic controllability of the parameters of interest.

Lastly, it must be kept in mind that although in this chapter we have discussed the applications in the context of bipolar CCCIIs, however, most of them are equally appropriate to be realized with CMOS CCCIIs though the dependence of the various parameters on the external bias currents will become somewhat different due to the fact that while in the former R_x is inversely proportional to I_B, in the latter, R_x is inversely proportional to the square root of I_B (see equation (9.6)).

9.2.1 Realization of Current Controlled Resistances

A current controlled resistance can be readily realized from CCCII+ by grounding terminal Y, and either leaving open or shorting to ground the terminal Z as shown in Fig. 9.6. Looking into terminal X, the equivalent input impedance is then found to be $R_{in} = \frac{V_T}{2I_B}$. If V_T is taken as 25mV and I_B is changed from 0.1uA to 1mA, the equivalent resistance will change from 12.5 Ω to 125 KΩ.

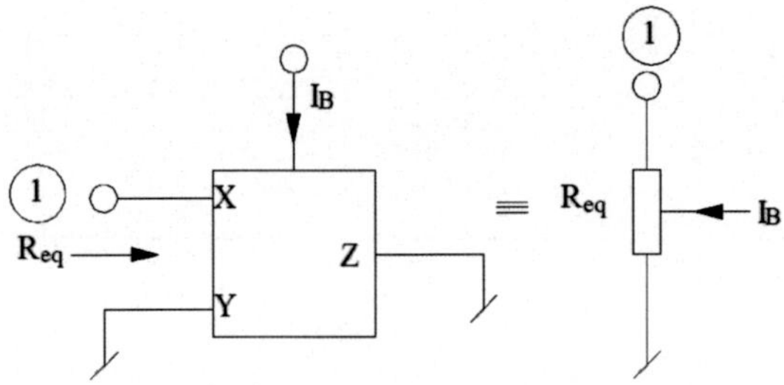

Fig. 9.6 Current-Controlled grounded resistance simulation using CCCII+ (adapted from [6] ©1996 IEE)

It may, however, be noted that the basic circuit responsible for creating a current controllable R_x is the four-transistor MTC composed of transistors Q_1 –Q_4. This MTC configured as grounded current- controlled resistance [6] is shown in Fig. 9.7.

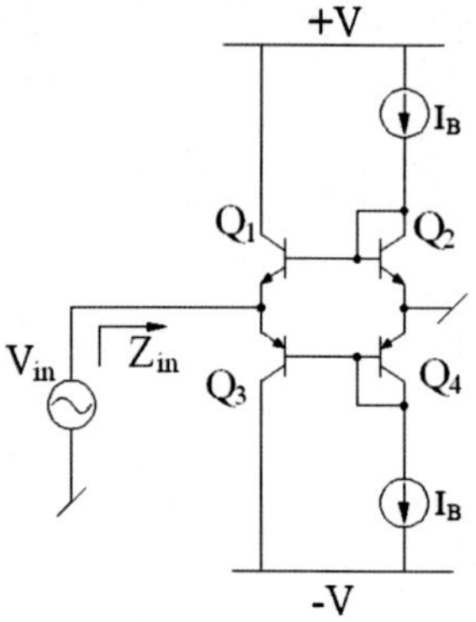

Fig. 9.7 The Mixed Translinear Cell as a grounded Current-Controlled Resistor (adapted from [6] ©1996 IEE)

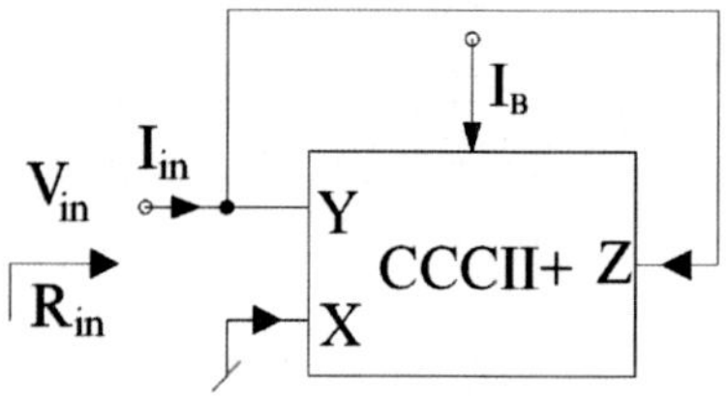

Fig. 9.8 A negative Current-controllable grounded resistance (adapted from [1] ©1996 IEEE)

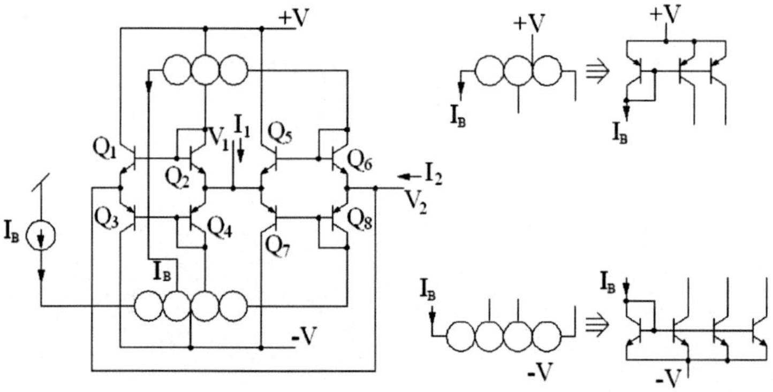

Fig. 9.9 A Floating Current Controlled Positive Resistance (adapted from [7] ©2004 IEEE)

A negative current controlled resistance can be realized from the CCCII+ by shorting terminals Y and Z, grounding terminal X and then looking into terminal Y; see Fig. 9.7. In this case, the realized resistance has a negative value given by

$$Z_{in} = R_{in} = -\frac{V_T}{2I_B}.$$

A floating current controlled positive resistance (FCCPR) circuit was proposed by Senani, Singh and Singh in [7] and is shown in Fig. 9.9. This circuit simulates a floating resistance of the same value i.e. $R_{1-2} = \frac{V_T}{2I_B}$.

9.2.2 *Current-Controlled Simulated Inductors*

A grounded inductance simulation circuit can be obtained from Sedra and Smith's two CCII-based gyrator [31] by using CCCII± instead of CCII± along with a capacitor C_0 to realize an integrator and then using a CCCII– as a V to I convertor as shown in Fig. 9.10.

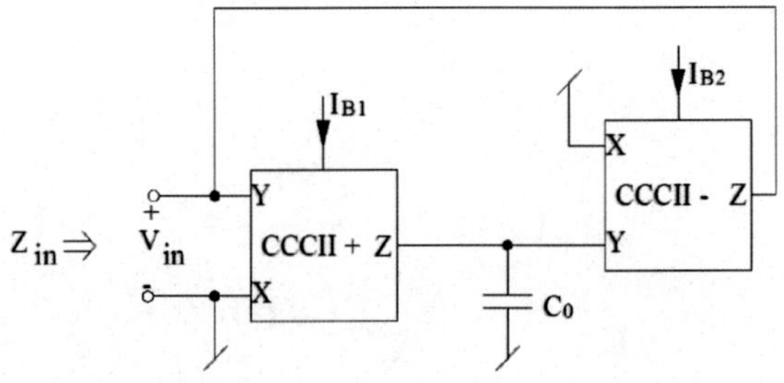

Fig. 9.10 Electronically-tunable grounded inductance

The circuit in Fig. 9.10 gives $Z_{in} = sC_0R_{x1}R_{x2}$ thereby realizing a grounded inductance of value

$$L = C_0R_{x1}R_{x2} = \frac{C_0V_T^2}{4I_{B1}I_{B2}} \tag{9.8}$$

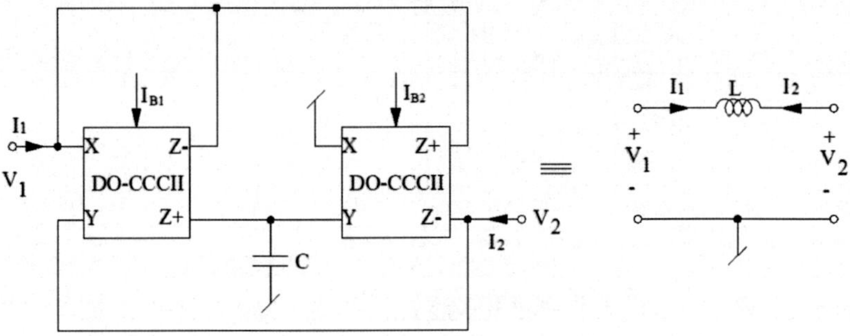

Fig. 9.11 Current-controlled floating inductance (adapted from [9] © 2008 Taylor & Francis)

That a floating inductance can be realized using only a single grounded capacitor along with four CCCII+ was shown by Kiranon and Pawarangkoon in [8]. Other dual output CCII- based circuits were subsequently presented in [9] and [10]. Here we show a circuit from [9] which employs only two dual outputs (DO)-CCCII to realize lossless floating inductance which is shown in Fig. 9.11.

The circuit is characterized by the following equations:

$$I_1 = -I_2 = (V_1 - V_2) / s\,C\,R_{x1}R_{x2} \tag{9.9}$$

so that the inductance value simulated by this circuit, between ports 1 and 2, is given by

$$L = C\,R_{x1}R_{x2} = \frac{CV_T^2}{4I_{B1}I_{B2}} \tag{9.10}$$

and is controllable by any or both of the two external DC bias currents $I_{B1,}$ and I_{B2}.

A negative floating inductor can also be realized from the above configuration (though not so recognized in [9]) by exchanging the Z+ and Z- terminals of second DO-CCCII as shown in Fig. 9.12 in which case the equivalent floating inductance is given by

$$L = -\,C\,R_{x1}\,(R_{x2} + R_{x3}) \tag{9.11}$$

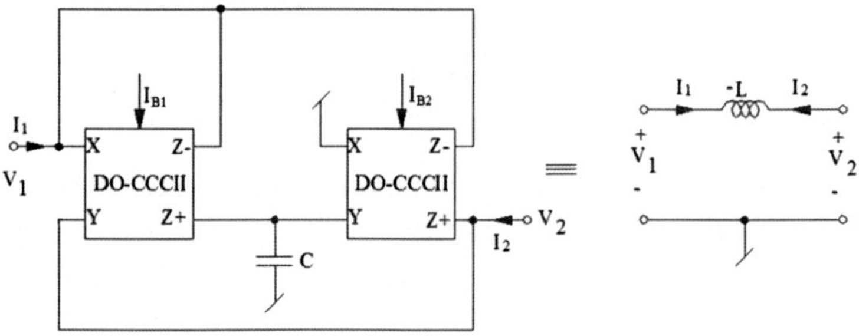

Fig. 9.12 Current-controlled floating negative inductance

9.2.3 *Active-Only Floating Inductance Simulator*

An interesting floating inductor circuit realizable with a CCCII along with consideration of the op-amp pole [11] is shown in Fig. 9.13, employing an internally compensated op-amp exhibiting a dominant pole (one pole roll off) characterized by A = B/s where B is the gain-bandwidth-product (unity gain bandwidth) of the op-amp employed. A straight forward analysis of the circuit shows that the circuit simulates a floating inductance of value $L = V_T / B\, I_B$ which can be electronically controlled by the bias current I_B.

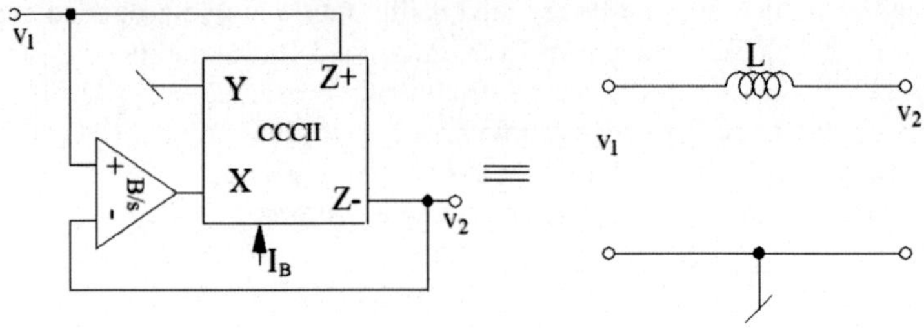

Fig. 9.13 Active-only Current-Controllable floating inductance (adapted from [11] ©2002 Taylor & Francis)

9.2.4 *Electronically-Controlled Capacitance-Multiplier and FDNR*

Consider the circuits shown in Fig. 9.14(a). A straight forward analysis of the circuit shows that the input impedance is given by

$$Z_{in} = R_{x1} + R_{x2} + \frac{Z_1 Z_2}{R_{x3}} \tag{9.12}$$

From this expression it can be easily seen that with $Z_1 = 1/sC_1$ and $Z_2 = R_2$, the circuit can realize a resistance in series with a grounded capacitor whose value is electronically-controllable by the external current I_{B3}. On the other hand, with Z_1 and Z_2 both taken as capacitors, the circuit would realize a resistance in series with a FDNR whose value would be controllable by I_{B3}.

A circuit which can simulate a floating capacitance multiplier while using a grounded capacitor is shown in Fig. 9.14(b). This circuit simulates a floating capacitance in series with a resistance with parameter values as $C_{eq} = C_1 \frac{R_{x4}}{R_{x3}}$ and

$$R_{eq} = (R_{x1} + R_{x2}) \tag{9.13}$$

9.2.5 *Current Controlled VM/CM Amplifiers*

The minimum number of resistors by which a VCVS can be realized through any active circuit building blocks is two. In view of this, it is apparent that with R_x of CCCII accounted for; only two CCCIIs should suffice for this purpose. Such a circuit is shown in Fig. 9.15(a) for which the voltage gain is given by

$$\frac{V_{out}}{V_{in}} = \frac{R_{x2}}{R_{x1}} = \frac{I_{B2}}{I_{B1}} \tag{9.14}$$

Thus, the voltage gain is electronically controllable by I_{B1} and/or I_{B2} and is temperature-insensitive since the parameter V_T is cancelled out!

A current amplifier with a current gain equal to ratio of the two external bias currents is also realizable by an appropriate connection of only two CCCIIs as shown in Fig. 9.15(b). Its current gain is given by

$$\frac{I_{out}}{I_{in}} = \frac{R_{x1}}{R_{x2}} = \frac{I_{B1}}{I_{B2}} \tag{9.15}$$

Note that this circuit also has the desirable properties of low input impedance, high output impedance and temperature-insensitive gain.

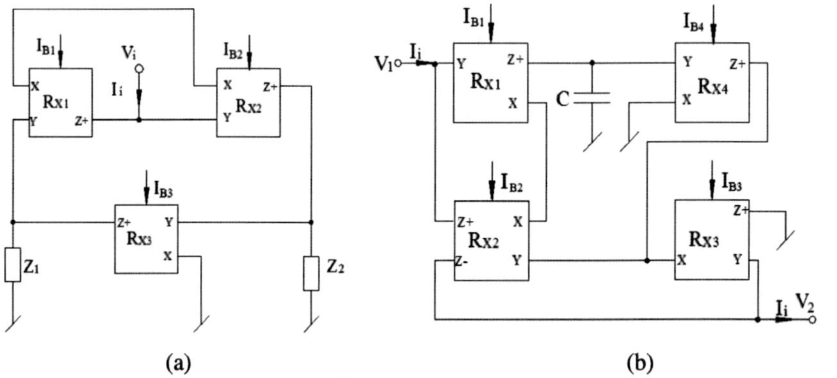

Fig. 9.14 Electronically-controllable Resistance, Capacitance and FDNR simulators (adapted from [12] ©1999 Elsevier): (a) Circuit for FDNR simulation, (b) Circuit for floating capacitance multiplier

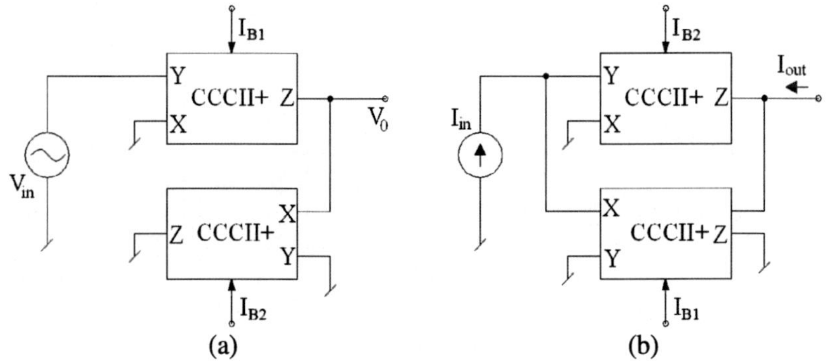

Fig. 9.15 Adjustable-gain Voltage and Current amplifiers: (a) current-controlled voltage amplifier, (b) Current-controlled current amplifier (adapted from [1] @ 1996 IEEE)

9.2.6 Sum and Difference Amplifiers

CCCIIs make it possible to realize sum and difference circuits using only active devices without requiring even a single resistor. One such circuit is shown in Fig. 9.16. The output voltage of this circuit is given by

$$V_o = -R_{x3}\left(\frac{V_1}{R_{x1}} \pm \frac{V_2}{R_{x2}}\right) \tag{9.16}$$

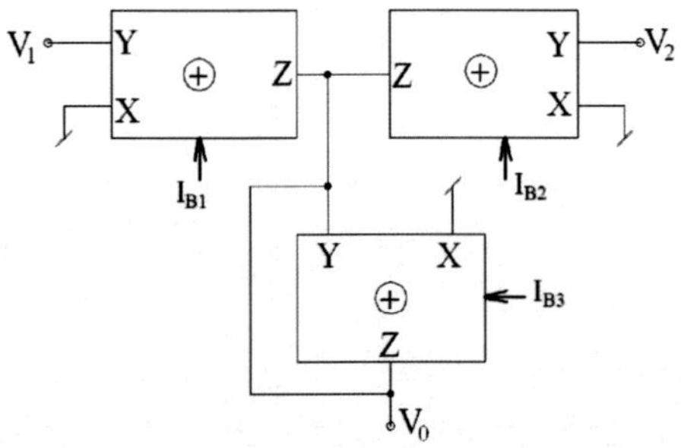

Fig. 9.16 Sum and Difference Amplifiers (adapted from [13] © 2005 Old city publishing. Inc.)

From the above equation it is clear that with the choice for CCCII as *positive*, a summing amplifier is realized and with the choice for CCCII as *negative*, a difference amplifier is realized.

9.2.7 Current-Tunable Active-Only Instrumentation Amplifier

It is well known that a traditional instrumentation amplifier typically requires as many as three op-amps along with seven resistors out of which, at least four are required to be identical. Such a circuit although provides a single resistance controllable variable gain with infinite input impedance at both the input terminals and zero output impedance at the output terminal, but suffers from the conflict between the maximum gain and the available bandwidth. This *gain-bandwidth-conflict* can be avoided by using either a three-CCII-based circuit or a three current feedback op-amp (CFOA) based circuit each of which provide the advantages possessed by the traditional op-amp based circuit and in addition, use a bare minimum of (only two) resistors for varying the voltage gain. Using CCCIIs however, an instrumentation amplifier can be made with only three active building blocks *without requiring any external resistors* and with the additional advantage that the voltage gain can now be controlled through the external DC bias currents as shown in Fig. 9.17.

$$V_0 = \frac{R_{x3}}{(R_{x1}+R_{x2})}(V_2 - V_1) = \frac{I_{B1}I_{B2}}{I_{B3}(I_{B1}+I_{B2})}(V_1 - V_2) \tag{9.17}$$

It is interesting to observe that the gain of this instrumentation amplifier is also temperature insensitive.

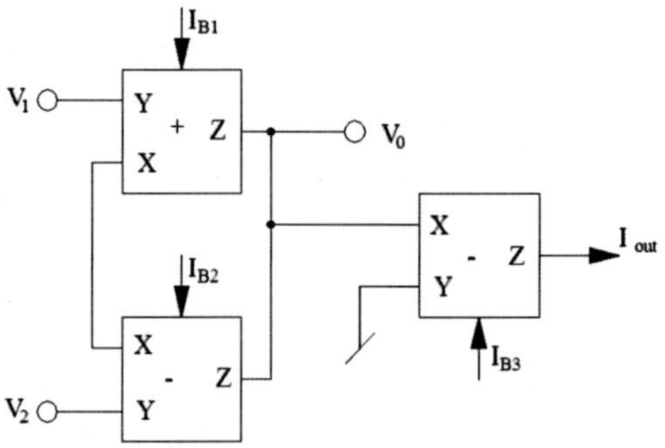

Fig. 9.17 *Active-only* instrumentation amplifier with electronically-controllable voltage gain (adapted from [14] © 2002 Taylor & Francis)

9.2.8 Electronically-Controllable Multifunction Voltage Mode Biquad

Although several researchers have proposed the realization of voltage-mode universal biquads using CCCIIs, here we show a very compact circuit realizable with only two CCCIIs along with a minimum number of capacitors as proposed by Parveen, Ahmed and Khan [15]. This circuit is shown in Fig. 9.18 and can realize all the five standard filter functions by taking different conditions/relationship between the four voltage inputs employed in the circuit shown in Fig. 9.18

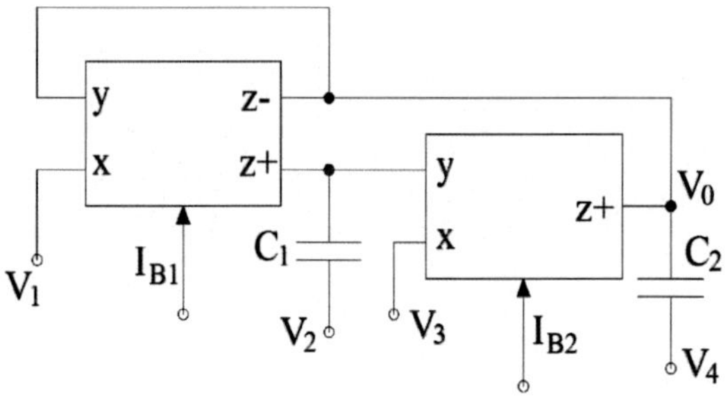

Fig. 9.18 Universal Voltage mode biquad (adapted from [15] ©2009 Old City Publishing Inc.)

An analysis of this circuit reveals that the output voltage of this circuit is given by

$$V_0 = \frac{s^2V_4 + s\left(\frac{1}{R_{x2}C_2}V_3 - \frac{1}{R_{x2}C_2}V_2 + \frac{1}{C_2R_{x1}}V_1\right) + \frac{V_1}{R_{x1}R_{x2}C_1C_2}}{s^2 + s\frac{1}{C_2R_{x1}} + \frac{1}{R_{x1}R_{x2}C_1C_2}} \tag{9.18}$$

From the above equation, the various filter responses can be obtained through the selection of the inputs as shown in Table 9.1.

Table 9.1 Realization of various filter responses

Type of Filter	Condition of Realization
HP	$V_4 = V_{in}$, $V_1 = V_2 = V_3 = 0$
Non-inverting BP	$V_3 = V_{in}$, $V_1 = V_2 = V_4 = 0$
inverting BP	$V_2 = V_{in}$, $V_1 = V_3 = V_4 = 0$
LP	$V_1 = V_2 = V_{in}$, $V_3 = V_4 = 0$ and $R_{x1} = R_{x2}$
Band stop (BS)	$V_1 = V_2 = V_4, = V_{in}$
AP	$V_1 = V_2 = V_4 = V_{in}$

It may be noted that the realization of LP, and BP responses does not require any matching constraints (see cases (i) to (iii)). On the other hand, the constraints in the case of LP, BS and AP cases are also simple to satisfy through design, particularly in monolithic technologies, where matching of the devices is not difficult to achieve. The pole frequency ω_o and the pole-Q of the filter are given by

$$\omega_o = \frac{1}{\sqrt{R_{x1}R_{x2}C_1C_2}}, \qquad Q = \sqrt{\frac{R_{x1}C_2}{R_{x2}C_1}} \tag{9.19}$$

9.2.9 Current-Mode Multifunction Filters

Although a large number of universal current-mode biquad filter circuits have been proposed by various authors using CCCII as active elements, an exemplary circuit which is capable of realizing all the five standard filter functions, namely low pass (LP), high pass (HP), band pass (BP), band stop (BS) and all pass (AP), is shown in Fig. 9.19.

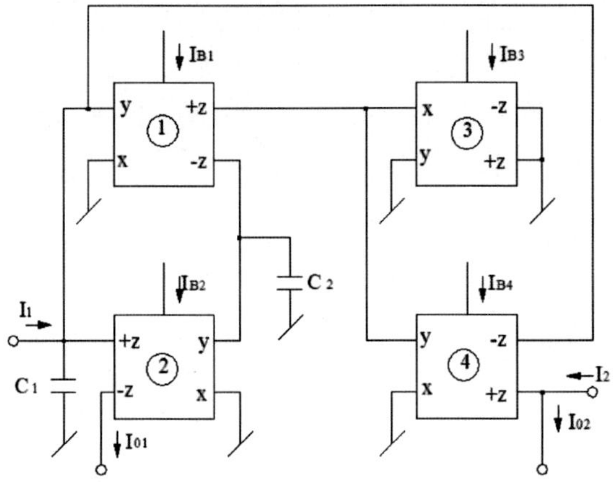

Fig. 9.19 A universal Current mode biquad (adapted from [16] ©2007 Elsevier)

By standard analysis, the two outputs of the circuit are given by

$$I_{01} = \left[\frac{\omega_0^2}{D(s)}\right] I_1 \tag{9.20}$$

$$I_{02} = \frac{\left(s^2 + \frac{\omega_o}{Q}s + \omega_o^2\right) I_2 + \left(\frac{\omega_o}{Q}s\right) I_1}{D(s)} \tag{9.21}$$

where

$$\mathrm{D(s)} = s^2 + \frac{\omega_0}{Q}s + \omega_o^2 \tag{9.22}$$

The various filter functions can be realized from this circuit as shown in Table 9.2.

Table 9.2 Realization of various filter functions

Type of Filter	Condition of Realization
LP	$I_1 = I_{in}, I_2 = 0$ and $I_{o1} = I_{out}$
BP	$I_1 = I_{in}, I_2 = 0$ and $I_{o2} = I_{out}$
HP	$-I_1 = I_2 = I_{in}$ and $I_{o1} + I_{02} = I_{out}$
BS	$-I_1 = I_2 = I_{in}$ and $I_{02} = I_{out}$
AP	$-I_1/2 = I_2 = I_{in}$ and $I_{02} = I_{out}$

Setting $C_1 = C_2 = C$ and $R_{x1} = R_{x2} = R_x = V_T/2I_B$ ($I_{B1} = I_{B2} = I_B$), the parameters ω_0 and Q_0 of all the realized filters are found to be

$$\omega_o = \frac{1}{R_x C} = I_B\left(\frac{2}{V_T C}\right) \text{ and } Q = \frac{I_{B3}}{I_{B4}} \tag{9.23}$$

Yet another current mode universal active filter with one input and multiple outputs was proposed by Senani, Singh, Singh and Bhaskar [17], which is reproduced in Fig. 9.20.

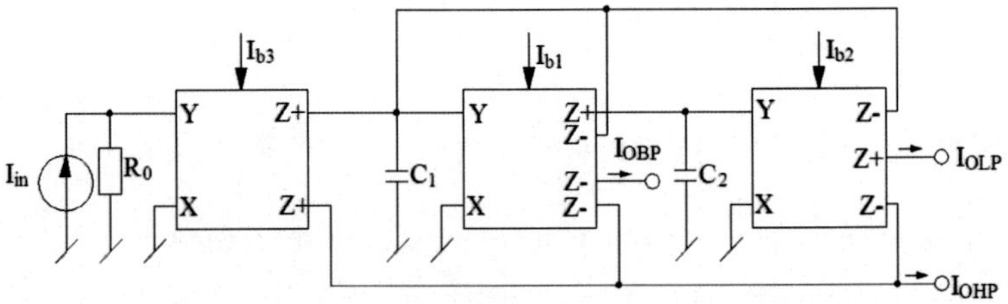

Fig. 9.20 Current-controlled CM universal biquad (adapted from [17] © 2004, IEICE)

By straight forward analysis of this circuit, the three responses are found to be:

$$\frac{I_{OLP}}{I_{in}} = \frac{H_{OLP}\omega_0^2}{D(s)};\frac{I_{OBP}}{I_{in}} = -\frac{H_{OBP}\left(\frac{\omega_0}{Q_0}\right)s}{D(s)} and \frac{I_{OHP}}{I_{in}} = \frac{H_{OHP}s^2}{D(s)} \tag{9.24}$$

$$H_{OHP} = -H_{OBP} = H_{OLP} = \frac{R_0}{R_{x3}} \tag{9.25}$$

and

$$\mathrm{D(s)} = s^2 + \frac{1}{R_{x1}C_1}s + \frac{1}{R_{x1}R_{x2}C_1C_2} \tag{9.26}$$

Using $R_{xi} = \dfrac{V_T}{2I_{Bi}}$; $i=1-3$, the parameters of the realized filters can be expressed as

$$\omega_0 = \frac{2}{V_T}\sqrt{\frac{I_{B1}I_{B2}}{C_1C_2}};\frac{\omega_0}{Q_0} = \frac{2I_{B1}}{V_T C_1};Q_0 = \sqrt{\frac{I_{B2}C_1}{I_{B1}C_2}} \tag{9.27}$$

$$H_{OHP} = -H_{OBP} = H_{OLP} = \frac{2R_0 I_{B3}}{V_T} \tag{9.28}$$

It may be seen that because of the ready availability of the three basic transfer functions (i.e. LP, BP and HP) with correct polarities as well as due to the equal values of all the three gains, a band stop (BS) function is realizable just by joining LP and HP outputs whereas an all-pass function is realizable by joining all the three output terminals. Also, note that a novel feature of this circuit is that no component-matching or equality constraint is needed in any of these two additional realizations.

9.2.10 Current Controlled Sinusoidal Oscillator

There have been a number of proposals for realizing current conveyor based sinusoidal oscillators, however, it is known that to realize such an oscillator, employing both grounded capacitors, at least two current conveyors along with two capacitors and three resistors are needed. Exploiting the current-controllable R_x of the CCCIIs it is interesting to note that an electronically controllable sinusoidal oscillator can be realized with only two CCCIIs along with both grounded capacitors as desirable for integrated circuit implementation. Such a circuit was proposed by Horng [18] and is shown in Fig. 9.21 for which, the condition of oscillation and frequency of oscillation are given by

$$C_1 = C_2 \text{ and } f_0 = \frac{1}{\pi V_T}\sqrt{\frac{I_{B1} I_{B2}}{C_1 C_2}} \tag{9.29}$$

with $I_{B1}=I_{B2}=I_B$, one obtains

$$f_0 = \frac{I_B}{\pi V_T}\sqrt{\frac{1}{C_1 C_2}} \tag{9.30}$$

Thus, the oscillation frequency is linearly tunable by I_B.

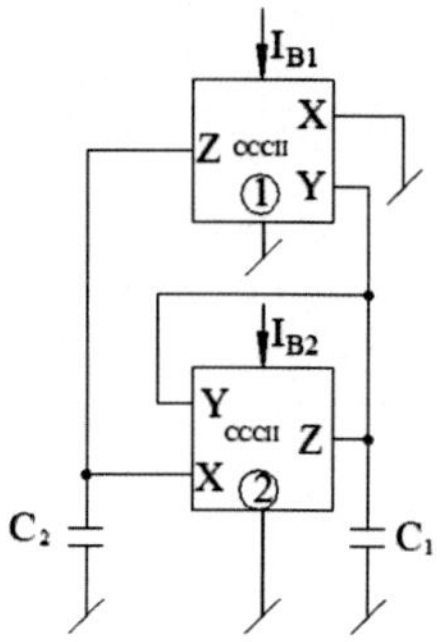

Fig 9.21 Current-controllable sinusoidal oscillator (adapted from [18] © 2001, Taylor & Francis)

9.2.11 CCCII-Based Precision Rectifier

There have been numerous circuits in literature proposing full wave precision rectifiers using a variety of building blocks such as op-amps, current conveyors and current feedback op-amps. A new circuit to realize a precision rectifier was presented in [19] which employs two CCCIIs, a pair of complimentary MOS transistors and two resistors. This circuit is shown in Fig. 9.22.

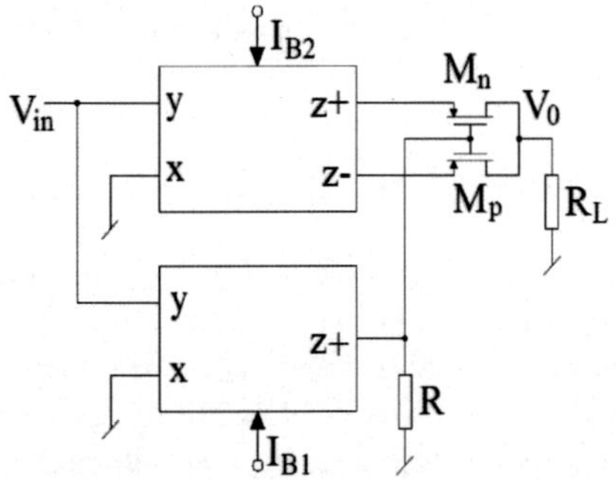

Fig. 9.22 CCCII-based precision rectifier (adapted from [19] © 2005 Old city Publishing Inc.)

The lower CCCII provides a current output proportion to the input voltage which creates a controlled voltage at the gates of the two MOSFETs in conjunction with the resistor R which is so selected that both the transistors remain in saturation even for small input signals. Depending upon whether V_{in} is positive or negative, the gate voltage V_G would be either +V or –V (this requires that the ratio R/R_{x1} is large which can be achieved by proper choice of I_{B1} and R). Here ± V denotes the DC biasing supply voltage of the CCCIIs. An analysis of this circuit shows that the circuit is characterized by the following equations.

$$V_{in} > 0, \quad V_G = +V, \quad V_o = \frac{2I_{B2}R_L}{V_T}V_{in} \tag{9.31}$$

$$V_{in} < 0: \ V_G = -V, \quad V_o = -\frac{2I_{B2}R_L}{V_T}V_{in} \tag{9.32}$$

An important merit of this circuit, over the previously known precision rectifiers using normal current conveyors, is that the output amplitude of this rectifier is electronically controllable by external bias current I_{B2}.

9.2.12 Frequency Doubler

Frequency doubler is widely used in analog signal processing, communication and instrumentation systems. Frequency doubler is often made from analog multipliers. However, Anuntahirunrat, Tangsrirat, Riewruga and Surkamponturn [20] have demonstrated that by imposing appropriate design conditions, the CCCII circuit shown in Fig. 9.23 can accomplish both these functions from the same structure. For an elaborate analysis (see [20]).

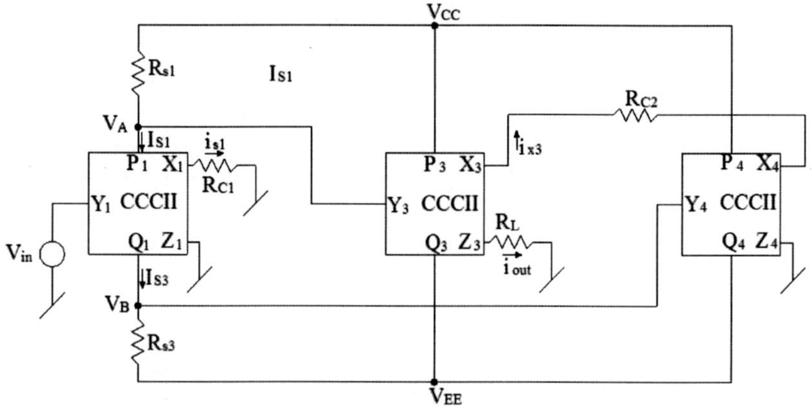

Fig. 9.23 Frequency doubler (adapted from [20] © 2004 Taylor & Francis)

It has been found that an approximate expression for the output current of this circuit is given by

$$i_{out} \cong k_5 - (I_{DC} + I_{2\omega} \cos 2\omega t) \tag{9.33}$$

$$k_5 = k_4(V_{CC} - V_{EE}) - 5I_1 k_4 R_S \;; k_4 = \frac{1}{R_{C2} + 2R_x} \tag{9.34}$$

$$I_{DC} \cong 4I_1 k_4 R_S (1 + \frac{1}{4} k_3);\; I_{2\omega} \cong -I_1 R_s K_3 K_4;\; K_3 = \frac{I_m^2}{(2I_1)^2} \tag{9.35}$$

From the above equations, it is clear that by appropriately selecting component values and parameters, the circuit can generate a cosine wave of frequency 2ω from an input current $i_{x1}=I_m \sin \omega t$ derived from the input V_m.

9.3 Other Current-Controlled Building Blocks and Some Exemplary Current-Controlled Function Realized from Them

An exhaustive review of various existing building blocks as well as several new possible building blocks was presented by Biolek Senani Biolkova-Kolka in [21]. In the following, we discuss a number of those building blocks whose front end consists of one or more MTC type structures thereby leading to the possibility of current controllability of the various functions realized from them.

9.3.1 CC-CFOA

Since the internal architecture of CFOA contains a CCII+ followed by a voltage buffer, it is, therefore, obvious that a CCCII+ followed by a voltage buffer would become a CC-CFOA. This concept has been studied by Siripruchyanun, Chanapromma, Silapan and Jaikla in [22] wherein using bipolar and MOS current mirrors several topologies of BiCMOS CCCIIs have been studied, ultimately resulting in Bi-CMOS CC-CFOA which has been studied in detail. In addition, a number of interesting applications have been proposed some of which are shown in Fig. 9.24.

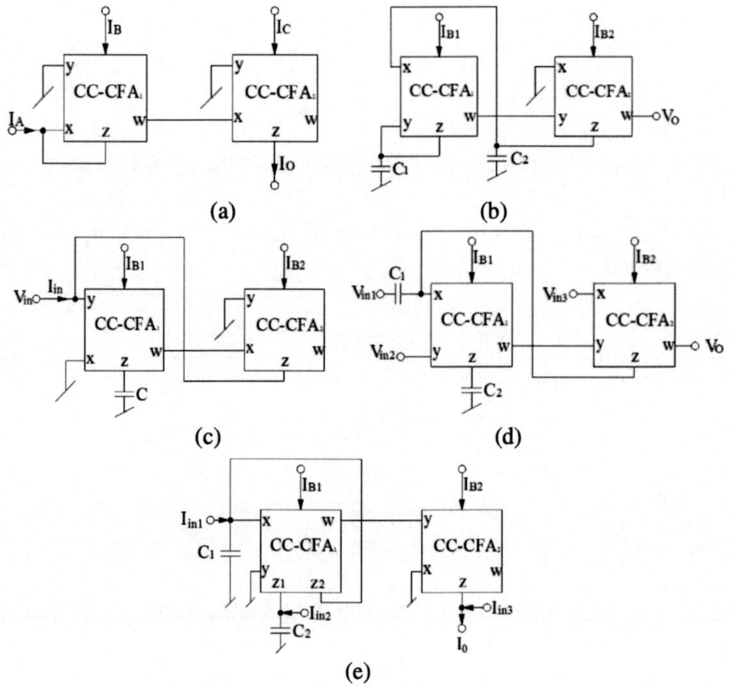

Fig. 9.24 Various applications of CC-CFAs (adapted from [22] ©WSEAS 2008)

For the circuit of Fig. 9.24(a), the I_0 is given by

$$I_o = -\frac{I_A I_C}{2I_B} \tag{9.36}$$

Hence, the circuit would function as current-mode multiplier/divider.

For the sinusoidal oscillator of the circuit of Fig. 9.24(b), the CO and FO are given by

$$C_1 = C_2; \qquad \omega_0 = \sqrt{\frac{1}{C_1 C_2 R_{x1} R_{x2}}} \tag{9.37}$$

In case of the grounded inductance circuit of Fig. 9.24(c), the value of simulated inductance is given by

$$L = CR_{x1} R_{X2} = \frac{CV_T^2}{4I_{B1} I_{B2}} \tag{9.38}$$

For the circuit of Fig. 9.24(d), the output voltage is given by

$$V_o = \frac{V_{in1} s^2 C_1 C_2 R_{x1} R_{x2} + V_{in2}(1 + sC_2 R_{x2}) - V_{in3} sC_2 R_{x1}}{s^2 C_1 C_2 R_{x1} R_{x2} + sC_2 R_{x2} + 1} \tag{9.39}$$

while in case of the circuit of Fig. 9.24(e), the output current is given by

$$I_o = \frac{I_{in1} sC_2 R_{x2} + I_{in2} - I_{in3}(s^2 C_1 C_2 R_{x1} R_{x2} + sC_2 R_{x2} + 1)}{s^2 C_1 C_2 R_{x1} R_{x2} + sC_2 R_{x2} + 1} \tag{9.40}$$

From the eqns (9.39) and (9.40) it is easily seen that by choosing inputs appropriately all the five standard filter functions can be realized in both multi-input-single-output-type VM and CM filters of Fig. 9.24(d) and 9.24(e) respectively.

9.3.2 Current Controlled Current Differencing Transconductance Amplifier (CC-CDTA)

A CC-CDTA is an extension of the concept of CDTA introduced by Biolek[2] and is characterized by $V_p = R_p I_p$, $V_n = R_n I_n$, $I_z = I_p - I_n$, $I_x = \pm g_m V_z$ where $R_p = R_n = V_T/2I_B$ and $g_m = I_C/2V_T$. A novel application of this implementation is the current mode multiplier/divider shown in Fig. 9.25 which has the output current given by

$$I_O = \frac{I_A I_C}{8I_B} \tag{9.41}$$

From the above it is seen that I_O is result of either multiplication of I_A and I_C, or dividing I_A or I_C by I_B. However, a limitation of this circuit is that it is only two quadrant multiplier/divider. On the other hand, a merit of this circuit is that the functions performed by this circuit are temperature insensitive since the terms involving V_T have been cancelled out and do not appear in the output equation.

[2] Biolek D, (2003) CDTA: Building block for current mode analog signal processing. Proc. ECCTD, 3:397-400.

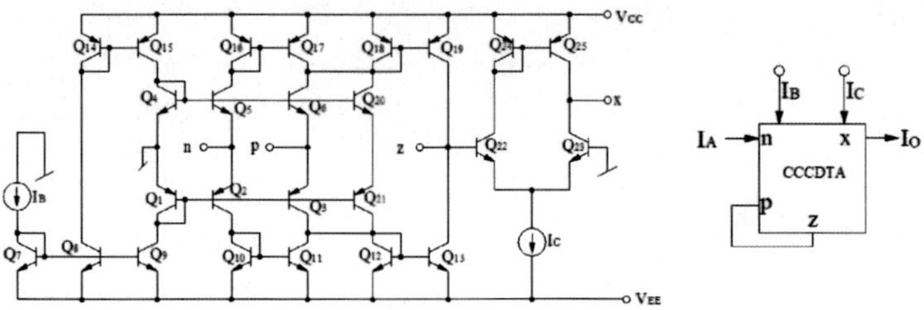

Fig. 9.25 Current controlled current differencing Transconductance amplifier CC-CDTA and CC-CDTA configured as a multiplier /divider (Adapted from [23] ©2008 Elsevier)

9.3.3 Current Controlled Current Conveyor Transconductance Amplifier (CCCC-TA)

This building block was proposed by Siripruchyanun and Jaikla [24] and is characterized by the following terminal equations: $I_y = 0$, $V_x = V_y + I_xR_x$, where $R_x = V_T/2I_{B1}$, $I_Z = I_x$ and $I_O = \pm g_m V_z$ where $g_m = I_{B2}/2V_T$.

An exemplary implementation of this building block is shown in Fig. 9.26 from where it can be seen that the input front end of this building block also happens to be the MTC which results in the current controllability of R_x.

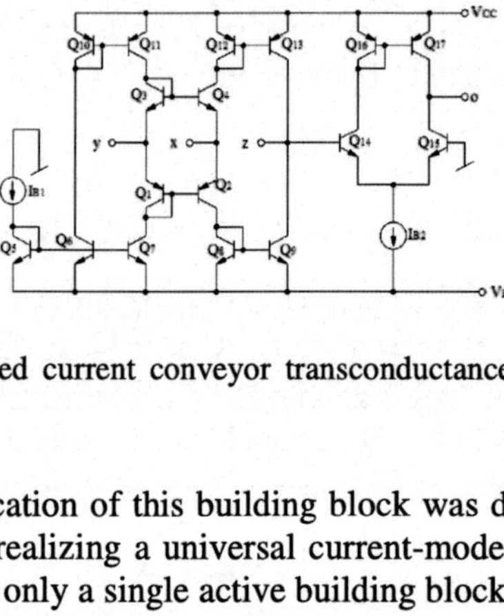

Fig. 9.26 Current controlled current conveyor transconductance amplifier (Adapted from [24] © 2007 Springer)

An interesting application of this building block was demonstrated by Siripruchyanun and Jaikla in realizing a universal current-mode biquad. The novelty of this circuit lies in using only a single active building block for realizing all the five standard filter responses with both grounded capacitors. This circuit can thus, be seen to be a minimum-component biquad exhibiting these properties.

The output current of this circuit is given by

$$I_o = \frac{I_{in1}sC_2 + g_m I_{in2} - I_{in3}(s^2C_1C_2R_x + sC_2 + g_m)}{s^2C_1C_2R_x + sC_2 + g_m} \tag{9.42}$$

from where one can easily work out that, subject to different input conditions /selections, it would be possible to realize all the five standard functions with tuning of ω_o.

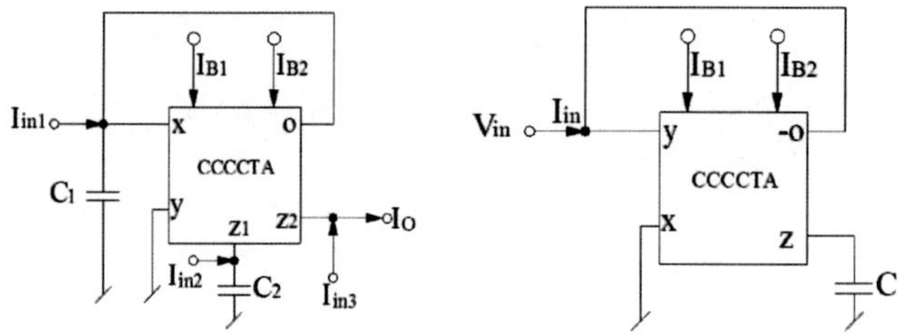

(a) (b)

Fig. 9.27 (a) Universal biquad filter based on the CCCC-TA, (b) Grounded inductance simulator (Adapted from [24] © 2007 Springer)

In Fig. 9.27(b), another novel application of CCCC-TA in realizing an electronically-controllable grounded inductor is shown, which employs only one CCCC-TA along with a grounded capacitor as advantageous for IC implementation. The realized inductance value is given by

$$L_{eq} = \frac{CV_T^2}{I_{B1}I_{B2}} \tag{9.43}$$

9.3.4 *CCCII with Negative Intrinsic Resistance and Its Applications*

Barthelemy and Fabre [25] proposed a second generation current controlled conveyor with negative intrinsic resistance whose value is given by $R_x = -\frac{V_T}{2I_B}$ for which the circuit implementation Fig. 9.28 was proposed. It was shown that with the circuit realized from ALA200 transistor arrays from ATT, with = ± 1.5 volt DC power supplies the circuit exhibit bandwidth higher than 40 MHz and a total power consumption only 1 mW for a bias current of 50μA. It was shown that this building block was particularly useful in realizing a floating negative controlled resistance.

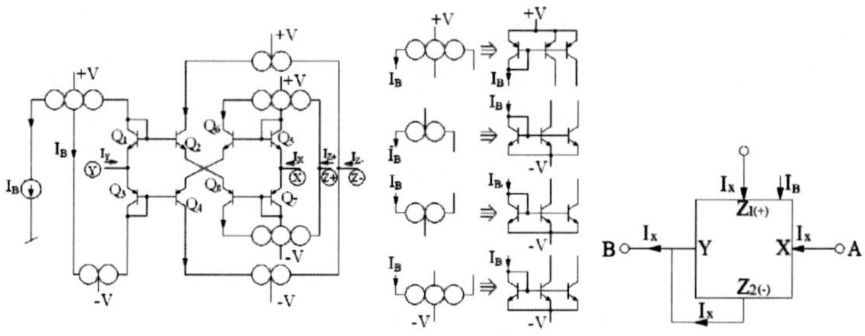

(a) (b)

Fig. 9.28 (a) Implementation of the CCCII± with negative intrinsic resistance, (b) circuit connections for realizing a floating negative resistance (Adapted from [25] 2002©IEEE)

9.3.5 Current-Controlled CDBA

The CDBA was introduced as a new building block by Acar and Ozoguz [26] and is characterized by equations $V_p=V_n=0$, $i_z=i_p - i_n$, $V_w=V_z$. An exemplary CDBA circuit exhibiting current controllable input resistances at its both p and n input terminals is shown here in Fig. 9.29. A CC-CDBA can be characterized by $V_p=R_p I_p$, $V_n=R_n I_n$, $I_z=I_p-I_n$, and $V_w=V_z$.

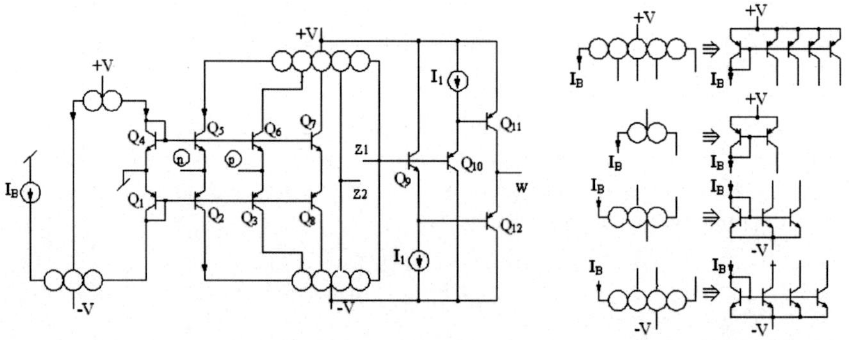

Fig. 9.29 Internal construction of a DO-CCCDBA (adapted from [27] 2006 © ECTI-CON)

An interesting application of CC-CDBA is in the realization of a very compact current mode analog multiplier/divider which is shown in Fig. 9.30.

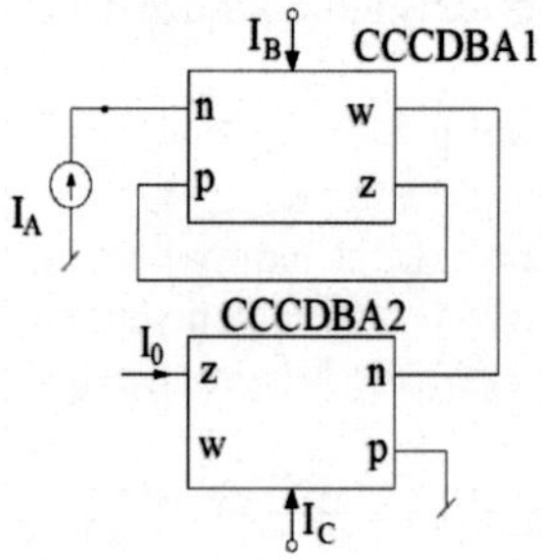

Fig. 9.30 Current mode multiplier/divider (adapted from [28] 2007©IEEE)

A routine analysis of the circuit revels that the output is given by

$$I_0 = \frac{I_A I_C}{2I_B} \tag{9.44}$$

From the above it is seen that one can perform either multiplication by taking I_A and I_C as the inputs or as a divider by taking I_A and I_B as inputs. However, it should be noted that in both cases, only two quadrant operations is possible. Two other novel applications of CCCDBA are in the realization of floating negative simulated grounded and floating inductors. These circuits are shown in Fig. 9.31

and have the novelty of employing only two active building blocks and a single grounded capacitor while providing electronically-controlled inductance value.

The circuit of Fig. 9.31 simulates a floating inductance of value $L = -C\,R_{n1}\,R_2$ where $R_{n2} = R_{p2} = R_2$. The circuit for grounded negative simulated inductor on the other hand realizes the negative inductance of the value $L = -C\,R_{p1}\,R_{p2}$. In both the cases, the value of the inductance can be adjusted electronically by I_{B1} or I_{B2}.

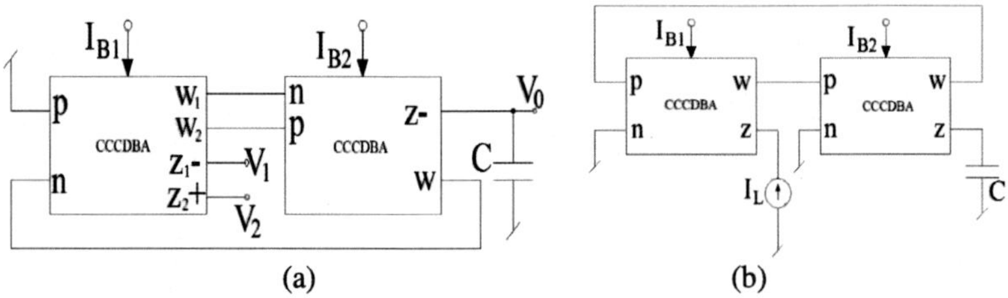

Fig. 9.31 (a) Floating negative simulated inductor, (b) grounded negative simulated inductor (adapted from [29] 2006©ECTI)

9.4 Concluding Remarks and Directions of Future Research

Although traditionally operational transconductance amplifiers (OTA) or CMOS transconductors with controllable transconductance values have been the main elements for designing electronically-controllable analog circuits, during the past 15 years, the so-called Current-Controlled Current Conveyors (CCCII) have also attracted considerable attention in devising various electronically-controllable linear as well as nonlinear functions. This chapter has presented, in a tutorial review format, some of the prominent works done in this area. Since the electronic-controllability of the input resistance looking into the X-terminal of the CCCII essentially comes from a four-transistor core known as Mixed Translinear-Cell (MTC), a number of other electronically-controlled building blocks have been proposed recently in which the front end of the hardware realizations is made from one or more MTCs. Thus, the newer building blocks such as CC-CFOA, CC-CDTA, CCCC-TA and CC-CDBA also belong to the same class. Due to this reason, some interesting applications of these building blocks have also been included. From a comprehensive survey of references, it has been revealed that the work on the various applications of these building blocks is far from complete. Lastly, it is worth mentioning that more recently, Fakhfakh, Pierzchala and Rodanski [30] have advanced a novel concept of electronically-controllable CCVS which appears to be yet another interesting building block for evolving electronically-controllable analog signal processing and signal generating circuits. It appears that numerous newer possibilities are yet to be unfolded which warrant further investigations and research on this exciting area.

Acknowledgement. The authors gratefully acknowledge the contribution of Ms. Shashi Rawat for her dedicated and meticulous help in the preparation of this manuscript.

References

[1] Fabre, A., Saaid, O., Wiest, F., Boucheron, C.: High frequency applications based on a new current controlled conveyor. IEEE Trans. on Circuits and Systems-I 43, 82–91 (1996)
[2] Chaisricharoen, R., Chipipop, B., Sirinaovakul, B.: CMOS CCCII: Structures, characteristics, and considerations. Int. J. Electron. Commun. (AEU) 64, 540–557 (2010)
[3] Minaei, S., Kaymak, D., Ibrahim, M.A., Kuntman, H.: New CMOS configurations for current-controlled conveyors (CCCIIs). In: Proc. of IEEE ICCSC, pp. 62–65 (2002)
[4] Zouaoui-Abouda, H., Fabre, A.: New high- value floating controlled resistor in CMOS technology. IEEE Trans. Instrum. Meas. 55, 1017–1020 (2006)
[5] Seguin, F., Fabre, A.: 2GHz controlled current conveyor in standard 0.8 um BiCMOS technology. Electron. Lett. 37, 329–330 (2001)
[6] Saaid, O., Fabre, A.: Class AB current controlled resistor for high performance current-mode applications. Electron. Lett. 32, 4–5 (1996)
[7] Senani, R., Singh, A.K., Singh, V.K.: A new floating current-controlled positive resistance using mixed translinear cells. IEEE Trans. on Circuits and Systems-II 51, 374–377 (2004)
[8] Kiranon, W., Pawarangkoon, P.: Floating inductance simulation based on current conveyors. Electron. Lett. 33, 1748–1749 (1997)
[9] Sedef, H., Sagbas, M., Acar, C.: Current-controllable fully-integrated inductor simulator using CCCIIs. Int. J. Electronics 95, 425–429 (2008)
[10] Yuce, E., Minaei, S., Cicekoglu, O.: Resistorless floating immittance function simulators employing current controlled conveyors and a grounded capacitor. Electrical Engg. 88, 519–525 (2006)
[11] Minaei, S., Cicekoglu, O., Kuntman, H., Turkoz, S.: Electronically tunable, active only floating inductance simulation. Int. J. Electron. 89, 905–912 (2002)
[12] Abuelma'atti, M.T., Tasadduq, N.A.: Electronically tunable capacitance multiplier and frequency-dependent negative-resistor simulator using the current-controlled current conveyor. Microelectronics J. 30, 869–873 (1999)
[13] Maheshwari, S.: Additional summing/difference amplifiers using CCCIIs. J. of Active and Passive Electronic Devices 1, 159–162 (2005)
[14] Maheshwari, S.: High CMRR wide bandwidth instrumentation amplifier using current controlled conveyors. Int. J. Electron. 89, 889–896 (2002)
[15] Parveen, T., Ahmed, M.T., Khan, I.A.: A Canonical voltage-mode universal CCCII-C filter. J. of Active and Passive Electron. Devices 4, 7–12 (2009)
[16] Tangsrirat, W.: Current-tunable current-mode multifunction filter based on dual-output current-controlled conveyors. Int. J. of Electron. Commun. (AEU) 61, 528–533 (2007)
[17] Senani, R., Singh, V.K., Singh, A.K., Bhaskar, D.R.: Novel electronically controllable current-mode universal biquad filter. IEICE Electron. Expr. 1, 410–415 (2004)
[18] Horng, J.W.: A sinusoidal oscillator using current controlled current conveyors. Int. J. Electron. 88, 659–664 (2001)
[19] Maheshwari, S., Khan, I.A.: An integrable precision rectifier with current controlled output. J. of Active and Passive Electronic Devices 1, 177–182 (2005)
[20] Anuntahirunrat, K., Tangsrirat, W., Riewruja, V., Surakampontorn, W.: Sinusoidal frequency doubler and full-wave rectifier based on translinear current-controlled current conveyors. Int. J. Electron. 91, 227–239 (2004)

[21] Biolek, D., Senani, R., Biolkova, V., Kolka, Z.: Active elements for analog signal processing: classification, Review and new proposals. Radioengineering J. 17, 15–32 (2008)
[22] Siripruchyanun, M., Chanapromma, C., Silapan, P., Jaikla, W.: BiCMOS current-controlled current feedback amplifier (CC-CFA) and its applications. WSEAS Trans. on Electron. 5, 203–219 (2008)
[23] Siripruchyanun, M., Jaikla, W.: A current-mode analog multiplier/divider based on CCCDTA. Int. J. Electron. and Commun. (AEU) 62, 223–227 (2008)
[24] Siripruchyanun, M., Jaikla, W.: Current controlled current conveyor Transconductance amplifier (CCCCTA): A building block for analog signal processing. In: Int. Symp. on Commun. and Inf. Tec. (ISCIT 2007), pp. 209–212. IEEE (2007)
[25] Barthelemy, H., Fabre, A.: A second generation current-controlled conveyor with negative intrinsic resistance. IEEE Trans. of CAS-I: Fundamental Theory and Applications 49, 63–65 (2002)
[26] Acar, C., Ozoguz, S.: A new versatile building block: Current differencing buffered amplifier suitable for analog signal processing filters. Microelectron. J. 30, 157–160 (1999)
[27] Jaikla, W., Siripruchyanun, M.: Novel current-mode four-quadrant Multiplier/divider based on current controlled CDBAs (CCCDBAs). In: ECTI International Conference, pp. 509–512 (2006)
[28] Silpan, P., Jaikla, W., Siripruchyanun, M.: High-performance BiCMOS Current Controled CDBA and application. In: Int. Symp. on Commun. and Inf. Tech. (ISCIT 2007), pp. 40–43. IEEE (2007)
[29] Jaikla, W., Siripruchyanun, M.: Current controlled CDBAs based-novel floating and grounded negative inductance simulators. The 21st Int. Technical Conference on Circuits/Systems, Computers and Commun. III, 701–704 (2006)
[30] Fakhfakh, M., Pierzchala, M., Rodanski, B.: On the design of active inductors with current-controlled voltage sources. Analog Integr. Circ. Sig. Process. (2011), doi:10.1007/s10470-011-9798
[31] Sedra, A., Smith, K.C.: A second-generation current conveyor and its applications. IEEE Trans. Circuits Theory 2, 132–134 (1970)
[32] Smith, K.C., Sedra, A.: The current conveyor- A new circuit building block. Proc. IEEE 56, 1368–1369 (1968)

Chapter 10
Synthesis of Generalized Impedance Converter and Inverter Circuits Using NAM Expansion

Ahmed M. Soliman

Abstract. The generalized impedance converter (GIC) is an active two port network in which the input impedance is equal to the load impedance times a conversion function of the complex frequency variable .There are two types of the GIC, the first is the voltage generalized impedance converter (VGIC) and the second is the current generalized impedance converter (CGIC). In this chapter the nodal admittance matrix (NAM) expansion is used to generate all possible VGIC and CGIC circuits. The realizations of two types of the generalized impedance inverter (GII) circuits using NAM expansion are also given.

10.1 Introduction

Antoniou introduced several realizations of the CGIC using the operational amplifier (Op Amp) as the active element and was used in active network synthesis of voltage transfer functions [1-2]. The GIC can also be used in the realization of grounded and floating inductances and frequency dependent negative resistances (FDNR) [3]. The singular network elements introduced in [4] are the nullator and norator and are shown symbolically in Figs. 10.1 (a) and (b).The nullator and norator are singular elements that possess ideal characteristics and are specified according to the constraints they impose on their terminal voltages and currents. For the nullator V = 0, I = 0; while the norator imposes no constraints on its voltage and current. A nullator-norator pair constitutes a universal active two-port network element called the nullor [4] and hence, nullator and norator are also called nullor elements.

The nullator and norator are very useful and powerful singular elements as demonstrated by the many circuit applications [5].Their use in the realization of active building blocks and active RC filters using Op Amp have been demonstrated by several examples [5].

Ahmed M. Soliman
Electronics and Communication Engineering Department, Faculty of Engineering,
Cairo University, Egypt 12613
e-mail: asoliman@ieee.org

M. Fakhfakh et al. (Eds.): *Analog/RF & Mixed-Signal Circuit Sys. Design*, LNEE 233, pp. 223–242.
DOI: 10.1007/978-3-642-36329-0_10

Since the introduction of the mirror elements a number of applications have been proposed and investigated [6-8] proving that the singular elements set is now complete [9].

The voltage mirror (VM), shown in Fig. 1(c), is a lossless two-port network element used to represent an ideal voltage reversing action and it is described by:

$$V_1 = -V_2 \tag{10.1.a}$$

$$I_1 = I_2 = 0 \tag{10.1.b}$$

The current mirror (CM), shown in Fig. 10.1(d), is a two-port network element used to represent an ideal current reversing action and it is described by:

$$V_1 \text{ and } V_2 \text{ are arbitrary} \tag{10.1.c}$$

$$I_1 = I_2, \text{ and they are also arbitrary} \tag{10.1.d}$$

Although the CM element shown in Fig. 10.1(d) has the same symbol as the regular current mirror, it is a bi-directional element and has a theoretical existence.

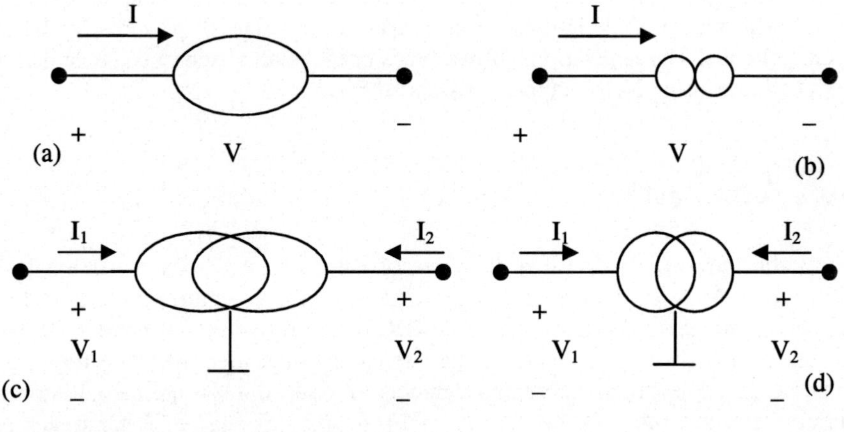

Fig. 10.1 Singular network elements: (a) Nullator, (b) Norator, (c) Voltage mirror, (d) Current mirror

The systematic synthesis method based on NAM expansion and using nullor elements [10-13] has been used to realize several active building blocks. The NAM expansion has been extended to accommodate mirror elements [14-20]. This results in a generalized framework encompassing all singular elements for ideal description of active elements. Accordingly, more alternative realizations are possible and a wide range of active devices can be used in the synthesis.

First a brief review of the pivotal expansion and the key steps in the NAM expansion are summarized.

The pivotal expansion is the reversal of Gaussian elimination [10,13] and is useful in circuit synthesis. The pivotal expansion has the following general form:

$$[\frac{ab}{d}] = \begin{bmatrix} \frac{ab}{d} & 0 \\ 0 & 0 \end{bmatrix} = \begin{bmatrix} 0 & \mp a \\ \pm b & d \end{bmatrix} \tag{10.2}$$

Next the key steps in the NAM expansion are summarized as follows:

A Nullator connected between two nodes (which is represented by bracket linking the corresponding columns in the NAM) moves a circuit element from one column to the other column with the same sign.

A Norator connected between two nodes (which is represented by bracket linking the corresponding rows in the NAM) moves a circuit element from one row to the other row with the same sign.

A VM connected between two nodes (which is represented by bracket linking the corresponding columns in the NAM) moves a circuit element from one column to the other column with opposite sign.

A CM connected between two nodes (which is represented by bracket linking the corresponding rows in the NAM) moves a circuit element from one row to the other row with opposite sign.

The nullator and norator are represented by straight brackets as in [10-13]. Curved brackets are used to represent the VM and CM as in [14-20].

10.2 Current Generalized Impedance Converters Using Op Amps

Before considering the generation of GIC using NAM expansion, it is important to review the history of the generation of GIC using Op Amps. The CGIC was introduced by Antoniou in [1-2] and defined by the following transmission matrix equation:

$$\begin{bmatrix} V_1 \\ I_1 \end{bmatrix} = \begin{bmatrix} 1 & 0 \\ 0 & \frac{Y_1 Y_3}{Y_2 Y_4} \end{bmatrix} \begin{bmatrix} V_2 \\ -I_2 \end{bmatrix} \tag{10.3}$$

The above T matrix can be expanded as follows:

$$T = \begin{bmatrix} 1 & 0 \\ 0 & -\frac{Y_1}{Y_2} \end{bmatrix} \begin{bmatrix} 1 & 0 \\ 0 & -\frac{Y_3}{Y_4} \end{bmatrix} \tag{10.4}$$

It is seen that the CGIC is realizable by two cascaded negative impedance converter (NIC) circuits and shown in Fig. 10.2(a) using nullor elements. The two Op Amp circuit realizing the two cascaded NIC is based on pairing the nullator to the left with the norator at node 3 and the nullator to the right with the norator at node 5. The realization in this case is shown in Fig. 10.2(b) and is known as the uncoupled CGIC [5]. The alternative nullator norator pairing results in the coupled CGIC well known circuit shown in Fig. 10.2(c) originally derived in [2] and detailed study on the number of possible Op Amp CGIC and stability analysis was given in details in [5].

It should be noted that there are no realizations for the VGIC using Op Amps, up to this author best knowledge. In the next section new approach for the realizations of the VGIC using current conveyors (CCII) [21] or inverting current conveyor (ICCII) [6] or combination of both is given.

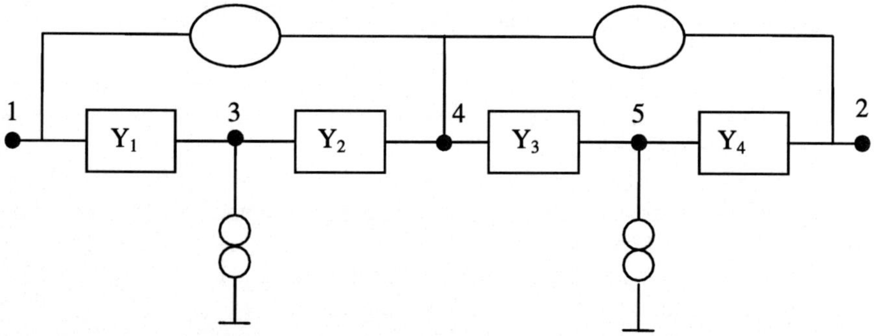

Fig. 10.2(a) Nullator norator realization of the CGIC

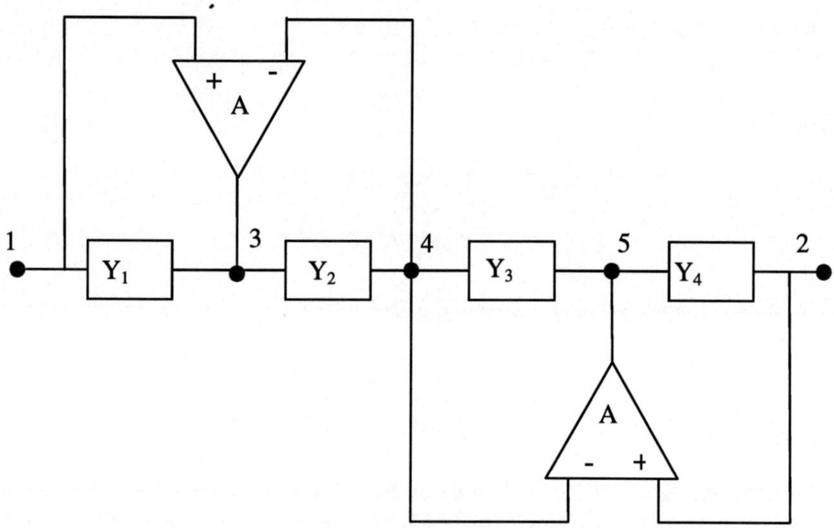

Fig. 10.2(b) Two Op Amp realization of the uncoupled CGIC

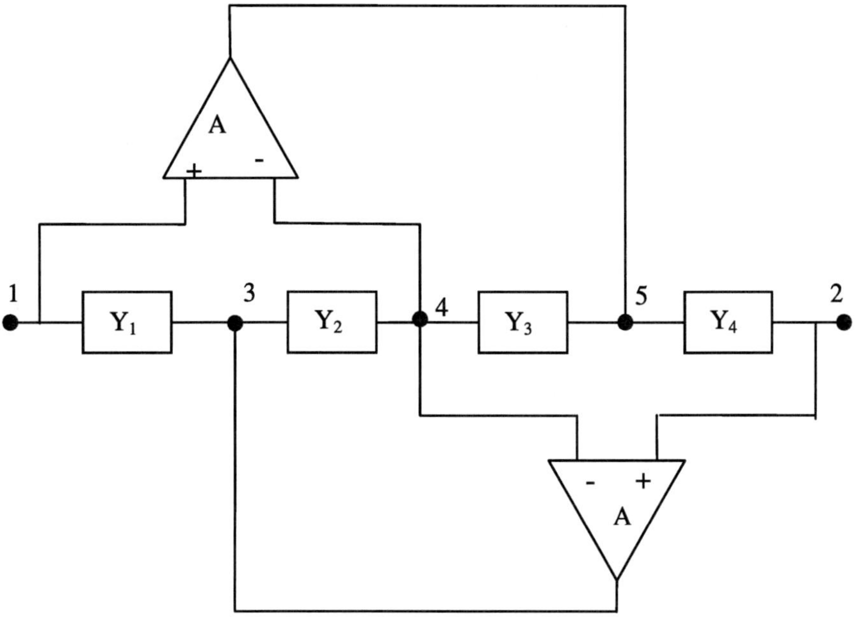

Fig. 10.2(c) Two Op Amp realization of the coupled CGIC

10.3 Voltage Generalized Impedance Converters

The VGIC is defined by the following T matrix:

$$T = \begin{bmatrix} \frac{Y_2 Y_4}{Y_1 Y_3} & 0 \\ 0 & 1 \end{bmatrix} \tag{10.5}$$

This equation is summarized in Table 10.1 together with the T matrix of all GIC and GII considered in this chapter. The above equation cannot be represented directly by an admittance matrix Y. In [19] the expanded Y matrix using infinity parameters introduced in [11] was used in order to obtain the circuit realizations.

Here an alternative approach is used to obtain the expanded NAM. The approach is based on considering the input admittance of the VGIC represented by Eq. (10.5) when terminated by load admittance Y_5.This approach simplifies the NAM expansion and was used before in the NIC realizations given in [22]. The input admittance is given by:

$$Y_{in} = \frac{Y_1 Y_3 Y_5}{Y_2 Y_4} \tag{10.6}$$

Adding a blank row and column to the above equation the following NAM is obtained:

$$Y = \begin{bmatrix} \frac{Y_1 Y_3 Y_5}{Y_2 Y_4} & 0 \\ 0 & 0 \end{bmatrix} \quad (10.7)$$

It is desirable to expand the above Y matrix to realize a one port circuit, then Y_5 will be removed from the circuit to create port 2.Thus in the expansion operation it is necessary to locate Y_5 at the diagonal position 2, 2. Also in order to maintain the current relation $I_1 = -I_2$ as given by Eq. (10.5) it is necessary to have a norator connected between nodes 1 and 2 to satisfy this condition on I_1 and I_2. This condition is summarized in Table 10.2 together with the four basic rules for realizing the four different types of the GIC.

Table 10.1 The T matrix of the GIC, GIC a, GII and GII a

GIC	T Matrix of VGIC	GII	T Matrix of GII
VGIC	$T = \begin{bmatrix} \frac{Y_2 Y_4}{Y_1 Y_3} & 0 \\ 0 & 1 \end{bmatrix}$	GII Type 1	$T = \begin{bmatrix} 0 & \frac{Y_2}{Y_1 Y_3} \\ Y_5 & 0 \end{bmatrix}$
VGICa	$T = \begin{bmatrix} -\frac{Y_2 Y_4}{Y_1 Y_3} & 0 \\ 0 & -1 \end{bmatrix}$	GII a Type 1	$T = \begin{bmatrix} 0 & -\frac{Y_2}{Y_1 Y_3} \\ -Y_5 & 0 \end{bmatrix}$
CGIC	$T = \begin{bmatrix} 1 & 0 \\ 0 & \frac{Y_1 Y_3}{Y_2 Y_4} \end{bmatrix}$	GII Type 2	$T = \begin{bmatrix} 0 & \frac{1}{Y_5} \\ \frac{Y_1 Y_3}{Y_2} & 0 \end{bmatrix}$
CGICa	$T = \begin{bmatrix} -1 & 0 \\ 0 & -\frac{Y_1 Y_3}{Y_2 Y_4} \end{bmatrix}$	GII a Type 2	$T = \begin{bmatrix} 0 & -\frac{1}{Y_5} \\ -\frac{Y_1 Y_3}{Y_2} & 0 \end{bmatrix}$

Table 10.2 Basic rules in the NAM expansion of the four different types of GIC

Type of GIC	Key step in the NAM expansion
VGIC	Norator between nodes 1 and 2
VGIC a	CM between nodes1 and 2
CGIC	Nullator between nodes 1 and 2
CGIC a	VM between nodes 1 and 2

Adding a third blank row and column to the above equation and using pivotal expansion [10] it follows that:

$$Y=\begin{bmatrix} 0 & 0 & -Y_5 \\ 0 & 0 & 0 \\ \frac{Y_1Y_3}{Y_2} & 0 & Y_4 \end{bmatrix} \tag{10.8}$$

In order to move $-Y_5$ to the diagonal position 2, 2 with a positive sign and with a norator connected between nodes 1 and 2 as mentioned above, it is necessary to use a VM to be connected between nodes 2 and 3, therefore:

$$Y=\begin{bmatrix} 0 & 0 & 0 \\ 0 & Y_5 & 0 \\ \frac{Y_1Y_3}{Y_2} & 0 & Y_4 \end{bmatrix} \tag{10.9}$$

Adding a fourth blank row and column to the above equation and apply pivotal expansion to the element in the 3, 1 position therefore:

$$Y=\begin{bmatrix} 0 & 0 & 0 & 0 \\ 0 & Y_5 & 0 & 0 \\ 0 & 0 & Y_4 & -Y_3 \\ Y_1 & 0 & 0 & Y_2 \end{bmatrix} \tag{10.10}$$

The above equation represents the 4x 4 Y matrix of the class I. Adding two blank rows and columns to Eq. (10.10), then a nullator is connected between nodes 4 and 5 to move $-Y_3$ to the position 3, 5, then a CM is connected between nodes 3, 5 to move $-Y_3$ to become Y_3 at the diagonal position 5, 5.Next a nullator is connected between nodes 1 and 6 to move Y_1 to the position 4, 6 and a norator is connected between nodes 4, 6 to move Y_1 to the diagonal position 6, 6. The expanded NAM is given by:

$$Y = \begin{bmatrix} 0 & 0 & 0 & 0 & 0 & 0 \\ 0 & Y_5 & 0 & 0 & 0 & 0 \\ 0 & 0 & Y_4 & 0 & 0 & 0 \\ 0 & 0 & 0 & Y_2 & 0 & 0 \\ 0 & 0 & 0 & 0 & Y_3 & 0 \\ 0 & 0 & 0 & 0 & 0 & Y_1 \end{bmatrix} \tag{10.11}$$

The above equation represents realization 1 of the class I VGIC shown in Fig. 10.3(a) after removing Y_5 at node 2 to create port 2.

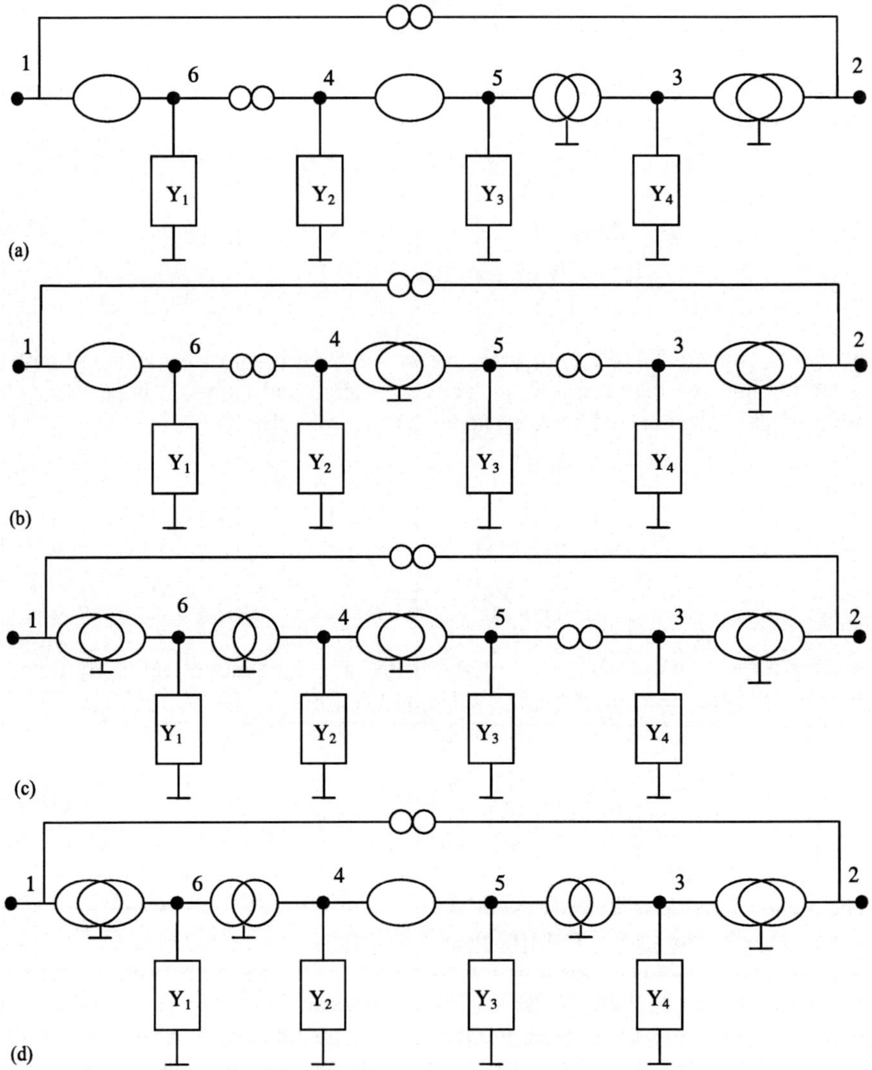

Fig. 10.3 Four nullor-mirror realizations of class I VGIC

The second realization is obtained by using a VM between nodes 4 and 5 to move $-Y_3$ to the position 3, 5 as Y_3, then a norator is connected between nodes 3, 5 to move Y_3 to the diagonal position 5, 5.The admittance Y_1 is moved to the diagonal position 6, 6 as in the previous case. The expanded NAM is given by:

$$Y = \begin{bmatrix} 0 & 0 & 0 & 0 & 0 & 0 \\ 0 & Y_5 & 0 & 0 & 0 & 0 \\ 0 & 0 & Y_4 & 0 & 0 & 0 \\ 0 & 0 & 0 & Y_2 & 0 & 0 \\ 0 & 0 & 0 & 0 & Y_3 & 0 \\ 0 & 0 & 0 & 0 & 0 & Y_1 \end{bmatrix} \qquad (10.12)$$

The above equation represents realization 2 of the Class I VGIC and shown in Fig. 10.3(b) after removing Y5 at node 2 to create port 2. Similarly realizations 3 and 4 shown in Figs. 10.3(c) and 10.3(d) can be obtained.

The 4x4 Y matrix of the class II VGIC is obtained from Eq. (10.9) by applying pivotal expansion in the alternative way from that given by Eq. (10.10) as follows:

$$Y = \begin{bmatrix} 0 & 0 & 0 & 0 \\ 0 & Y_5 & 0 & 0 \\ 0 & 0 & Y_4 & Y_3 \\ -Y_1 & 0 & 0 & Y_2 \end{bmatrix} \qquad (10.13)$$

The above matrix can be expanded in four alternative ways to obtain the 6x6 NAM resulting in additional four realizations of the class II VGIC.

The class III and IV of the VGIC are obtainable from Eq. (10.7) by adding a third blank row and column and using pivotal expansion as follows:

$$Y = \begin{bmatrix} 0 & 0 & Y_5 \\ 0 & 0 & 0 \\ -\dfrac{Y_1 Y_3}{Y_2} & 0 & Y_4 \end{bmatrix} \qquad (10.14)$$

The 6x6 NAM of the class III and IV can be obtained as in the case of class I. Each of these two classes includes four alternative realizations and is not included to limit the chapter length.

Each of the four nullor-mirror realizations for the VGIC shown in Fig. 10.3 can be realized in two alternative forms resulting in eight generalized conveyor (GC) circuits using CCII or ICCII or combination of both depending on the pairing of the nullor-mirror elements. The first GC realization of Fig. 10.3(a) is shown in Fig. 10.4(a) and the second realization of Fig. 10.3(a) is shown in Fig. 10.4(b). It is worth noting that Fig. 10.4(a) includes the type 2b L-C mutator as a special case [18, 23].

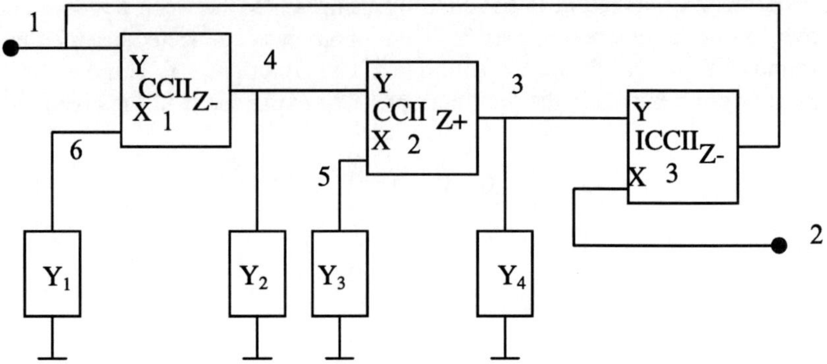

Fig. 10.4(a) Realization 1of the VGIC of Fig. 3(a)

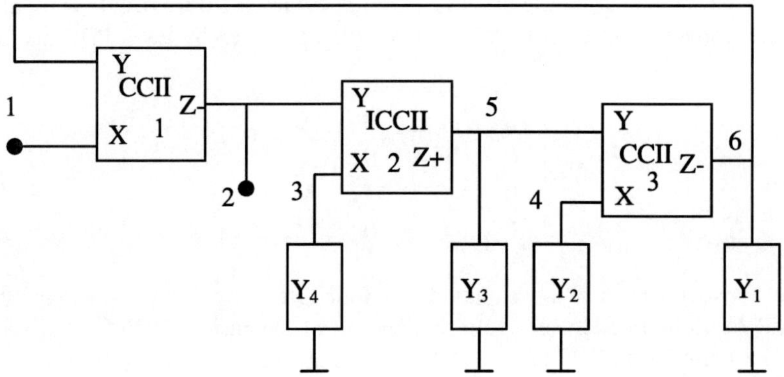

Fig. 10.4(b) Realization 2 of the VGIC of Fig. 3(a)

10.4 The VGIC with Negative A and D Coefficients

The VGIC a (where a stands for additional) is defined by the following T matrix:

$$T = \begin{bmatrix} -\frac{Y_2 Y_4}{Y_1 Y_3} & 0 \\ 0 & -1 \end{bmatrix} \tag{10.15}$$

The above T matrix with negative A and D was first defined in [24] and was realized using coefficient sign relations in [25] and using NAM expansion in [20]. In order to maintain the current relation $I_1 = I_2$ as given by Eq. (10.15) it is necessary to have a CM connected between nodes 1 and 2 as summarized in the right column of Table 10.2.

From Eq. (10.8) which applies also to the VGIC a, and in order to move $-Y_5$ to the diagonal position 2, 2 with a positive sign and using a CM connected between rows 1 and 2, it is necessary to use a nullator to be connected between columns 2 and 3, therefore:

$$Y=\begin{bmatrix} 0 & 0 & 0 \\ 0 & Y_5 & 0 \\ \frac{Y_1Y_3}{Y_2} & 0 & Y_4 \end{bmatrix} \tag{10.16}$$

Following similar steps as in the previous case the 6x6 NAM is obtained. The nullor-mirror realization 1 of the class I VGIC a, is shown in Fig. 10.5(a) after removing Y_5 at node 2 to create port 2. The first GC realization of Fig. 10.5(a) is shown in Fig. 10.5(b) and the second realization is shown in Fig. 10.5(c).

5 Current Generalized Impedance Converters

The CGIC is defined by the T matrix given by Eq. (10.3). It is desirable to expand the Y matrix in Eq. (10.7) which applies also to the CGIC, to realize a one port cir cuit, then Y_5 will be removed from the circuit to create port 2.Thus in the expansion operation it is necessary to locate Y_5 at the diagonal position 2, 2. Also in order to maintain the voltage relation $V_1 = V_2$ as given by Eq. (10.3), it is necessary to have a nullator connected between nodes 1 and 2 to satisfy this condition.

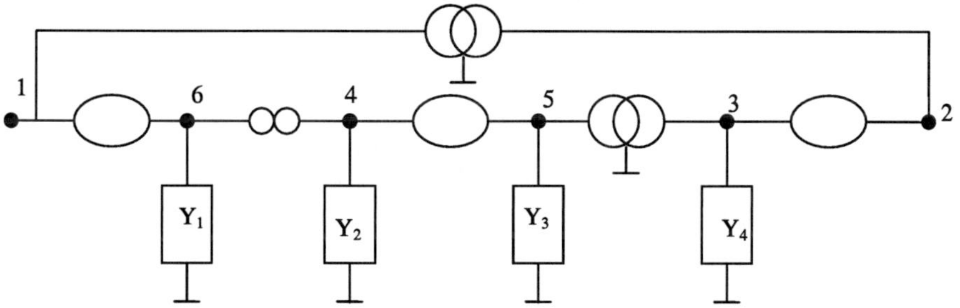

Fig. 10.5(a) Nullor-mirror realization 1 of class I VGIC a

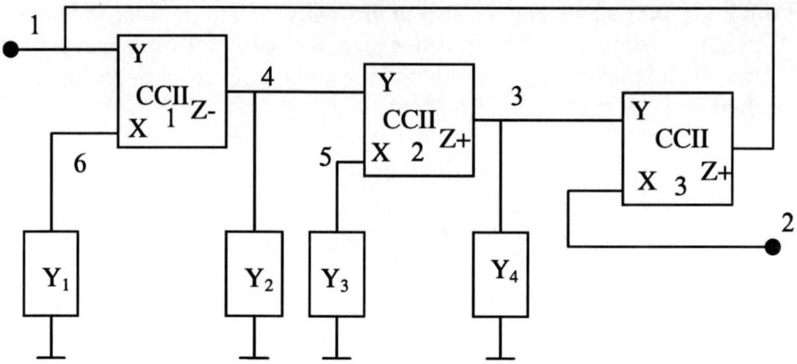

Fig. 10.5(b) Realization 1 of the VGIC a of Fig. 5(a)

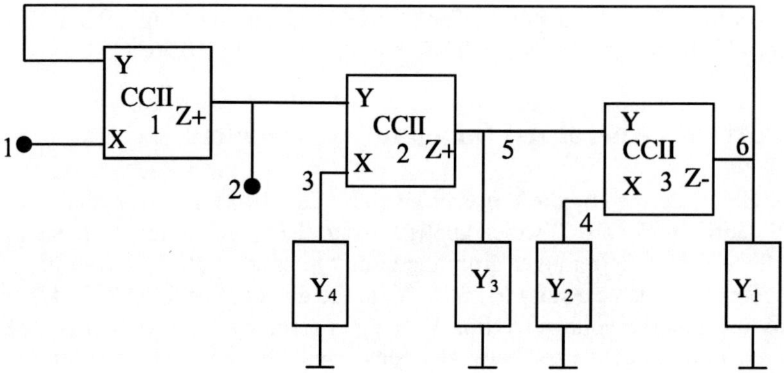

Fig. 10.5(c) Realization 2 of the VGIC a of Fig. 5(a)

Adding a third blank row and column to Eq. (10.7) and using pivotal expansion [10] it follows that:

$$Y=\begin{bmatrix} 0 & 0 & \frac{Y_1Y_3}{Y_2} \\ 0 & 0 & 0 \\ -Y_5 & 0 & Y_4 \end{bmatrix} \qquad (10.17)$$

In order to move $-Y_5$ to the diagonal position 2, 2 with a positive sign with a nullator connected between nodes 1 and 2, it is necessary to use a CM to be connected between nodes 2 and 3, therefore:

$$Y=\begin{bmatrix} 0 & 0 & \frac{Y_1Y_3}{Y_2} \\ 0 & Y_5 & 0 \\ 0 & 0 & Y_4 \end{bmatrix} \qquad (10.18)$$

The above matrix can be expanded to a 6x6 NAM as class I CGIC. The nullor-mirror realization 1 of the class I CGIC is shown in Fig. 10.6.

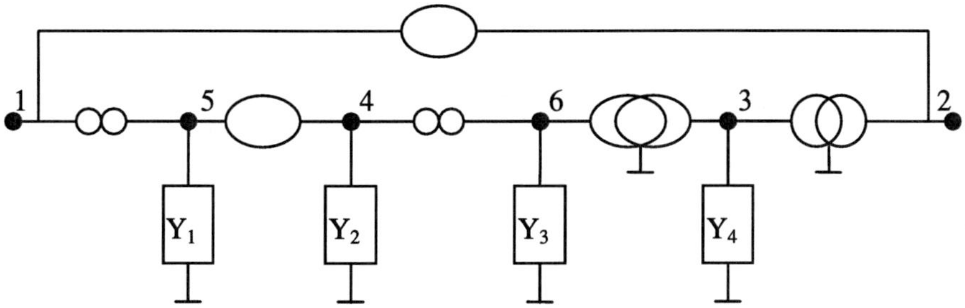

Fig. 10.6 Nullor-mirror realization 1 of class I CGIC

There are four nullor-mirror realizations for the CGIC each of them can be realized in two alternative forms resulting in eight GC circuits using CCII or ICCII or combination of both depending on the pairing of the nullor-mirror elements. The first GC realization of Fig. 10.6 is shown in Fig. 10.7(a) and the second realization is shown in Fig. 10.7(b). It should be noted that these realizations are the adjoints of the VGIC realizations given in Fig. 10.4 [26-28]. It is worth noting that the CCII- is self adjoint and the CCII+ is adjoint to ICCII-.

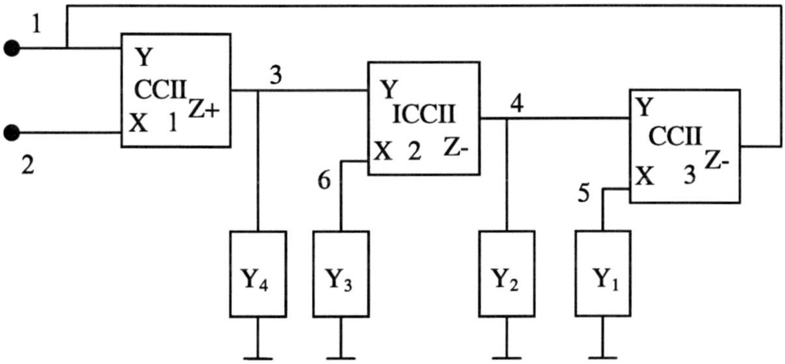

Fig. 10.7(a) Realization 1of the CGIC of Fig. 6

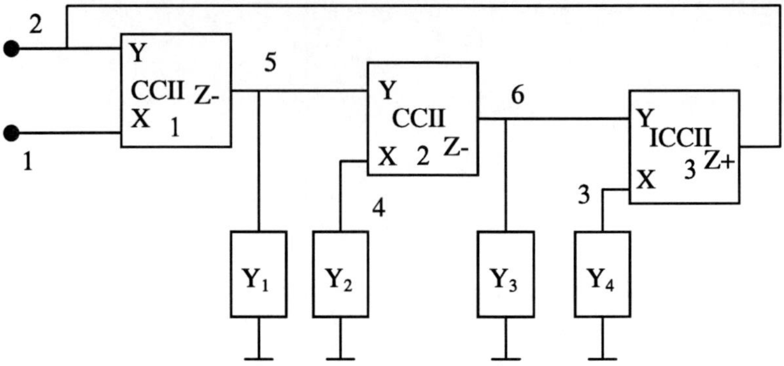

Fig. 10.7(b) Realization 2 of the CGIC of Fig. 6

10.6 The CGIC with Negative A and D Coefficients

The CGIC a, is defined by the following T matrix:

$$T = \begin{bmatrix} -1 & 0 \\ 0 & -\dfrac{Y_1 Y_3}{Y_2 Y_4} \end{bmatrix} \tag{10.19}$$

In order to maintain the voltage relation $V_1 = -V_2$ as given by the above equation, it is necessary to have a VM connected between nodes 1 and 2 to satisfy this condition. Fig. 10.8 represents realization 1 of the CGIC a.

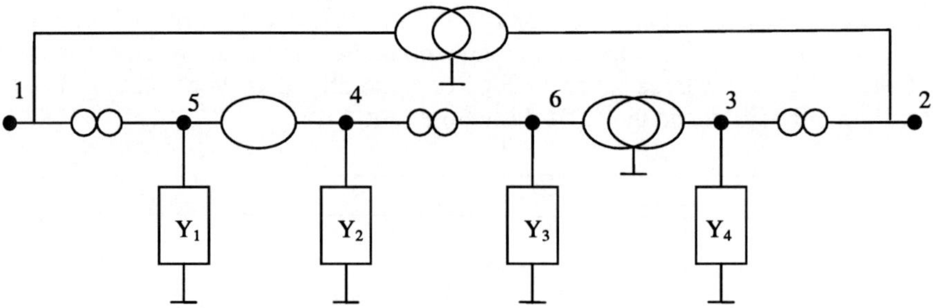

Fig. 10.8 Nullor-mirror realization 1 of class I CGIC a

10.7 Generalized Impedance Inverter Type 1

The GII type 1 is defined by the following T matrix:

$$T = \begin{bmatrix} 0 & \dfrac{Y_2}{Y_1 Y_3} \\ Y_5 & 0 \end{bmatrix} \tag{10.20}$$

This is a generalization to the type 2a-LR mutator considered in [18]. The notations in the above equation are chosen such that when port 2 is terminated by Y_4 the same expression for Y_{in} given by Eq. (10.6) is obtained. The corresponding Y matrix of Eq. (10.20) is given by:

$$Y = \begin{bmatrix} 0 & Y_5 \\ -\frac{Y_1 Y_3}{Y_2} & 0 \end{bmatrix} \tag{10.21}$$

It is desirable to locate the admittances Y_1, Y_2, Y_3 and Y_5 at diagonal positions leaving the diagonal positions 1, 1 and 2, 2 open as they represent the two ports.Adding a third blank row and column to the above equation and connecting a nullator between nodes 2 and 3 and a norator between nodes 1 and 3 in order to move Y_5 to the diagonal position 3, 3 therefore:

$$Y = \begin{bmatrix} 0 & 0 & 0 \\ -\frac{Y_1 Y_3}{Y_2} & 0 & 0 \\ 0 & 0 & Y_5 \end{bmatrix} \tag{10.22}$$

Applying pivotal expansion to the term in the 2, 1 position, thus:

$$Y = \begin{bmatrix} 0 & 0 & 0 & 0 \\ 0 & 0 & 0 & Y_3 \\ 0 & 0 & Y_5 & 0 \\ Y_1 & 0 & 0 & Y_2 \end{bmatrix} \tag{10.23}$$

The nullor-mirror realization 1 of the class I GII is shown in Fig. 10.9. There are 16 different GC realizations realizing Eq. (10.20) and are obtained using coefficient conditions and are tabulated in [18, 25]

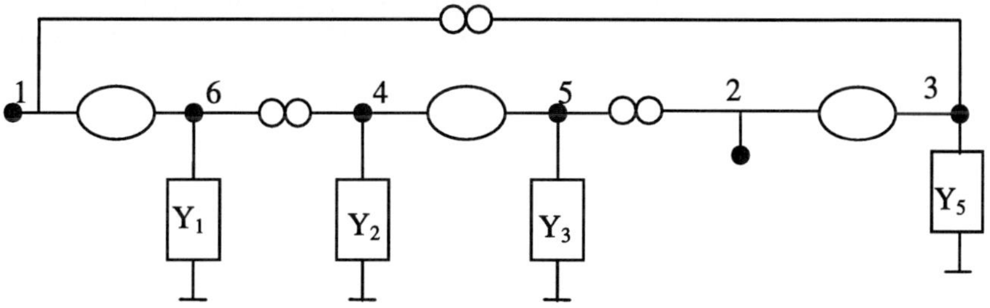

Fig. 10.9 Nullor-mirror realization 1 of class I type 1 GII

10.8 Generalized Impedance Inverter a Type 1

The GII a type 1 is defined by the following T matrix:

$$T = \begin{bmatrix} 0 & -\dfrac{Y_2}{Y_1 Y_3} \\ -Y_5 & 0 \end{bmatrix} \tag{10.24}$$

The corresponding Y matrix of Eq. (10.24) is given by:

$$Y = \begin{bmatrix} 0 & -Y_5 \\ \dfrac{Y_1 Y_3}{Y_2} & 0 \end{bmatrix} \tag{10.25}$$

Adding a third blank row and column to the above equation and connecting a nullator between nodes 2 and 3 and a CM between nodes1 and 3 in order to move Y_5 to the diagonal position 3, 3 therefore:

$$Y = \begin{bmatrix} 0 & 0 & 0 \\ \dfrac{Y_1 Y_3}{Y_2} & 0 & 0 \\ 0 & 0 & Y_5 \end{bmatrix} \tag{10.26}$$

Applying pivotal expansion to the term in the 2, 1 position, thus:

$$Y = \begin{bmatrix} 0 & 0 & 0 & 0 \\ 0 & 0 & 0 & Y_3 \\ 0 & 0 & Y_5 & 0 \\ -Y_1 & 0 & 0 & Y_2 \end{bmatrix} \tag{10.27}$$

The nullor-mirror realization 1 of the class I GII a, is shown in Fig. 10.10. There are 16 different GC realizations realizing Eq. (10.24) and are obtained using coefficient conditions as given in [25].

10.9 Generalized Impedance Inverter Type 2

The GII type 2 is defined by the following T matrix:

$$T = \begin{bmatrix} 0 & \dfrac{1}{Y_5} \\ \dfrac{Y_1 Y_3}{Y_2} & 0 \end{bmatrix} \tag{10.28}$$

This is a generalization to the type 2b-LR mutator considered in [18]. The corresponding Y matrix of Eq. (10.28) is given by:

$$Y = \begin{bmatrix} 0 & \frac{Y_1 Y_3}{Y_2} \\ -Y_5 & 0 \end{bmatrix} \tag{10.29}$$

The nullor-mirror realization 1of the class I GII type 2, is shown in Fig. 10.11. There are 16 different GC realizations realizing Eq. (10.28) and are obtained using coefficient conditions and are tabulated in [18].

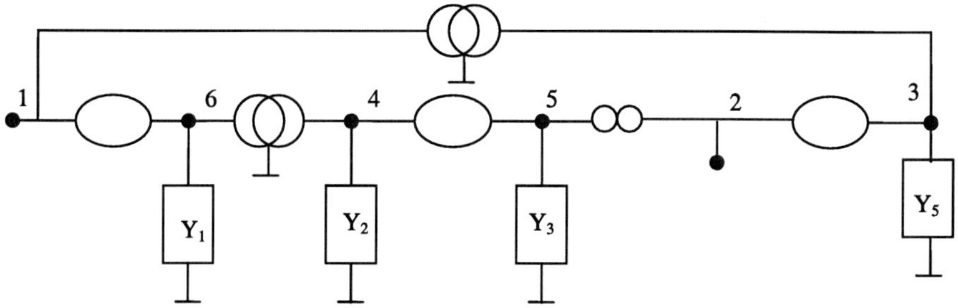

Fig. 10.10 Nullor-mirror realization 1 of class I type 1 GII a

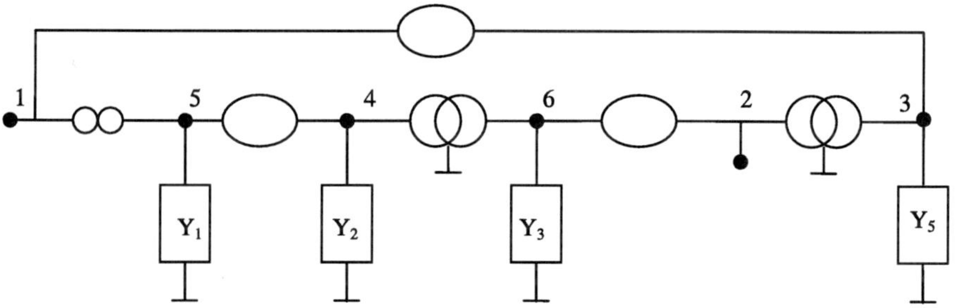

Fig. 10.11 Nullor-mirror realization 1 of class I type 2 GII

10.10 Generalized Impedance Inverter a Type 2

The GII a type 2 is defined by the following T matrix:

$$T = \begin{bmatrix} 0 & -\frac{1}{Y_5} \\ -\frac{Y_1 Y_3}{Y_2} & 0 \end{bmatrix} \tag{10.30}$$

The corresponding Y matrix of Eq. (10.30) is given by:

$$Y = \begin{bmatrix} 0 & -\frac{Y_1 Y_3}{Y_2} \\ Y_5 & 0 \end{bmatrix} \tag{10.31}$$

The nullor-mirror realization 1of the class I GII a type 2, is shown in Fig. 10.12.

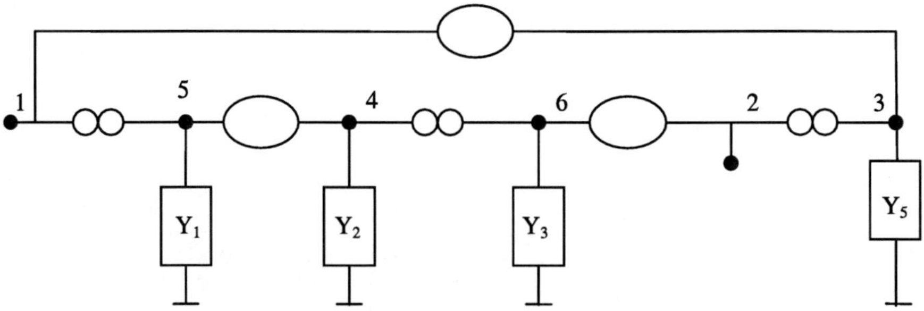

Fig. 10.12 Nullor-mirror realization 1 of class I type 2 GII a

10.11 Conclusions

The generation of the GIC and GII circuits using NAM expansion is explained in details. It is seen that there are 16 nullor-mirror realizations of each of the VGIC, VGIC a, CGIC, CGIC a, GII type 1, GII a type1, GII type 2 and the GII a type 2. Each nullor-mirror realization can be implemented in two alternative ways using GC, one version is suitable for inductance simulation and the other one is suitable for FDNR realization.

It is worth noting that the CGIC is the adjoint to the VGIC, CGIC a is the adjoint to the VGIC a. Also the GII type 1 is the adjoint to the GII a type 2 and the GII a type 1 is the adjoint to the GII type 2.

References

[1] Antoniou, A.: Novel RC-active-network synthesis using generalized immittance converters. IEEE Transactions on Circuit Theory 17, 212–217 (1970)

[2] Antoniou, A.: Realization of gyrators using operational amplifiers and their use in RC-active network synthesis. Proc. of IEE 116, 1838–1850 (1969)

[3] Bruton, L.T.: Network transfer functions using the concept of frequency dependant negative resistance. IEEE Transactions on Circuit Theory 16, 406–408 (1969)

[4] Carlin, H.J.: Singular network elements. IEEE Transactions on Circuit Theory 11, 67–72 (1964)

[5] Bruton, L.T.: RC Active Circuits Theory and Design. Prentice Hall, Inc., New Jersey (1980)

[6] Awad, I.A., Soliman, A.M.: Inverting second-generation current conveyors: the missing building blocks, CMOS realizations and applications. International Journal of Electronics 86, 413–432 (1999)
[7] Awad, I.A., Soliman, A.M.: On the voltage mirrors and the current mirrors. Analog Integrated Circuits and Signal Processing 32, 79–81 (2002)
[8] Awad, I.A., Soliman, A.M.: A new approach to obtain alternative active building blocks realizations based on their ideal representations. Frequenz 54, 290–299 (2000)
[9] Soliman, A.M., Saad, R.A.: The voltage mirror–current mirror pair as a universal element. International Journal of Circuit Theory and Applications 38, 787–795 (2010)
[10] Haigh, D.G., Tan, F.Q., Papavassiliou, C.: Systematic synthesis of active-RC circuit building-blocks. Analog Integrated Circuits Signal Processing 43, 297–315 (2005)
[11] Haigh, D.G., Clarke, T.J.W., Radmore, P.M.: Symbolic framework for linear active circuits based on port equivalence using limit variables. IEEE Transactions Circuits Systems I 53, 2011–2024 (2006)
[12] Haigh, D.G., Radmore, P.M.: Admittance matrix models for the nullor using limit variables and their application to circuit design. IEEE Transactions Circuits Systems I 53, 2214–2223 (2006)
[13] Haigh, D.G.: A method of transformation from symbolic transfer function to active-RC circuit by admittance matrix expansion. IEEE Transactions Circuits Systems. I 53, 2715–2728 (2006)
[14] Saad, R.A., Soliman, A.M.: Use of mirror elements in the active device synthesis by admittance matrix expansion. IEEE Transactions Circuits Systems I 55, 2726–2735 (2008)
[15] Saad, R.A., Soliman, A.M.: Generation, modeling, and analysis of CCII-based gyrators using the generalized symbolic framework for linear active circuits. International Journal of Circuit Theory and Applications 36, 289–309 (2008)
[16] Saad, R.A., Soliman, A.M.: A new approach for using the pathological mirror elements in the ideal representation of active devices. International Journal of Circuit Theory and Applications 38, 148–178 (2010)
[17] Saad, R.A., Soliman, A.M.: On the systematic synthesis of CCII based floating simulators. International Journal of Circuit Theory and Applications 38, 935–967 (2010)
[18] Soliman, A.M.: New CCII and ICCII based realizations of L-C and L-R mutators. Journal of Circuits Systems and Signal Processing 29, 1181–1191 (2010)
[19] Soliman, A.M.: Generation of generalized impedance converter circuits using NAM expansion. Journal of Circuits Systems and Signal Processing 20, 1091–1114 (2011)
[20] Soliman, A.M.: A Note on the generation of generalized impedance converter circuits using NAM expansion. Journal of Circuits Systems and Signal Processing 31, 1147–1157 (2012)
[21] Sedra, A.S., Smith, K.C.: A second generation current conveyor and its applications. IEEE Transactions Circuit Theory 17, 132–134 (1970)
[22] Soliman, A.M.: Generation and classification of CCII and ICCII based negative impedance converter circuits using NAM expansion. International Journal of Circuit Theory and Applications 39, 835–847 (2011)
[23] Soliman, A.M.: Two new LC mutators and their realizations. IEEE Transactions Circuit Theory 19, 371–372 (1972)
[24] Soliman, A.M.: Realizations of ideal FDNC and FDNR elements using new types of mutators. International Journal of Electronics 44, 317–323 (1978)

[25] Swamy, M.N.S.: Mutators, generalized impedance converters and inverters, and their realization using generalized current conveyors. Circuits, Systems and Signal Process 30, 209–232 (2011)
[26] Bhattacharyya, B.B., Swamy, M.N.S.: Network transposition and its application in synthesis. IEEE Transactions Circuit Theory 18, 394–397 (1971)
[27] Swamy, M.N.S.: Transpose of a multi-terminal element and applications. IEEE Transactions Circuit and Systems II 57, 696–700 (2010)
[28] Soliman, A.M.: Adjoint network theorem and floating elements in the NAM. Journal of Circuits Systems and Computers 18, 597–616 (2009)

Chapter 11
Fractional Step Analog Filter Design

Todd Freeborn, Brent Maundy, and Ahmed Elwakil

Abstract. Using the fractional Laplacian operator, s^{α}, this chapter outlines the process to design, analyze, and implement continuous-time fractional-step lowpass, highpass, and bandpass filters of order $(n+\alpha)$, where α is the fractional-step between the integer orders with value $0 < \alpha < 1$. The design of these filters is done using transfer functions in the $s-$domain without solving fractional-order differential equations in the time domain. The design process, stability analysis, PSPICE simulations, and physical realization of these filters are presented based on minimum-phase error approximations of the operator s^{α}. Four methods of implementation, using fractional capacitors in the Tow-Thomas biquad, Single Amplifier Biquads (SABs), Field Programmable Analog Array (FPAA) hardware and Frequency Dependent Negative Resistor (FDNR) topologies to realize decomposed transfer functions are demonstrated.

Keywords: Fractional calculus, Fractional filters, Analog circuits.

11.1 Introduction

The design of continuous-time analog filters for signal processing has traditionally involved the use of the Laplacian operator, s, raised to an integer order — i.e. s, s^2, s^n. However, the recent import of concepts from fractional calculus, the branch of mathematics concerned with differentiation and integration to non-integer orders,

Todd Freeborn · Brent Maundy
University of Calgary, 2500 University Dr. N.W., Calgary, Canada
e-mail: {tjfreebo,bmaundy}@ucalgary.ca

Ahmed Elwakil
University of Sharjah, P.O. Box 27272, Sharjah, United Arab Emirates
e-mail: elwakil@ieee.org

M. Fakhfakh et al. (Eds.): *Analog/RF & Mixed-Signal Circuit Sys. Design*, LNEE 233, pp. 243–267.
DOI: 10.1007/978-3-642-36329-0_11

offers attenuation characteristics not possible using integer order filters [15, 16] and applications in many interdisciplinary fields [6].

A fractional derivative of order α with initial condition a is given by the Grünwald-Letnikov approximation [4] as

$$ {}_aD^{\alpha}f(x) = \lim_{h\to 0}\frac{1}{h^{\alpha}}\sum_{m=0}^{\left[\frac{x-a}{h}\right]}(-1)^m\frac{\Gamma(\alpha+1)}{m!\Gamma(\alpha-m+1)}f(x-mh) \tag{11.1}$$

where $\Gamma(\cdot)$ is the gamma function. Applying the Laplace transform to the general fractional derivative of (11.1) with zero initial conditions yields

$$\mathscr{L}\{{}_0D^{\alpha}f(t)\} = s^{\alpha}F(s) \tag{11.2}$$

The fractional Laplacian operator is especially useful in the design of filters with fractional step stopband characteristics, as the design of transfer functions can be done algebraically rather than through solving the difficult time domain representations of fractional derivatives. The stopband attenuation of integer order filters has been limited to increments based on the order, n, but using s^{α} attenuations between integer orders n and $(n+1)$, where $0 \le \alpha \le 1$, are possible.

In the subsequent sections we consolidate the recent progress in fractional filters to present a process to design and implement these filters. In Section 11.2 we present the design of lowpass, highpass, and bandpass fractional filters using transfer functions in the s-domain, with Section 11.3 presenting the method to analyze the stability of these designed filters and implement higher order stable fractional filters. Section 11.4 outlines the methods and design equations for the physical realization of these filters using fractional capacitors in the Tow-Thomas biquad, as well as using Single Amplifier Biquads (SABs), Field Programmable Analog Array (FPAA) hardware and Frequency Dependent Negative Resistor (FDNR) topologies to realize approximated fractional step filters using integer-order approximations of s^{α}.

11.2 Design of Fractional Filters

11.2.1 Fractional Lowpass Filters (FLPFs)

Consider the $(1+\alpha)$ order transfer function

$$T_{1+\alpha}^{FLPF}(s) = \frac{k_1}{s^{1+\alpha}+k_2s^{\alpha}+k_3} \tag{11.3}$$

where $k_{1,2,3}$ are positive constants and $0 < \alpha < 1$. Using (11.3) yields a lowpass filter response with a fractional step of $-20(1+\alpha)$ dB/dec through the stopband while it is possible to maintain a flat passband, based on the selection of $k_{2,3}$, for the desired α [7, 8]. The values of $k_{2,3}$ when $k_1 = 1$ for a flat passband response are given, respectively, as

$$k_2 = 1.1796\alpha^2 + 0.16765\alpha + 0.21735 \tag{11.4}$$
$$k_3 = 0.19295\alpha + 0.81369 \tag{11.5}$$

The -3 dB frequency, $\omega_{3\text{dB}}$, can be calculated by solving for the positive real roots of the equation

$$\omega_{3\text{dB}}^{2+2\alpha} - 2\omega_{3\text{dB}}^{1+\alpha} k_3 \sin\left(\frac{\alpha\pi}{2}\right) + \omega_{3\text{dB}}^{2\alpha} k_2^2 + 2\omega_{3\text{dB}}^{\alpha} k_2 k_3 \cos\left(\frac{\alpha\pi}{2}\right) - k_3^2 = 0 \tag{11.6}$$

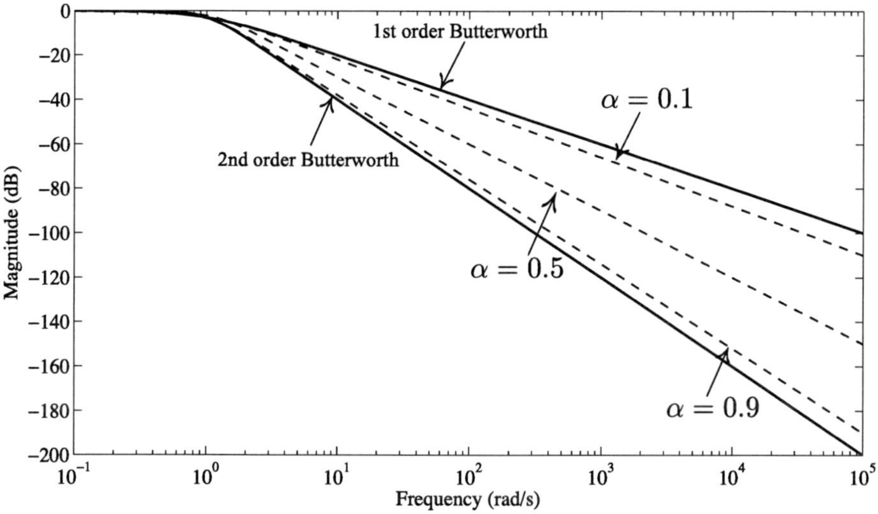

Fig. 11.1 MATLAB simulated magnitude response of FLPFs of order $(1+\alpha) = 1.1$, 1.5, and 1.9 when $k_1 = 1$ and $k_{2,3}$ selected for flat passband response.

MATLAB simulations of (11.3) for $\alpha = 0.1$, 0.5, and 0.9 when $k_1 = 1$ and $k_{2,3}$ selected using (11.4) and (11.5), respectively, are shown in Fig. 11.1. We note the fractional steps of -22, -30, and -38 dB/dec in the stopband, between the 1st and 2nd order Butterworth responses, not possible using traditional integer order filters, and -3 dB frequencies of 0.6723, 0.9961, and 0.9281 rad/s, respectively.

11.2.2 Fractional Highpass Filters (FHPFs)

To obtain a FHPF transfer function we apply the LP-to-HP transformation, replacing s with $1/s$, to the FLPF of (11.3) yielding:

$$T_{1+\alpha}^{FHPF}(s) = \frac{k_1}{k_3} \frac{s^{1+\alpha}}{s^{1+\alpha} + \frac{k_2}{k_3} s + \frac{1}{k_3}} \tag{11.7}$$

where $k_{1,2,3}$ are positive constants and $0 < \alpha < 1$. Using (11.7) yields a highpass filter response with fractional step of $20(1+\alpha)$ dB/dec through the stopband with a flat passband when $k_1 = 1$ and (11.4) and (11.5) are used for $k_{2,3}$, respectively. MATLAB simulations of the magnitude response of (11.7) for $\alpha = 0.1, 0.5$, and 0.9

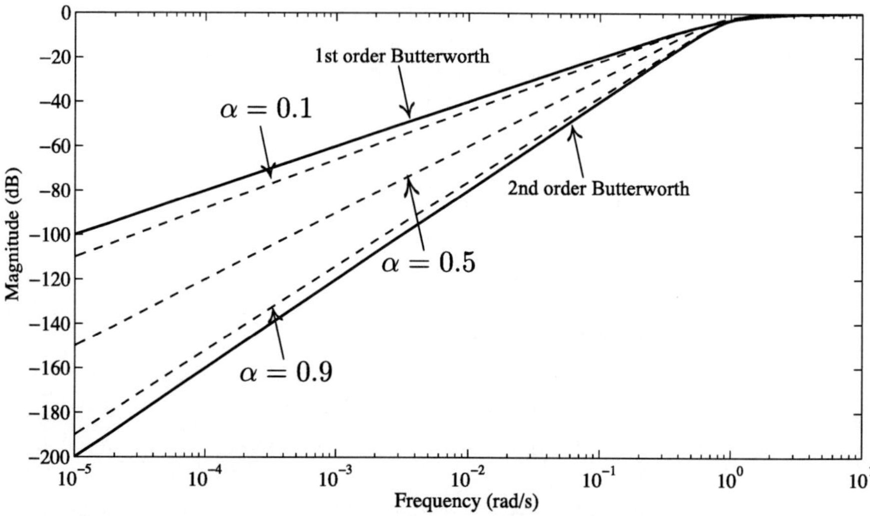

Fig. 11.2 MATLAB simulated magnitude response of FHPFs of order $(1+\alpha) = 1.1, 1.5$, and 1.9 when $k_1 = 1$ and $k_{2,3}$ selected for flat passband response.

when $k_1 = 1$ and $k_{2,3}$ values selected using (11.4) and (11.5), respectively, are shown in Fig. 11.2. Like the FLPFs of the previous section, fractional steps of 22, 30, and 38 dB/dec in the stopband, between the 1st and 2nd order Butterworth responses are measured. The -3 dB frequency, $\omega_{3\text{dB}}$, of these filters can be calculated by solving for the positive real roots of the equation

$$1 + 2\omega_{3\text{dB}}^{2+\alpha} k_2 k_3 \cos\left(\frac{\alpha\pi}{2}\right) - \omega_{3\text{dB}}^{2(1+\alpha)} k_3^2 - 2\omega_{3\text{dB}}^{1+\alpha} k_3 \sin\left(\frac{\alpha\pi}{2}\right) = 0 \tag{11.8}$$

Solving (11.8) for the -3 dB frequencies of the responses in Fig. 11.2 yields 1.487, 1.004, and 1.077 rad/s for filters of order 1.1, 1.5, and 1.9, respectively.

11.2.3 Fractional Bandpass Filters (FBPFs)

The use of s^α in the design of bandpass filters presents a new method for the realization of bandpass filters with asymmetric stopband characteristics and high quality factors. Consider the $(\alpha_1 + \alpha_2)$ order transfer function

$$T_{\alpha_1+\alpha_2}^{FBPF}(s) = \frac{k_1 s^{\alpha_2}}{s^{\alpha_1+\alpha_2} + k_2 s^{\alpha_2} + k_3} \tag{11.9}$$

where $k_{1,2,3}$ are positive constants and $0 < \alpha_{1,2} < 1$. Using (11.9) yields a bandpass filter response with fractional steps of $20\alpha_2$ and $-20\alpha_1$ dB/dec for frequencies lower and higher, respectively, than the center frequency. Therefore, this fractional transfer function can realize bandpass filters with asymmetric stopband characteristics when $\alpha_1 \neq \alpha_2$. MATLAB simulations of the normalized magnitude response of

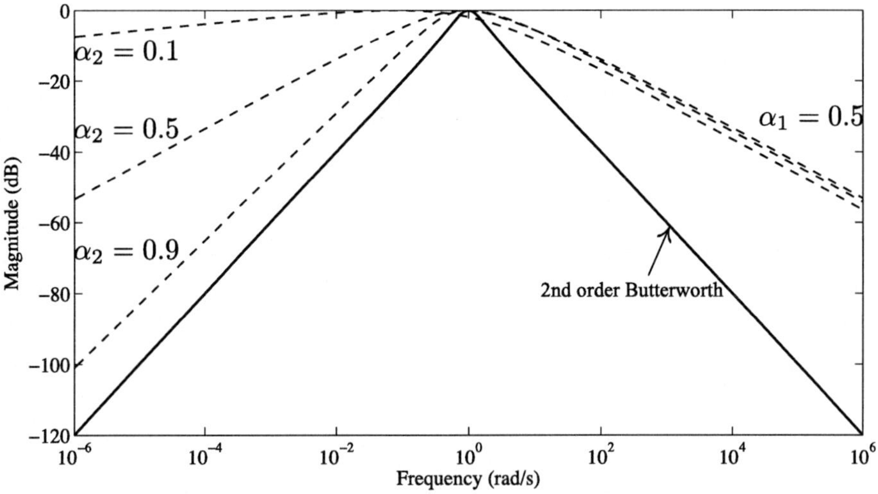

Fig. 11.3 MATLAB simulated magnitude response of normalized FBPFs of order $(\alpha_1 + \alpha_2) = 0.6$, 1.0, and 1.4 when $\alpha_1 = 0.5$ and $k_{2,3}$ selected for flat passband response

(11.9) for $\alpha_2 = 0.1, 0.5$, and 0.9 when $\alpha_1 = 0.5$ and $k_{2,3}$ values selected using (11.4) and (11.5), respectively, are shown in Fig. 11.3. We note the fractional steps of 2, 10, and 18 dB/dec in the low frequency stopband while maintaining a fractional step of -10 dB/dec in the high frequency stopband, providing an asymmetric stopband response not easily realizable using traditional integer order filters. The frequency at which the FBPF reaches a maxima, ω_m, can be calculated by solving for the positive real root of the equation

$$0 = \omega_m^{2\alpha_2+\alpha_1} k_1 k_3 (\alpha_2 - \alpha_1) \cos\left(\frac{(\alpha_1+\alpha_2)\pi}{2}\right) + \omega_m^{\alpha_2} k_1 k_3 \alpha_2 \left(\omega_m^{\alpha_2} \cos\left(\frac{\alpha_2 \pi}{2}\right) + k_3\right)$$
$$-\omega_m^{3\alpha_2+\alpha_1} k_1 \alpha_1 \left(\omega_m^{\alpha_1} + k_2 \cos\left(\frac{\alpha_1 \pi}{2}\right)\right) \tag{11.10}$$

Knowing ω_m the -3 dB frequencies, $\omega_{1,2}$, can be calculated numerically by solving for ω the equation

$$\frac{\left|T^{FBPF}_{\alpha_1+\alpha_2}(j\omega_m)\right|}{\sqrt{2}} = \frac{k_1}{\sqrt{x_0\omega^{\alpha_1-\alpha_2} + x_1\omega^{-\alpha_2} + x_2\omega^{\alpha_1} + x_3 + x_4\omega^{-2\alpha_2} + \omega^{2\alpha_1}}} \tag{11.11}$$

where $x_0 = 2k_3\cos((\alpha_1+\alpha_2)\pi/2)$, $x_1 = 2k_2k_3\cos(\alpha_2\pi/2)$, $x_2 = 2k_2\cos(\alpha_1\pi/2)$, $x_3 = k_2^2$, and $x_4 = k_3^2$. The quality factors, maxima frequencies, and -3 dB frequencies of the FBPF responses in Fig. 11.3 are given in Table 11.1.

Table 11.1 Quality factors, maxima frequencies, and −3 dB frequencies of simulated FBPFs in Fig. 11.3 for $\alpha_2 = 0.1$, 0.5, and 0.9 when $\alpha_1 = 0.5$ and $k_{2,3}$ selected for flat passband response

α_2	Q	ω_m (rad/s)	ω_1 (rad/s)	ω_2 (rad/s)
0.1	0.0473	0.0839	0.0003	1.775
0.5	0.1950	0.9102	0.1712	4.839
0.9	0.2296	0.9450	0.3287	4.445

11.2.3.1 High-Q Asymmetric Bandpass Filters

In addition to asymmetric stopband characteristics, fractional filters provide a method for obtaining bandpass filters with high quality factors using fractional transfer functions [1, 2]. Two transfer functions which realize high-Q asymmetric bandpass filters are given as

$$T_I^{FBPF}(s) = k_1 \frac{k_2 s^\alpha}{s^2 + k_2 s^\alpha + k_3} \tag{11.12}$$

$$T_{II}^{FBPF}(s) = k_1 \frac{k_2 s^{1+\alpha}}{s^2 + k_2 s^{1+\alpha} + k_3} \tag{11.13}$$

where (11.12) and (11.13) are referred to as the Type I and Type II transfer functions, respectively. These transfer functions realize attenuations of 20α and $20(1+\alpha)$ dB/dec in the low frequency stopbands and $-20(2-\alpha)$ and $-20(1-\alpha)$ dB/dec in the high frequency stopbands for the Type I and Type II transfer functions, respectively. The maxima frequency, ω_m, and −3 dB frequencies, $\omega_{1,2}$, for the Type I transfer function can be calculated numerically by solving the equations

$$0 = \omega_m^2 - k_2 \omega_m^\alpha \cos\left(\frac{\alpha\pi}{2}\right) - k_3 \tag{11.14}$$

$$0 = \omega_1^2 - \sqrt{2} k_2 \omega_1^\alpha \cos\left(\frac{\alpha\pi}{2} + \frac{\pi}{4}\right) - k_3 \tag{11.15}$$

$$0 = \omega_2^2 - \sqrt{2} k_2 \omega_2^\alpha \sin\left(\frac{\alpha\pi}{2} + \frac{\pi}{4}\right) - k_3 \tag{11.16}$$

for ω_m, ω_1, and ω_2, respectively. The quality factor, Q, of these filters can then be calculated as

$$Q = \frac{\omega_m}{\omega_2 - \omega_1} \tag{11.17}$$

and the center frequency gain (CFG) at ω_m can be calculated as

$$\text{CFG} = \frac{k_1}{\sin\left(\frac{\alpha\pi}{2}\right)} \tag{11.18}$$

Equations (11.14)-(11.16) and (11.18) can be used for the Type II transfer function by replacing α with $(1+\alpha)$. MATLAB simulations of the magnitude response of (11.12) for $\alpha = 0.1, 0.5$, and 0.9 when $k_{1,2,3} = 1, 0.01$, and 1, respectively, are shown in Fig. 11.4. The characteristics of these filters calculated using (11.14) to (11.17) are given in Table 11.2. Note that these filters can be normalized to have a CFG = 1 by setting $k_1 = \sin(\alpha\pi/2)$.

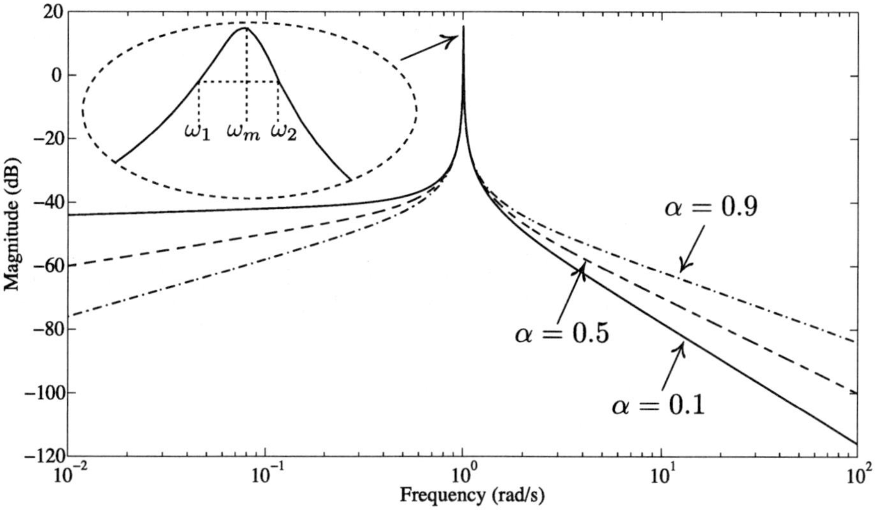

Fig. 11.4 MATLAB simulated magnitude response of high-Q asymmetric Type I FBPF when $\alpha = 0.1, 0.5$, and 0.9 when $k_1 = 1$, $k_2 = 0.01$, and $k_3 = 1$

Table 11.2 Quality factors, CFG, maxima frequencies, and −3 dB frequencies of simulated Type I FBPFs in Fig. 11.4 for $\alpha = 0.1, 0.5$, and 0.9 when $k_{1,2,3} = 1, 0.01$, and 1, respectively

α	Q	CFG	ω_m (rad/s)	ω_1 (rad/s)	ω_2 (rad/s)
0.1	644.9	6.393	1.005	1.004	1.005
0.5	141.9	1.414	1.004	1.000	1.007
0.9	101.2	1.013	1.001	0.9959	1.006

11.3 Stability Analysis

To analyze the stability of fractional filters requires conversion of the s-domain transfer functions to the W-plane defined in [17]. This transforms the transfer function from a fractional one to an integer order one which can be analyzed using traditional integer order analysis methods. The process for this analysis can be done using the following steps:

1. Convert the fractional transfer function to the W-plane using the transformations $s = W^m$ and $\alpha = k/m$ [17],
2. Select k and m for the desired α value,
3. Solve the transformed transfer function for all poles in the W-plane and if any of the absolute pole angles, $|\Theta_W|$, are less than $\frac{\pi}{2m}$ rad/s then the system is unstable, otherwise if all $|\Theta_W| > \frac{\pi}{2m}$ then the system is stable.

Example. Applying the analysis process on the FLPFs of Fig. 11.1 yields

1. The FLPF transfer function, (11.3), after transformation to the W-plane becomes:

$$T_{1+\alpha}^{FLPF}(W) = \frac{k_1}{W^{m+k} + k_2 W^k + k_3} \tag{11.19}$$

2. For $\alpha = 0.1, 0.5$, and 0.9 values of $k = 1, 5$, and 9 when $m = 10$ are selected.
3. Solving for the poles of (11.19) yields minimum pole angles of 0.2916, 0.2421, and 0.2404 rad/s when $k = 1, 5$, and 9, respectively, for $m = 10$ and $k_{2,3}$ selected using (11.4) and (11.5), respectively. The minimum pole angles for the FLPFs are all greater than $\frac{\pi}{2m} = 0.1571$ rad/s and therefore are all stable.

11.3.1 Higher Order Fractional Filters

Expanding (11.3) to the general case, that is a $(n+\alpha)$ order filter, yields the transfer function:

$$T_{n+\alpha}^{FLPF}(s) = \frac{k_1}{s^{n+\alpha} + k_2 s^{\alpha} + k_3} \tag{11.20}$$

The highest order filter that (11.20) can implement while maintaining stability is $(n+\alpha) \leq 2$ when $n < 2$ [8]. Therefore, this transfer function is not able to realize stable higher-order fractional step filters. To overcome this limitation $T_{1+\alpha}^{FLPF}(s)$, which is always stable when $0 < \alpha < 1$ is divided by higher order normalized Butterworth polynomials [12] creating stable higher-order fractional step filters of order $(n+\alpha)$ written as [8, 12]

$$T_{n+\alpha}^{FLPF}(s) \approx \frac{T_{1+\alpha}^{FLPF}(s)}{B_{n-1}(s)}; n \geq 2 \tag{11.21}$$

where $B_n(s)$ is a standard Butterworth polynomial of order n [5]. MATLAB simulations of the magnitude response of (11.21) for $(4+\alpha) = 4.1, 4.5$, and 4.9 order filters when $k_1 = 1$ and $k_{2,3}$ selected using (11.4) and (11.5), respectively, are shown in Fig. 11.5. We note the fractional steps of -82, -90, and -98 dB/dec in the stopband, between the 4th and 5th order Butterworth responses, providing a stable higher order fractional filter not possible using (11.20). Note, that this same method can be applied to create stable higher order FHPFs and FBPFs as well.

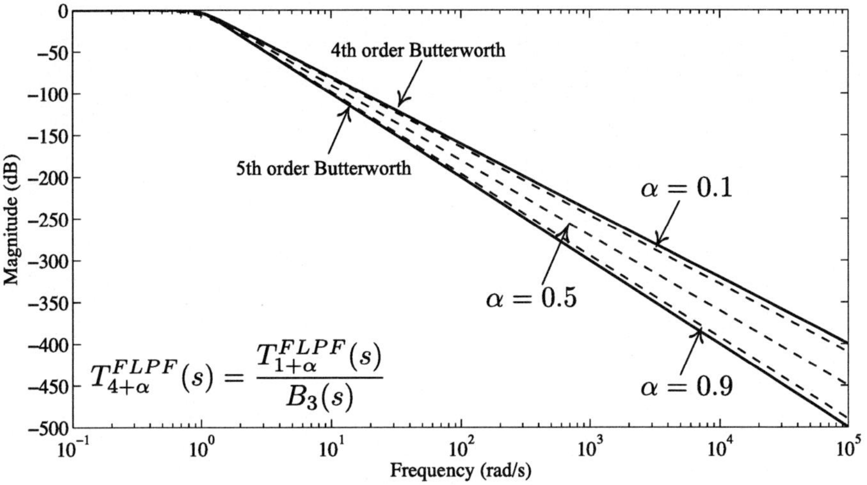

Fig. 11.5 MATLAB simulated magnitude response of FLPFs of order $(4+\alpha) = 4.1$, 4.5, and 4.9 when $k_1 = 1$ and $k_{2,3}$ selected for flat passband response

11.4 Simulation and Realization

Until the commercial availability of fractance devices with impedances of $Z = Fs^{\alpha}$ are available, the simulation and physical realization of fractional filters will require the use of integer order approximations of s^{α}. Using these integer order approximations, three methods have been presented for the realization of fractional filters, using fractional capacitors in traditional filter topologies [16] and using SABs [8] or FPAAs [9] to realize approximated fractional transfer functions.

11.4.1 Fractional Tow-Thomas Biquad

While the traditional Tow-Thomas biquad, shown in Fig. 11.6, uses standard capacitors, the available filter responses can be further generalized by replacing traditional capacitors with fractional capacitors [10]. This approach has also been investigated for both the Sallen-Key filter and the Kerwin-Huelsman-Newcomb biquad [16] as well as in the design multivibrator circuits [13]. By replacing C_1 with a fractional capacitor the filter output at the lowpass node yields a FLPF with transfer function

$$T_{1+\alpha}^{FLPF}(s) = -\frac{\frac{R_6}{R_1R_4R_5C_1C_2}}{s^{1+\alpha} + \frac{s^{\alpha}}{R_3C_1} + \frac{R_6}{R_2R_4R_5C_1C_2}} \tag{11.22}$$

while replacing both C_1 and C_2 with fractional capacitors the filter output at the bandpass node yields a FBPF with transfer function

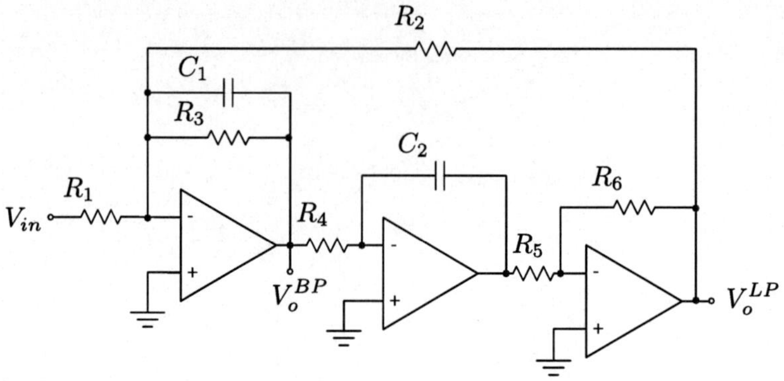

Fig. 11.6 Tow-Thomas biquad topology

$$T^{FBPF}_{\alpha_1+\alpha_2}(s) = -\frac{s^{\alpha_2}\frac{1}{R_1C_1}}{s^{\alpha_1+\alpha_2} + \frac{s^{\alpha_2}}{R_3C_1} + \frac{R_6}{R_2R_4R_5C_1C_2}} \tag{11.23}$$

Comparing the coefficients of (11.3) to (11.22) and (11.9) to (11.23) while ignoring the signs shows 3 design equations and 8 variables yielding 5 degrees of freedom in our selection of the component values to realize $k_{1,2,3}$. Therefore, setting $C_1 = C_2 =$ 1 F and $R_2 = R_4 = R_5 = 1\ \Omega$ our design equations for the remaining components to realize the FLPF response become

$$R_1 = \frac{k_3}{k_1}\frac{R_2R_5}{R_4} = \frac{k_3}{k_1} \tag{11.24}$$

$$R_3 = \frac{1}{k_2}\frac{1}{C_1} = \frac{1}{k_2} \tag{11.25}$$

$$R_6 = k_3C_1C_2R_2R_4R_5 = k_3 \tag{11.26}$$

and the design equations to realize the FBPF become

$$R_1 = \frac{1}{k_1C_1} = \frac{1}{k_1} \tag{11.27}$$

$$R_3 = \frac{1}{k_2}\frac{1}{C_1} = \frac{1}{k_2} \tag{11.28}$$

$$R_6 = k_3C_1C_2R_2R_4R_5 = k_3 \tag{11.29}$$

The component values to realize the FLPFs of Fig. 11.1, magnitude scaled by a factor of 1000 and frequency shifted to 1 kHz, are given in Table 11.3. The frequency and magnitude scaling factors for fractional elements are different than traditional scaling such that

$$C_{\text{new}} = \frac{C_{\text{old}}}{K_fK_m} \tag{11.30}$$

Table 11.3 Component values to realize FLPFs of orders 1.1, 1.5, and 1.9 using the Fractional Tow-Thomas biquad

Order $(1+\alpha)$	C_1 (μF)	C_2 (μF)	R_1 (Ω)	R_3 (Ω)	R_6 (Ω)	$R_{2,4,5}$ (Ω)
$(1+0.1)$	0.159	417	833	4067	833	1000
$(1+0.5)$	0.159	12.6	910	1678	910	1000
$(1+0.9)$	0.159	0.382	987	755	987	1000

$$R_{\text{new}} = R_{\text{old}} K_m \tag{11.31}$$

where K_m is the desired magnitude scaling factor, $K_f = \omega^\alpha$ is the frequency scaling factor [16], ω is the desired frequency to be shifted to, and α is the order of the capacitor to frequency shift.

11.4.1.1 PSPICE Simulations

While most capacitors do exhibit fractional behaviour [18] and should be modeled with an impedance $Z_C = \frac{1}{s^\alpha C}$, the value of α is very near to 1 preventing their use in implementing fractional filters such as the fractional Tow-Thomas Biquad. Therefore, until commercial fractance devices become available to physically realize circuits that make use of s^α, integer order approximations have to be used. There are many methods to create an approximation of s^α that include Continued Fraction Expansions (CFEs) as well as rational approximation methods [14]. These methods present a large array of approximations with varying order and accuracy, with the accuracy and approximated frequency band increasing as the order of the approximation increases. Here, a CFE method [11] was selected to model the fractional capacitors for PSPICE simulations. Collecting eight terms of the CFE yields a 4th order approximation of the fractional capacitor that can be physically realized using the RC ladder network in Fig. 11.7.

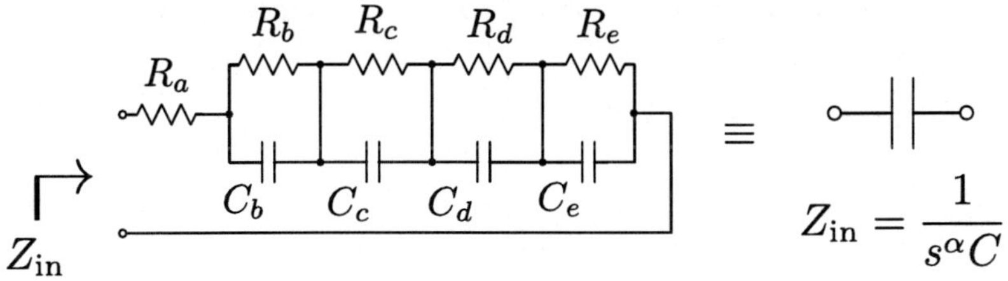

Fig. 11.7 RC ladder network to realize a 4th order approximated fractional capacitor

The component values required for the 4th order approximation of C_2 with values from Table 11.3 and orders of 0.1, 0.5, and 0.9 using the RC ladder network in Fig. 11.7, shifted to a center frequency of 1 kHz, are given in Table 11.4.

Table 11.4 Component values to realized 4th order approximations of fractional capacitors of 417, 12.6, and 0.382 μF with orders of 0.1, 0.5, 0.9, respectively, centered at a frequency of 1 kHz

Component	$C = 417$ μF $\alpha = 0.1$	$C = 12.6$ (μF) $\alpha = 0.5$	$C = 0.382$ (μF) $\alpha = 0.9$
R_a (Ω)	658.7	111.1	6.8
R_b (Ω)	196.3	251.7	43.3
R_c (Ω)	134.6	378.7	130.7
R_d (Ω)	159.0	888.9	670.4
R_e (Ω)	369.5	7.369 k	146.2 k
C_b (nF)	68.9	83.8	705
C_c (μF)	0.627	0.296	1.13
C_d (μF)	2.18	0.537	1.03
C_e (μF)	6.64	0.695	0.207

The magnitude and phase of the ideal (solid line) and 4th order approximated (dashed) fractional capacitor with capacitance 12.6 μF and order $\alpha = 0.5$, shifted to a center frequency of 1 kHz, are presented in Fig. 11.8. From this figure we observe that the approximation is very good over almost 4 decades, from 200 Hz to 70 kHz, for the magnitude and almost 2 decades, from 200 Hz to 6 kHz, for the phase. In these regions, the deviation of the approximation from ideal does not exceed 1.23 dB and 0.23° for the magnitude and phase, respectively. PSPICE simulations of the low-pass response of the approximated fractional Tow-Thomas biquad, shown in Fig. 11.9, compared to the MATLAB simulations of (11.3) for filters of order $(1+\alpha) =$ 1.1, 1.5, and 1.9 are given in Fig. 11.10 as dashed and solid lines, respectively. The component values to realize the fractional Tow-Thomas biquad and approximated fractional capacitors are given in Tables 11.3 and 11.4, respectively.

11.4.2 SAB Realization

The FLPF transfer function of (11.3) can be realized using SABs when a 2nd order approximation of s^α, given for any order α as

$$s^\alpha \approx \frac{(\alpha^2+3\alpha+2)s^2+(8-2\alpha^2)s+(\alpha^2-3\alpha+2)}{(\alpha^2-3\alpha+2)s^2+(8-2\alpha^2)s+(\alpha^2+3\alpha+2)} \tag{11.32}$$

when substituted into (11.3) yielding the integer order transfer function

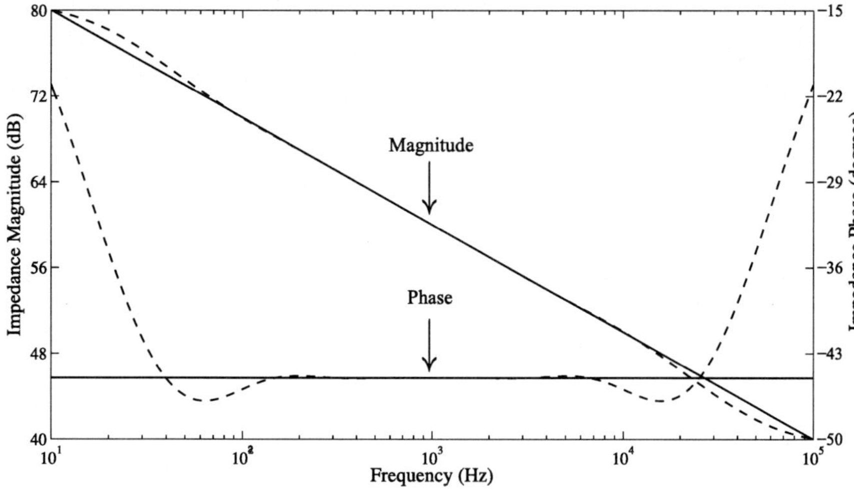

Fig. 11.8 Magnitude and phase response of the approximated fractional capacitor (dashed) compared to the ideal (solid) with capacitance of 12.6 μF and order 0.5 after scaling to a center frequency of 1 kHz

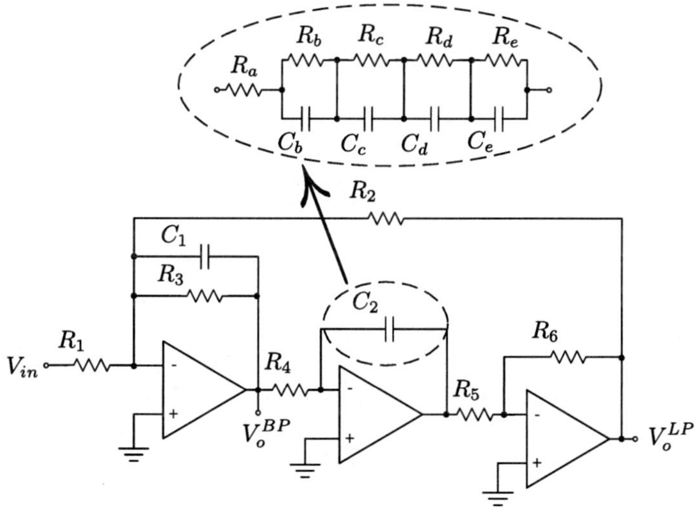

Fig. 11.9 Fractional Tow-Thomas biquad with the RC ladder network to realize a 4th order approximation of the fractional capacitor C_2

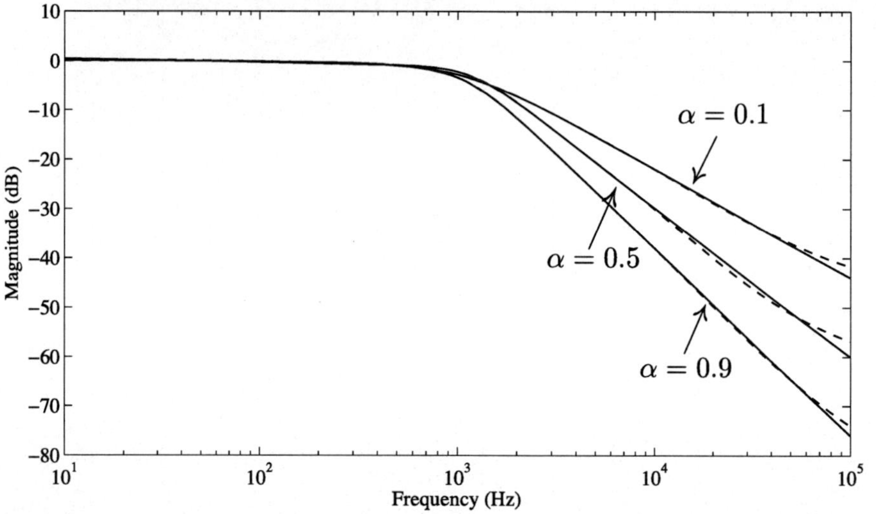

Fig. 11.10 PSPICE simulations of FLPF responses of the approximated Fractional Tow-Thomas biquad compared to the theoretical simulations of (11.3) when $\alpha = 0.1$, 0.5, and 0.9

$$H_{1+\alpha}^{FLPF}(s) \approx \frac{k_1}{a_0}\frac{a_2 s^2 + a_1 s + a_0}{s^3 + b_0 s^2 + b_1 s + b_2} \tag{11.33}$$

where $a_0 = \alpha^2 + 3\alpha + 2$, $a_1 = 8 - 2\alpha^2$, $a_2 = \alpha^2 - 3\alpha + 2$, $b_0 = (a_1 + a_0 k_2 + a_2 k_3)/a_0$, $b_1 = (a_1(k_2 + k_3) + a_2)/a_0$, and $b_2 = (a_0 k_3 + a_2 k_2)/a_0$. The integer order approximation, (11.33), can be physically realized by decomposing it into 1st and 2nd order transfer functions given as

$$H_{1+\alpha}^{FLPF}(s) \approx \frac{1}{s + d_0}\frac{e_0 s^2 + e_1 s + e_2}{s^2 + d_1 s + d_2} \tag{11.34}$$

Coefficients $d_{0,1,2}$ and $e_{0,1,2}$ are determined through the solution of the system of equations by equating like terms of (11.33) to (11.34) yielding

$$d_0 + d_1 = \frac{a_1 + a_0 k_2 + a_2 k_3}{a_0} \tag{11.35}$$

$$d_0 d_1 + d_2 = \frac{a_1(k_2 + k_3) + a_2}{a_0} \tag{11.36}$$

$$d_0 d_2 = \frac{a_0 k_3 + a_2 k_2}{a_0} \tag{11.37}$$

$$e_0 = k_1 \frac{a_2}{a_0} \tag{11.38}$$

$$e_1 = k_1 \frac{a_1}{a_0} \tag{11.39}$$

$$e_2 = k_1 \tag{11.40}$$

Using (11.35)-(11.40) to approximate the FLPFs of Fig. 11.1 yields the coefficient values in Table 11.5. These values can be realized using the circuit in Fig. 11.11

Table 11.5 Coefficients $d_{0,1,2}$ and $e_{0,1,2}$ values for decomposed 1st and 2nd order transfer functions to realize approximated FLPFs of orders 1.1, 1.5, and 1.9

Order $(1+\alpha)$	d_0	d_1	d_2	e_0	e_1	e_2
(1+0.1)	0.3174	4.000	3.1978	0.7403	3.4545	1.0000
(1+0.5)	0.4938	2.2843	2.0844	0.2000	2.0000	1.0000
(1+0.9)	0.7141	1.7872	1.4200	0.0200	1.1579	1.0000

which is a cascade of 1st and 2nd order sections realizable via a parallel RC network and a SAB [5], as shown in Fig. 11.11. The resistor values to approximate FLPFs of orders 1.1, 1.5, and 1.9 when all time constants were shifted to 0.1 ms using unit resistors of 1 kΩ and 0.1 μF capacitors are given in Table 11.6.

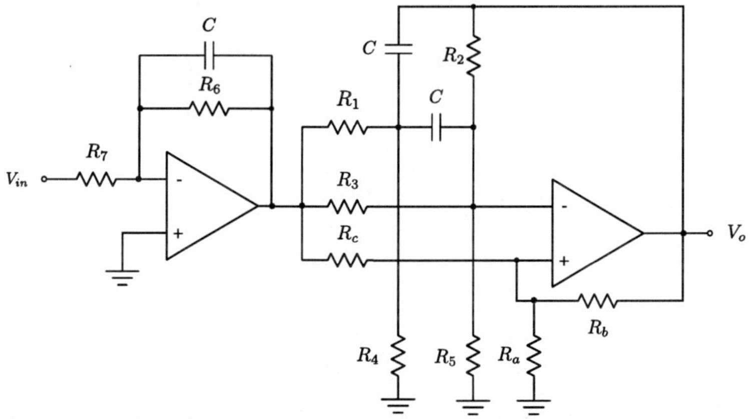

Fig. 11.11 Circuit topology to approximate FLPFs of order $(1+\alpha)$

PSPICE simulation results of Fig. 11.11 compared to the theoretical simulations of (11.3) are given in Fig. 11.12 for FLPFs of order $(1+\alpha) = 0.1$, 0.5, and 0.9.

11.4.3 FPAA Realization

Anadigm FPAAs are *analog signal processors* consisting of fully configurable analog modules (CAMs) surrounded by programmable interconnect and analog input/output cells. The signal processing occurs in the CAMs using fully differential

Table 11.6 Resistor values to realize approximated FLPFs of orders $(1+\alpha) = 1.1$, 1.5, and 1.9 using the circuit in Fig. 11.11

Resistor (Ω)	$(1+0.1)$	$(1+0.5)$	$(1+0.9)$
R_1	52.16 k	6.327 k	4.707 k
R_2	78.21	102.7	251.2
R_3	1.304 k	7.335 k	∞
R_4	1.652 k	1.803 k	796.4
R_5	∞	814.9	33.18
R_6	3.150 k	2.025 k	1.400 k
R_7	3.198 k	2.084 k	1.419 k
R_a	38.50 k	6.250 k	102
R_b	1.000 k	1.000 k	1.000 k
R_c	13.51 k	25.00 k	5.009 k

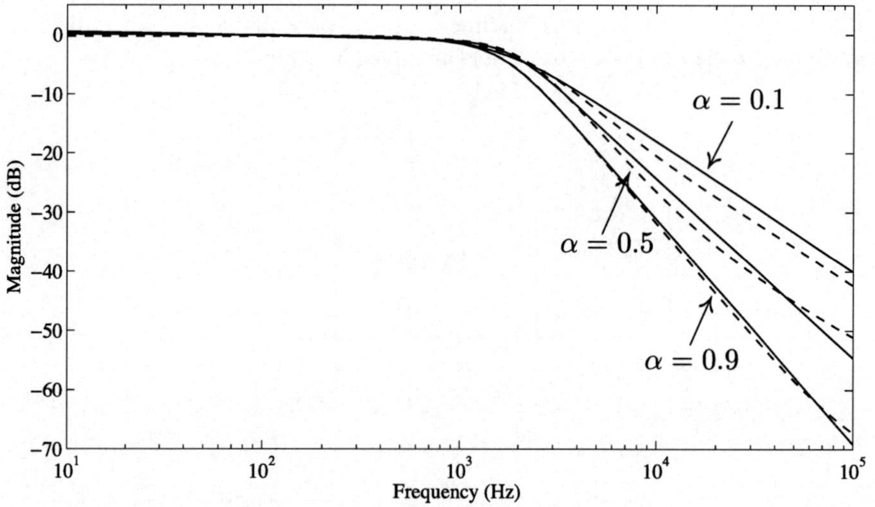

Fig. 11.12 PSPICE simulations of FLPF responses of Fig. 11.11 compared to the theoretical simulations of (11.3) when $\alpha = 0.1$, 0.5, and 0.9

switched capacitor circuitry, which provide specialized behaviours such as filtering, gain, sample and hold, summing, rectification and more. This provides a very flexible architecture that can be easily reconfigured using the AnadigmDesigner tools. These tools are a graphical design environment to build circuits using the design CAMs. In this design environment CAMs can be dropped in, wired together and configured for the desired design requirements. From the graphical implementation of a circuit, the AnadigmDesigner tools generates the configuration data file to program the FPAA.

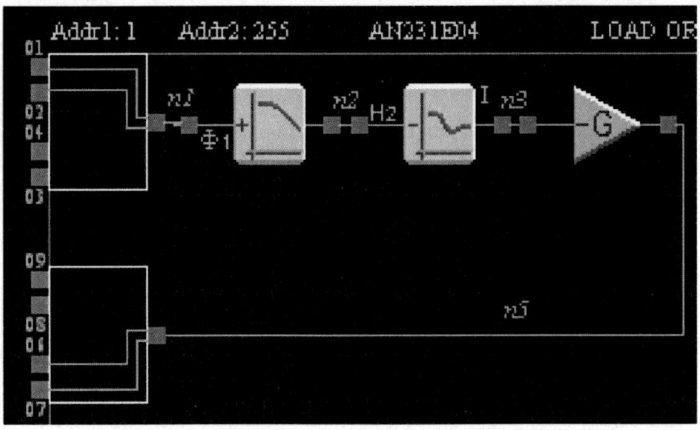

Fig. 11.13 FLPF implementation using the bilinear filter, biquadratic filter, and inverting gain CAMs of the AnadigmDesigner2 tools

Two CAMs that are particularly useful in the implementation of approximated fractional step filters are the bilinear and biquadratic filter CAMs. These CAMs realize bilinear and biquadratic transfer functions given the pole and zero frequencies and quality factors making them ideal for the realization of filters that have been decomposed into biquadratic and bilinear sections.

11.4.3.1 FLPF Design Equations

To realize (11.34) using the FPAA requires the use of bilinear filter, biquadratic filter, and gain CAMs, shown in Figure 11.13. However, we must apply the frequency transformation ($s = s/2\pi f_0$) to (11.34), where f_0 is the denormalized frequency, before using the CAMs. The FPAA design equations to implement the approximated denormalized FLPF of order $(1+\alpha)$ can be summarized as

$$f_1 = d_0 f_0 \tag{11.41}$$

$$f_{2_z} = f_0 \sqrt{\frac{e_2}{e_0}} \tag{11.42}$$

$$f_{2_p} = f_0 \sqrt{d_2} \tag{11.43}$$

$$Q_{2_z} = \frac{\sqrt{e_0 e_2}}{e_1} \tag{11.44}$$

$$Q_{2_p} = \frac{\sqrt{d_2}}{d_1} \tag{11.45}$$

$$G = \frac{e_0}{d_0} \tag{11.46}$$

where f_1 is the pole frequency of the bilinear CAM, $f_{2_{p,z}}$ and $Q_{2_{p,z}}$ are the pole and zero frequencies and quality factors of the biquadratic CAM, respectively, and G is the DC gain of (11.34). As examples, the theoretical values of pole and zero frequencies and quality factors for both the bilinear and biquadratic CAMs, for approximated FLPFs of orders $(1+\alpha) = 1.1$, 1.5, and 1.9, when $f_0 = 1$ kHz, are given in Table 11.7. Note that these values are calculated for $k_1 = 1$ and $k_{2,3}$ using

Table 11.7 Theoretical biquad and bilinear CAM values for physical implementation of approximated FLPFs of orders 1.1, 1.5, and 1.9.

Order $(1+\alpha)$	f_1 (kHz)	f_{2_z} (kHz)	f_{2_p} (kHz)	Q_{2_z}	Q_{2_p}	G
$(1+0.1)$	0.3174	1.1623	1.7882	0.2491	0.4471	2.3322
$(1+0.5)$	0.4938	2.2361	1.4437	0.2236	0.6320	0.4050
$(1+0.9)$	0.7141	7.0775	1.1915	0.1220	0.6667	0.0280

(11.4) and (11.5), respectively. The experimental magnitude and phase results of the $(1+\alpha) = 1.1$, 1.5, and 1.9 order FLPFs implemented with an Anadigm AN231E04 FPAA with the values in Table 11.7 FPAA compared to the theoretical simulations of (11.3) are given in Fig. 11.14. From Fig. 11.14 we see major deviations from the theoretical phase response above 2 kHz by the experimental FPAA results. This results from using a 2nd order approximation of s^α with the FPAA over the 4th order approximation used in Section 11.4.1.1 and non-idealities of the FPAA.

It should be mentioned that the realized FPAA values will differ from theoretical due to limitations on the values that can be implemented with the FPAA. The biquadratic and bilinear filter CAMs cannot realize all possible values because of hardware limits as a result of the design parameters being interrelated to other parameters as well as the sample clock frequency. As a result of these interrelations and the finite number of capacitor values implemented on silicon, the AnadigmDesigner tools select the capacitor values with the best ratios to satisfy the input design parameters (pole and zero frequencies, quality factors, and DC gain). However, these best ratios do not always meet the exact parameters which results in minor deviations between the theoretical and realized values. While an FPAA has the advantages of quickly realizing fractional filters and simplifying the design process its frequency range is limited by the bandwidth of the FPAAs, where the AN231E04 has a typical bandwidth of 2 MHz, lower than those realizable with other topologies.

FPAAs present the possibility to modify the fractional order of a filter by dynamically reconfiguring the FPAA using a connected microprocessor. The bilinear and biquadratic CAMs can be adjusted to modify the approximated α changing the stop band attenuation. This modification is not possible using other topologies, as changing α would require a complete new set of passive components.

11.4.3.2 FHPF Design Equations

Approximated FHPFs can also be realized using the FPAA with the same pole and zero frequency and quality factor design equations, (11.41) to (11.46), when the bilinear and biquadratic filter CAMs are set to the high-pass configuration. However, while the same design equations can be utilized, the values of $d_{0,1,2}$ and $e_{0,1,2}$ are different from their low-pass counterparts and must be calculated from the following system of equations

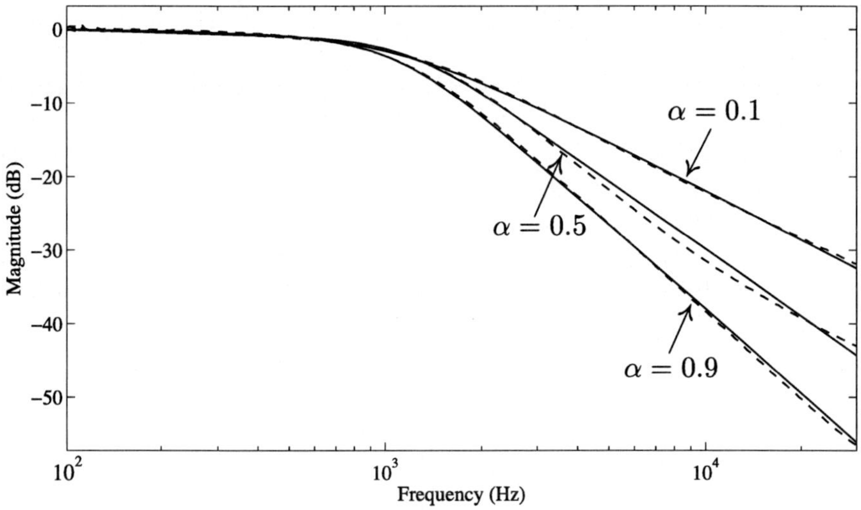

(a)

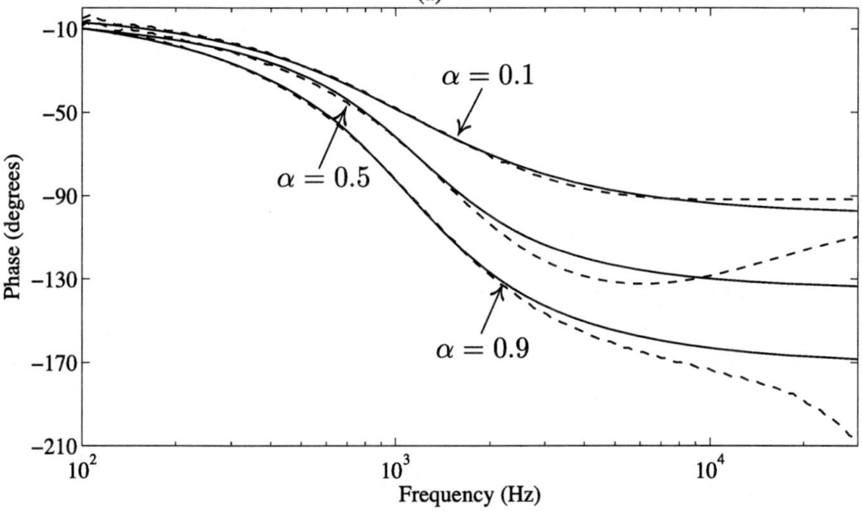

(b)

Fig. 11.14 Experimental FPAA (a) magnitude and (b) phase results of implemented FLPFs of order $(1+\alpha) = 1.1$, 1.5, and 1.9 compared to the theoretical simulations of (11.3)

$$d_0 + d_1 = \frac{a_1(k_2 + k_3) + a_2}{a_0 k_3 + a_2 k_2} \tag{11.47}$$

$$d_0 d_1 + d_2 = \frac{a_0 k_2 + a_1 + a_2 k_3}{a_0 k_3 + a_2 k_2} \tag{11.48}$$

$$d_0 d_2 = \frac{a_0}{a_0 k_3 + a_2 k_2} \tag{11.49}$$

$$e_0 = k_1 \tag{11.50}$$

$$e_1 = k_1 \frac{a_1}{a_0} \tag{11.51}$$

$$e_2 = k_1 \frac{a_2}{a_0} \tag{11.52}$$

The theoretical values of pole and zero frequencies and quality factors for both the bilinear and biquadratic CAMs, for approximated FHPFs of orders $(1+\alpha) = 1.1$, 1.5, and 1.9, when $f_0 = 1$ kHz, are given in Table 11.8, calculated for $k_1 = 1$ and $k_{2,3}$ using (11.4) and (11.5), respectively. It should be noted that the accuracy of the approximated fractional step filters compared to the theoretical can be improved by using a higher order approximation of s^α rather than the 2nd order approximation of (11.32).

Table 11.8 Theoretical biquad and bilinear CAM values for physical implementation of approximated FHPFs of orders 1.1, 1.5, and 1.9

Order $(1+\alpha)$	f_1 (kHz)	f_{2_z} (kHz)	f_{2_p} (kHz)	Q_{2_z}	Q_{2_p}	G
$(1+0.1)$	0.3454	0.8604	1.6889	0.2491	0.4164	2.8951
$(1+0.5)$	2.025	0.4472	0.6926	0.2236	0.6320	0.4938
$(1+0.9)$	1.400	0.1413	0.8393	0.1220	0.6667	0.7141

11.4.3.3 Higher Order Implementations

Higher order implementations of fractional filters using (11.21) are very easy to implement on an FPAA. Requiring cascading further bilinear and biquadratic filter CAMs, designed to realize the appropriate Butterworth response, with those previously designed in Section 11.4.3.1 to realize approximated $(1+\alpha)$ order filters. Therefore, to realize approximated FLPFs of order $(4+\alpha)$ requires cascading a single bilinear and biquadratic filter CAM to realize a 3rd order Butterworth response, with the bilinear and biquadratic CAMs to implement the $(1+\alpha)$ FLPF. The experimental results from implementing higher order FLPFs of orders $(4+\alpha) = 4.1$, 4.5, and 4.9, using the previously calculated design parameters for $(1+\alpha)$ filters in Table 11.7, compared to the theoretical simulations of (11.21) are given in Fig. 11.15. However, the highest order fractional filter that can be realized by a single FPAA requires $(n+m) \leq N$ where n is the integer order of the filter, m is the order of the s^α approximation, and N is the number of CAMs; where $N = 8$ for the Anadigm AN231E04. Filters with orders $(n+m) > N$ can be realized by cascading multiple FPAAs, increasing the number of CAMs to Nx where x is the number of FPAAs; providing $8x$ CAMs when cascading multiple AN231E04s.

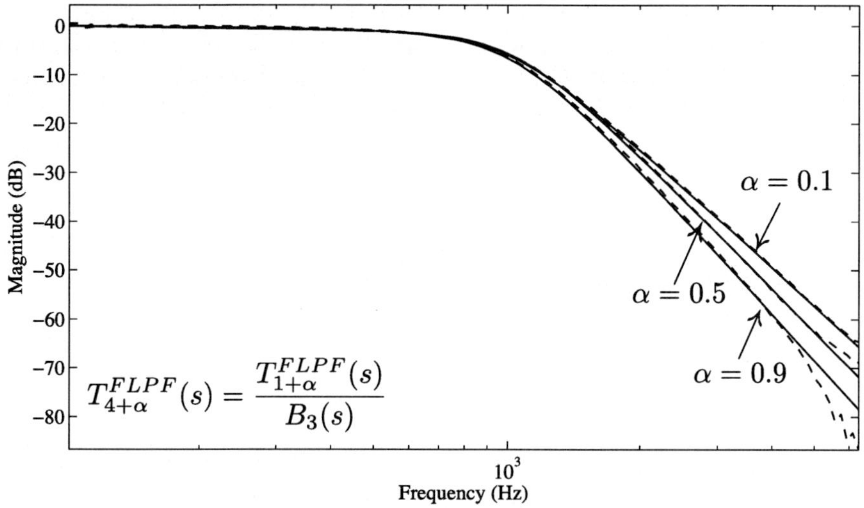

Fig. 11.15 Experimental results FPAA implemented FLPFs of order $(4+\alpha) = 4.1$, 4.5, and 4.9 compared to the theoretical simulations of (11.21)

11.4.4 Application of a Fractional Step Filter

To highlight the precise filtering achieved by a fractional step filter two tones, at 3 kHz and 10 kHz with peak-to-peak voltages of 500 mV, are applied to approximated FHPF of orders $(4+\alpha) = 4.1$ to 4.9 in steps of 0.2 shifted to a frequency of $f_0 = 10$ kHz. The peak value of the two tones for each filter are presented in Table 11.9. Note

Table 11.9 Signal power of tones at 3 and 10 kHz after application to approximated FHPFs of orders $(4+\alpha) = 4.1$ to 4.9 in steps of 0.2

Order $(4+\alpha)$	Power @ 3kHz (dBm)	Power @ 10kHz (dBm)
$(4+0.1)$	−32.7	3.94
$(4+0.3)$	−35.9	3.37
$(4+0.5)$	−38.2	3.5
$(4+0.7)$	−38.4	5.01
$(4+0.9)$	−40.7	4.45

that the use of the approximated fractional Laplacian operator results in the deviation of the linear spacing between the powers of the tone at 3 kHz as α increases. This control of the attenuation is not possible using integer order filters. Highpass 4th and 5th order Butterworth filters, frequency shifted to $f_0 = 10$ kHz, result in signals of −32.5 and −43.4 dBm for a 3 kHz tone. This precise control is also shown in Fig. 11.16. The spectrum of the 4th and 5th order Butterworth filters shown as dotted and solid lines, respectively, are compared to that of the 4.5 order FHPF, shown as a dashed line. All of the filters maintain the tone at 10 kHz with the attenuation of the 4.5 order filter at 3 kHz clearly between those of the standard Butterworth filters. These results reinforce the precise control of the attenuation characteristics that fractional filters offer.

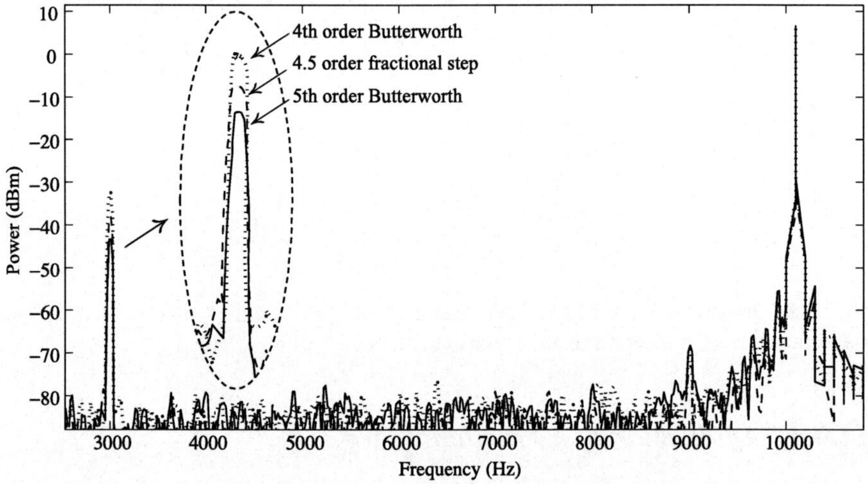

Fig. 11.16 Frequency spectrum of the approximated 4.5 order FHPF (dashed) compared to 4th (dotted) and 5th (solid) order highpass Butterworth filters

11.4.5 High-Q FBPF Realizations

The Type I and Type II FBPFs can be physically realized using the passive prototypes shown in Figs. 11.17(a) and (b), respectively, where C_α is a fractional capacitor of order $0 < \alpha \leq 1$. In the case of the Type I FPBF, the element D is a FDNR. The Type I passive prototype can be easily realized by replacing the FDNR with

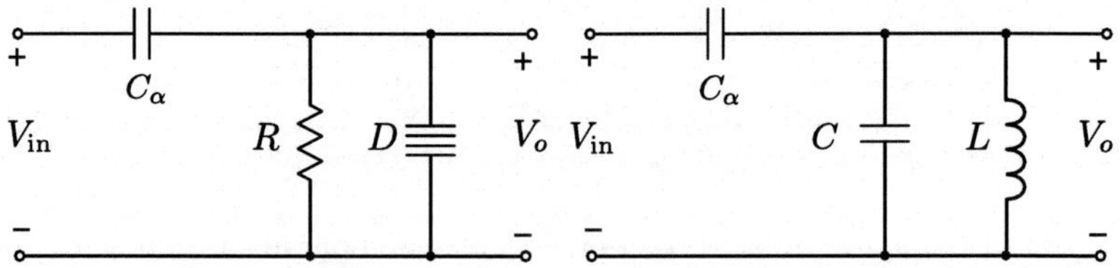

Fig. 11.17 Passive prototype models of (a) Type I and (b) Type II asymmetric-slope fractional bandpass filters

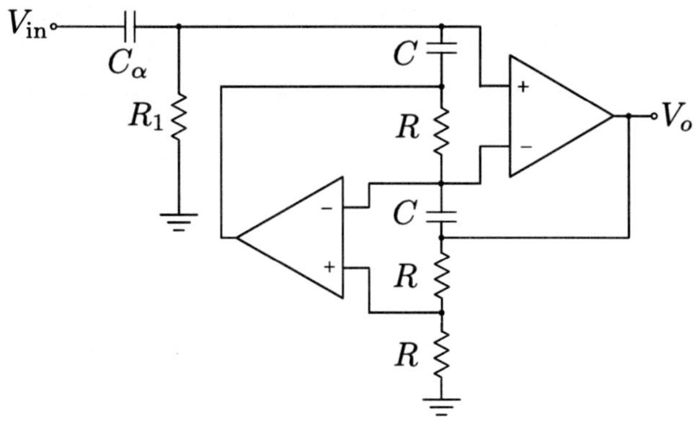

Fig. 11.18 Possible realization of Type I FBPF using a FDNR

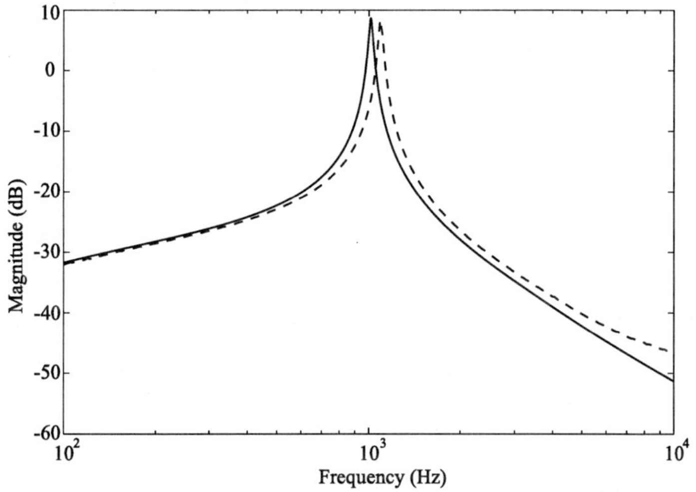

Fig. 11.19 PSPICE (solid) and experimental (dashed) results of approximated Type I FBPF realized using the FDNR topology in Fig. 11.18 with an approximated fractional capacitor

its active realization. If realized using operational amplifiers one implementation of the FBPF is given in Fig. 11.18. This circuit realizes the transfer function of (11.12) with $k_1 = 2$, $k_2 = C_\alpha/RC^2$, and $k_3 = 1/R_1RC^2$. The FBPF of Fig. 11.18 was implemented when $C_\alpha = 1$ μF with order $\alpha = 0.5$, approximated using Carlson's method [3] and realized with the RC ladder approximation of Fig. 11.7 centered around 1 kHz. The component values required to realize this approximated fractional capacitor are $(R_a, R_b, R_c, R_d, R_e) = (1.4, 3.2, 4.77, 11.21, 92.97)$ kΩ and $(C_b, C_c, C_d, C_e) = (6.64, 23.45, 42.57, 55)$ nF. To realize a Type I FBPF with $Q = 33$ and $f_0 = 1$ kHz using this fractional capacitor requires $R_1 = 531$ Ω, $R = 4.7$ kΩ, and $C = 0.1$ μF. The PSPICE simulations and experimental results of this FBPF are given in Fig. 11.19 as solid and dashed lines, respectively. The experimental results deviate from the designed response with $Q = 31.65$ and $f_0 = 1.087$ kHz due to the use of an approximated fractional capacitor and tolerances of the components to realize the circuit. The measured slopes both lower and higher than the center frequency are 10 and -30 dB/dec, respectively, confirming the asymmetric stop band characteristics possible using these filters.

11.5 Conclusion

In this chapter we have consolidated the recent progress in the design of analog filters with fractional step stop band attenuations. Presenting the design of $1 < (1+\alpha) < 2$ order fractional filters with lowpass and highpass responses; $0 < (\alpha_1 + \alpha_2) < 2$ order fractional bandpass responses with asymmetric stop bands and high quality factors; and the stable higher order fractional step filters. Finally, presenting three methods and design equations for the physical realization of these filters using fractional capacitors, SABs, FPAA hardware, and FDNR topologies.

References

1. Ahmadi, P.: Asymmetric-slope band-pass filters. M.Sc. thesis, Dept. Electr. and Comput. Eng., University of Calgary, Canada (2011)
2. Ahmadi, P., Maundy, B., Elwakil, A.S., Belostotski, L.: High-quality factor asymmetric-slope band-pass filters: a fractional-order capacitor approach. IET Circuits Devices Syst. 6, 187–197 (2012)
3. Carlson, G., Halijak, C.: Approximation of fractional capacitors of $(1/s)^{1/n}$ by regular Newton process. Trans. on Circuit Theory CT-11, 210–213 (1964)
4. Das, S.: Functional Fractional Calculus for System Identification and Controls. Springer, Heidelberg (2010)
5. Deliyannis, T.L., Sun, Y.Y., Fidler, J.: Continuous Time Active Filter Design. CRC Press LLC, New York (1999)
6. Elwakil, A.S.: Fractional-Order Circuits and Systems: An Emerging Interdisciplinary Research Area. IEEE Circuits Syst. Mag. 10, 40–50 (2010)
7. Freeborn, T.J.: Design and implementation of fractional step filters. M.Sc. thesis, Dept. Electr. and Comput. Eng., University of Calgary, Canada (2010)
8. Freeborn, T.J., Maundy, B., Elwakil, A.S.: Towards the realization of fractional step filters. In: IEEE Int. Symp. on Circuits and Systems (ISCAS), pp. 1037–1040 (2010)
9. Freeborn, T.J., Maundy, B., Elwakil, A.S.: Field programmable analogue array implementation of fractional step filters. IET Circuits Devices Syst. 4, 514–524 (2010)
10. Freeborn, T.J., Maundy, B., Elwakil, A.S.: Fractional-step Tow-Thomas biquad filters. Nonlinear Theory and its Applications, IEICE (NOLTA) 3, 357–374 (2012)
11. Krishna, B., Reddy, K.: Active and passible realization of fractance device of order 1/2. Act. Passive Electron. Compon. (2008), doi:10.1155/2008/369421
12. Maundy, B., Elwakil, A.S., Freeborn, T.J.: On the practical realization of higher-order filters with fractional stepping. Signal Process. 91, 484–491 (2011)
13. Maundy, B., Elwakil, A.S., Gift, S.: On a multivibrator that employs a fractional capacitor. Analog Integr. Circuits Signal Process. 62, 99–103 (2010)

14. Podlubny, I., Petras, I., O'Leary, P., Dorcak, L.: Analogue realizations of fractional-order controllers. Nonlinear Dyn. 29, 281–296 (2002)
15. Radwan, A., Elwakil, A., Soliman, A.: First-order filters generalized to the fractional domain. J. Circuit Syst. Comp. 17, 55–66 (2008)
16. Radwan, A., Elwakil, A., Soliman, A.: On the generalization of second-order filters to the fractional-order domain. J. Circuit Syst. Comp. 18, 361–286 (2009)
17. Radwan, A., Soliman, A., Elwakil, A., Sedeek, A.: On the stability of linear systems with fractional-order elements. Chaos, Solitons and Fractals 40, 2317–2328 (2009)
18. Westerlund, S., Ekstam, L.: Capacitor theory. IEEE Trans. Dielectr. Electr. Insul. 1, 826–839 (1994)

Chapter 12
The Flipped Voltage Follower: Theory and Applications

Jaime Ramirez-Angulo, Maria Rodanas Valero-Bernal, Antonio Lopez-Martin, Ramon G. Carvajal, Antonio Torralba, Santiago Celma-Pueyo, and Nicolás Medrano-Marqués

Abstract. The "flipped voltage follower (FVF)" is an enhanced buffer cell widely employed for low-power and/or low voltage operation. The applications of the FVF and its variations in analog and mixed signal (AMS) circuit design have increased continuously over the last few years. These are specially promising in deep submicrometer CMOS technology. In 2004 the paper "The Flipped Voltage Follower: A useful cell for low-voltage low-power circuit design" was published [1], where the most important information about this versatile cell was compiled. Since then, several modifications of the FVF and a wide variety of new applications have been reported. No other article with a similar purpose has been published since then. The aim of this chapter is to make a compendium summarizing some of the most relevant information published to date on the FVF. In order to facilitate AMS designers the utilization of this cell in this chapter we revisit, classify and summarize most applications of the basic FVF and its variations. Although it is tutorial in nature, it can be helpful for those who want to introduce themselves in the study of this cell or for experienced designers who want to become familiar with the principle of operation, improved versions and applications, so that both can exploit to its maximum the potential of the FVF in a wide number of applications or also to develop new ones.

Jaime Ramirez-Angulo
Klipsch School of Electrical Eng., New Mexico State University, Las Cruces NM, USA

Maria Rodanas Valero-Bernal · Santiago Celma-Pueyo · Nicolás Medrano-Marqués
Electronic Design Group, University of Zaragoza, Zaragoza, Spain

Antonio Lopez-Martin
Dept. Electrical and Electronic Eng., Public University of Navarra, Pamplona, Spain

Antonio Torralba · Ramon G. Carvajal
Dept. of Electronic Eng., University of Seville, Sevilla, Spain

M. Fakhfakh et al. (Eds.): *Analog/RF & Mixed-Signal Circuit Sys. Design*, LNEE 233, pp. 269–287.
DOI: 10.1007/978-3-642-36329-0_12

12.1 Introduction

The interest in low voltage analog cells has increased in recent years since the downscaling of CMOS processes has forced analog circuits to operate with continuously decreasing supply voltages. The demand for low voltage systems has been driven mainly by the need to reduce power consumption of the digital circuitry in mixed-signal very large-scale integrated (VLSI) systems, to prevent oxide breakdown with decreasing gate-oxide thickness and also to satisfy the requirements of the portable electronic equipment market. The reduction of supply voltages has forced to reconsider or modify many of the existing CMOS analog building blocks. In fact, some of them are not functional at all in the reduced supply environment of current CMOS technology.

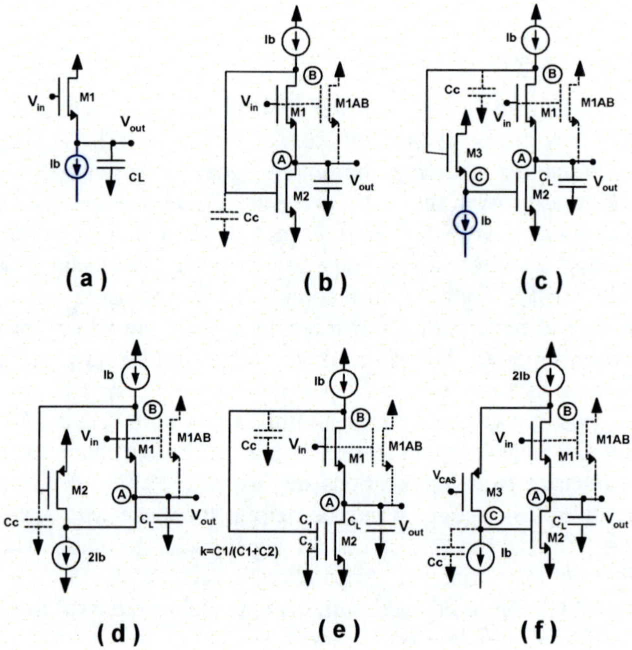

Fig. 12.1 Voltage followers: (a) Conventional Voltage follower (VF), (b) Flipped voltage follower (FVF), (c) Flipped voltage follower with VF level shifter (LSFVF), (d) Folded flipped voltage follower (FFVF), (e) Flipped voltage follower using floating gate level shifter (FGFVF), (f) Cascoded flipped voltage follower (CASFVF)

The voltage follower (VF, Fig. 12.1(a)), also known as source follower, is one of the basic building blocks of analog VLSI systems. The VF is biased on the source side with a constant current source I_b which ideally keeps the gate-source voltage V_{GS} of M1 constant. If body effect is neglected, the output follows the input voltage with a DC level shift, i.e., $V_0 = V_{in} - V_{GS}$. This feature, together with its high input impedance, relatively low output impedance $R_{out}=1/g_{m1}$ (in the range of few kΩ) high bandwidth $BW=g_{m1}/(2\pi C_L)$ and large input/output swing $V_{inpp}=V_{DD}-(V_T+2V_{DSsat})$ are the main reasons why the VF is used as a voltage buffer. Following notation is used throughout this chapter: V_{DSsat} is the drain-source saturation voltage, V_{GS} the gate-source voltage, V_T and g_m the transistor's threshold voltage and small signal transconductance gain.

Nevertheless, in practice, the VF suffers from various problems: a) In many applications the output resistance is not low enough. It can be decreased only by increasing the transconductance gain g_m. This requires a large bias current and/or large W/L dimensions. b) The output current is given by $I_{out}(\omega)=V_{out}\omega C_L$ and the drain current is $I_{D1}(\omega)= I_b-I_{out}(\omega)$. Given that I_{out} is a function of the input signal, the gate-source voltage of M1 varies with the input signal, which leads to distortion that increases at high frequencies. c) The slew rate is non-symmetrical since the sourcing capability is very large, while the sinking capability is limited by the bias current I_b. The Flipped Voltage Follower (FVF) is an enhanced voltage follower cell that overcomes some of the limitations of the VF [1]. In this chapter we summarize its characteristics, implementations and applications.

This chapter is structured as follows: In section 2 the basic single input, single branch FVF is presented and discussed together with six improved modifications, which overcome some of its limitations. In section 3, a detailed study and classification of low-voltage/low-power CMOS circuits based on the FVF is shown. In section 4, new FVFs with differential input are presented and discussed. In each section we revisit the applications of the different FVF cells in linear and nonlinear circuits that were reported originally and summarize new applications that have been reported since then. Although for the sake of space some remarks are just made on the applications of the different cells, the authors have made their best to compile a comprehensive (but not exhaustive) list of references. The chapter is finished with the conclusions drawn in section 5 where we summarize the most important facts of the FVF and hint at possible future applications.

12.2 Single Input FVF Structures

12.2.1 Flipped Voltage Follower

The basic FVF (Fig. 12.1(b)) corresponds to a voltage follower with improved characteristics [1]. In fact, it can be described as a cascode amplifier with negative feedback where the gate terminal of M1 is used as the input and its source as output voltage. It is characterized by very low output impedance $R_{out} = 1/(g_{m2}g_{m1}r_{o1})$ (tens of Ωs), high current sinking capability, very low supply requirements ($V_{DDmin}=V_{GS2}+V_{DSsat}$) close to a transistor's threshold voltage, low static power dissipation and high gain-bandwidth $GB=g_{m2}/(2\pi C_c)$. Output current variations are absorbed by the current sensing transistor M2, keeping the current in M1 essentially constant. Neglecting body effect and channel modulation, the gate-source voltage of M1 remains also constant. As a result, distortion remains low even at high frequencies. In the basic FVF of Fig. 12.1b, the output swing of M1 is 'strangled' by the gate-source voltage V_{GS2} of M2. This leads to a small input/output peak-peak swing $V_{inpp} = V_T - V_{DSsat}$. Unlike the case for most circuits, swing does not increase with the supply voltage V_{DD}. This is a severe limitation in modern deep sub-micrometer CMOS technology with V_T~0.4V peak-to-peak input/output swing is very small, typically $V_{inpp} < 300mV$.

12.2.2 FVF with Level Shifter

A possible solution to overcome the problem of the small input/output swing of the FVF is to include a DC level shifter between the drain of M1 and the gate of M2 [2]. Fig. 12.1(c) shows a modified version of the FVF that includes a DC level shifter (transistor M3) using a conventional voltage follower (LSFVF). In this circuit the input/output swing is increased by the DC level shift of the follower (V_{GS}) to a value $V_{inpp}=2V_T$, which is still independent of V_{DD} and might still be small in modern CMOS technologies. Bandwidth is moderately degraded due to the introduction of an additional high frequency pole at node C in the negative feedback loop formed by M1-M2 and M3. In addition, the introduction of M3 increases the quiescent power consumption and the minimum supply requirements of the circuit to the relatively high value $V_{DDmin}=2V_{GS}+V_{DSsat}$.

12.2.3 Folded Flipped Voltage Follower

In the folded flipped voltage follower (FFVF) or 'super source follower' [3] (Fig. 12.1(d)), the current sensing transistor M2 is folded (the NMOS is replaced by a PMOS transistor) by introducing an additional biasing source $2I_b$, making the FFVF to require additional power dissipation. The gain-bandwidth product is given by $GB=g_{m2}/(2\pi C_B)$, where C_B is the capacitance at node B. Given that all FVF structures are negative feedback structures, stability requires a dominant pole which should limit the gain bandwidth product (GB) to a value typically one half to one third of the effective high frequency pole of the open loop.

The effective high frequency open loop pole for the FVF, the LSFVF and the FFVF is given by $f_{phf}=f_{pout}=g_{m1}/(2\pi C_L)$. In general the additional high frequency pole introduced by M3 at node C is at much higher frequency than the pole at the output node. This assumption is valid if the load capacitance C_L is much greater than the parasitic capacitance at node C. In order to ensure stability in all FVF circuits (including the ones to be discussed below) a small compensation capacitor C_c can be used at node B to introduce a dominant pole that limits the gain bandwidth product to a value $GB=g_{m2}/(2\pi C_c)$.

12.2.4 Flipped Voltage Follower with Floating Gate Level Shifter

In the Flipped voltage follower with floating gate level shifter (Fig. 12.1(e)), denoted as FGFVF [2], the DC level shifting is achieved by using for M2 a two input floating gate transistor with capacitors C_1 and C_2 connected to the drain of M1 and to V_{SS} (ground) respectively. Capacitors C_1 and C_2 form a capacitive voltage divider, and based on charge conservation [4] the voltage at the gate of M2 is given by $V_{GS2} = kV_{D1}$ where $k = C_1/(C_1+C_2)$. Therefore, the voltage at the drain of M1 will be $V_{D1} = V_{GS2}\,(1+C_2/C_1)$. In this circuit V_{D1} is level shifted by $(C_2/C_1)V_{GS1}$ with respect to V_{GS1}. The minimum supply requirements of this circuit are given by $V_{DDmin} = V_{GS2}\,(1+C_2/C_1) + V_{DSsat}$.

12.2.5 Cascoded FVF

In the cascoded FVF [2] or CASFVF (Fig. 12.1(f)) a PMOS cascoding transistor M3 is introduced between the gate of M2 and the drain of M1. This transistor MC provides additional gain to the FVF negative feedback loop, leading to an extremely low output resistance $R_{out}=1/[g_m(g_m r_o)^2]$ (tenths of Ω). The quiescent voltage at the drain of M1 is set by the cascode biasing voltage of MC, with a value $V_{D1}=V_{CAS}+V_{SG,MC}$ that can be close to V_{DD}, in order to maximize the signal swing. The minimum supply requirements are $V_{DDmin}=V_{GS}+2V_{DSsat}$. The peak-to-peak output swing is dependent on V_{DD} and given by $V_{outpp}=V_{DD}-3V_{DSsat}$. An advantage of the CASFVF with respect to other FVF configurations is that due to the additional loop gain provided by MC, the voltage variations at the drain of M1 are reduced by the factor $g_m r_o$ (the gain of MC) with respect to those at the gate of M2. In this case lambda effect on M1 and on the transistor acting as biasing current source $2I_b$ are minimized.

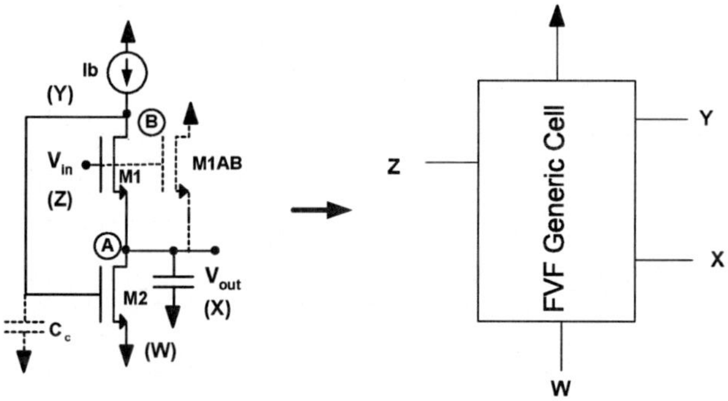

Fig. 12.2 Generic FVF cell. Note the identification of nodes: X is the low impedance node corresponding to the source terminal of the input transistor M1 (Node A), Z is a high impedance node corresponding to the gate terminal of the input transistor M1 (V_{in}), Y is a moderate low impedance node corresponding to the gate terminal of the current.

12.2.6 Other Improvements of Single Input FVF Structures

As can be easily identified from the FVF cell and its variations presented above all these cells are characterized by the presence of three terminals X, Y and Z (Fig. 12.2), where Z is a very low impedance node, Y is a moderate low impedance node (with value $1/g_{m2}$) and Z is a high impedance node. In this way, in node X current may be taken as the input or output variable, while voltage is the input or output variable for nodes Y and Z. An additional low impedance node W can be identified, which is the source terminal of the current sensing transistor M2, and that is usually connected to a rail. For simplicity in notation and generality, from here onwards, we will refer to the FVF as the FVF generic cell with four terminals X, Y, Z and W presented in Fig. 12.2. The following techniques can be applied both separately or combined.

12.2.6.1 Class AB Operation of Single Input FVF Structures

Similar to the conventional voltage follower, all FVF structures are characterized by nonsymmetrical slew rate. They all have large current sinking capability but their current sourcing capability is limited to the bias current I_b. A simple modification of the basic FVF [5]-[6] consists of including an additional input transistor M1AB with its drain terminal connected to V_{DD}, that provides class AB operation with both large current sourcing and sinking capabilities. This simple solution can be used in all FVF structures discussed above as shown in Figs. 12.1 and 12.2 in discontinuous trace. Although M1AB increases quiescent current dissipation by I_b (assuming same W/L for M1 and M1AB), it allows class AB operation with large positive and negative output currents $I_{out}{>>}I_b$. Table 12.1 shows a comparison summarizing the characteristics of the single input FVFs discussed and those of the conventional voltage follower.

Table 12.1 Comparison of buffer performance characteristics

Single Input Buffer comparison					
Circuit	**R_{out}**	**V_{DD}^{min}**	**V_{outPP}**	**GB**	**I_{DD}**
VF	$1/g_m$	$V_{GS}+V_{Dssat}$	$V_{DD}-V_{DD}^{min}$	g_{m1}/C_L	Ib
FVF	$1/(g_m(g_m r_o))$	$V_{GS}+V_{Dssat}$	V_T-V_{Dssat}	g_{m2}/C_c	Ib
LSFVF	$1/(g_m(g_m r_o))$	$2V_{GS}+V_{Dssat}$	$2V_T$	g_{m2}/C_c	2Ib
FGFVF	$1/(kg_m(g_m r_o))$	$V_{GS}/k+V_{DSsat}$	$V_{GS}/k-2V_{DSsat}$	kg_{m2}/C_c	Ib
FFVF	$1/(g_m(g_m r_o))$	$V_{GS}+2V_{Dssat}$	$V_{DD}-V_{DD}^{min}$	g_{m2}/C_c	2Ib
CASFVF	$1/g_m(g_m r_o)^2$	$V_{GS}+2V_{Dssat}$	$V_{DD}-V_{DD}^{min}$	g_{m2}/C_c	2Ib

12.2.6.2 Non Trapped Current Single Input FVFs

In all FVF cells discussed above, the current that flows in the current sensing transistor M2, I_{FVF}, is 'trapped', that is, it is not available as an output current. It can be replicated using mirroring techniques by tying the gate of an additional mirroring transistor to V_B. However, current mirroring inevitably introduces distortion and degrades the circuit's frequency response. This is especially critical in the implementation of highly linear OTAs [7].

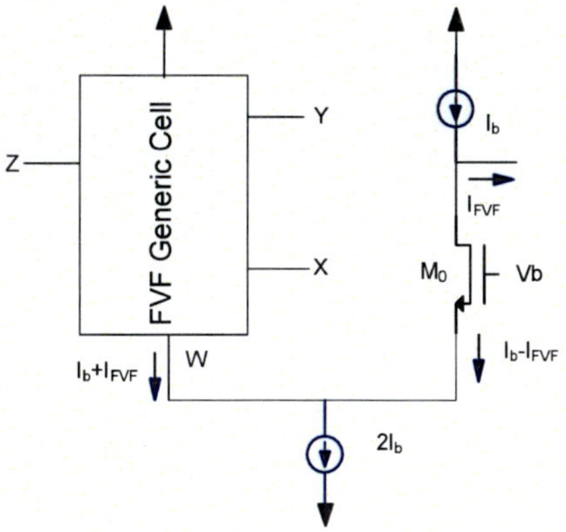

Fig. 12.3 Current 'de-trapping' technique

A solution to overcome this problem is to use the scheme depicted in Fig. 12.3 for NMOS input transistor. As it can be seen, now node W (the source of the current sensing transistor M2 in Fig. 12.2) is connected to a constant current source instead of being tied to a rail. In this way, the current I_{FVF} is transferred to transistor M_0 whose current equals $I_0 = I_b - I_{FVF}$. The additional component I_b is subtracted at the drain of M_0, leading to an output current $I_{out} = I_b - I_0 = I_{FVF}$, which is free of the distortion and the degradation in frequency behavior introduced by current mirroring. The output node is a high impedance, high swing node. In addition to its simplicity, this technique can be applied to any of the FVF cells discussed above.

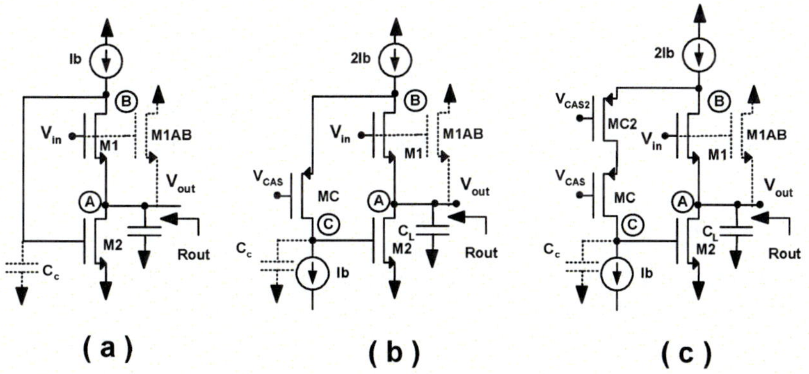

Fig. 12.4 Illustration of multiple cascoding technique: (a) Basic FVF, (b) Cascoded FVF, (c) double cascoded FVF using additional cascode transistor MC2

12.2.6.3 Multiple Cascoding Techniques

Note that the performance of FVF cells can be improved at no cost in power consumption by including additional cascode transistors in the negative feedback loop of the FVF. This leads to an increase in the open loop gain of the FVF cell with the consequent further reduction of the output resistance at node A by a factor $A_i = g_{mi}r_{oi}$ for each additional cascoding stage. This has very little effect on the phase margin since cascode transistors introduce very high frequency poles in the feedback loop. Fig. 12.4 illustrates how the conventional FVF (Fig. 12.4(a) with $R_{out}=1/[g_m(g_m r_o)]$) is transformed into the cascoded FVF (Fig. 12.4(b) with $R_{out}=1/[g_m(g_m r_o)^2]$) by adding a cascode transistor MC. Fig. 12.4(c) shows the addition of another cascode transistor, MC2, to achieve an extremely low output resistance with value $R_{out}=1/[g_m(g_m r_o)^3]$.

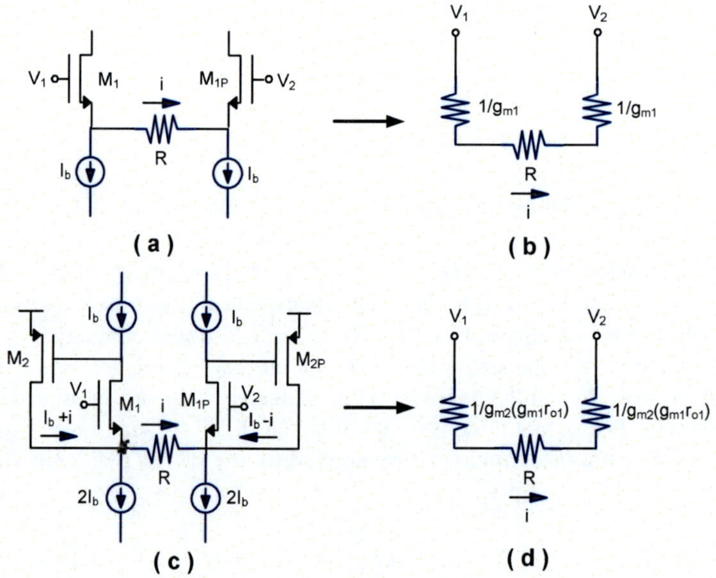

Fig. 12.5 (a) Source degeneration technique, (b) Equivalent circuit, (c) FVF CMOS linear transconductor employing source degeneration technique, (d) equivalent circuit of (c)

12.2.7 Applications of Single Input FVF Structures

As said before, the FVF can be seen as a high performance buffer. Therefore the basic applications of the FVF are those in which a high performance buffer is required. One example is in the implementation of linear CMOS transconductors. Linearization techniques have to be used to reduce the total harmonic distortion (related to the quadratic relation of the output current into the input voltage in a MOS transistor operating in saturation). A commonly used linearization technique is based on source degeneration [8], in which a resistor R is connected between the sources of a differential pair (Fig. 12.5(a)). Fig. 12.5(b) shows the equivalent circuit of Fig. 12.5(a), in which we have neglected the output impedance of the bias currents I_b and those of the input transistors. The input voltage $V_{i\ (i=1,2)}$ of each input transistor is translated into its source $V_{Si\ (i=1,2)}$ with a level shift. In this way, $V_{Si} = V_i - V_{shift}$, where V_{shift} depends inversely on their transconductance g_{mi}.

If $g_{m1} = g_{m2} = g_m$, the current which flows in the linear resistor R is given by $i = (V_1 - V_2)/((2/g_m) + R)$. If $R >> 2/g_m$, then $G \approx i/(V_1-V_2) = 1/R$, and the transconductance G is linear since it only depends on R and not in the non-linear resistor $r_{nl} = 1/g_m$, (whose value depends on the current i and input voltage V_i). In this way, linearity increases, but the price paid is that the transconductance G decreases inversely. Note that the gain bandwidth product (GB) and signal to noise ratio (SNR) of a circuit with OTAs is proportional to the transconductance gain. A possible solution to improve GB and SNR would be to reduce the value of r_{nl}, so R can be lower (and G higher). This requires a larger bias current and/or large transistor dimensions in the differential pair. An alternative is shown in Fig. 12.5(c) [9] the input pairs have been replaced by FVFs (actually FFVFs). In this case $r_{nl} = 1/(g_{m2}g_{m1}r_{01})$, which is much smaller than in the previous case, and $G = 1/[2/(g_m^2 r_0) + R]$ where we have assumed $g_m=g_{m1}=g_{m2}$, $r_{o2}=r_o$. In addition, the frequency response of the system is unaltered, since the bandwidth of the FVF is the same as a conventional voltage follower. Note that in this case it is possible to use a much smaller R value corresponding to a much higher linear transconductance gain $G\approx 1/R$ (with correspondingly higher GB and SNR).

In fine line CMOS technologies, the transistor's intrinsic gain $A_i=g_m r_o$ has decreased drastically which has resulted in decreased open loop gain of conventional Op-Amp architectures. For example, in one stage Op-Amps the gain bandwidth product (GB) and A_{OL} are given by $GBW \approx g_{mDP}/C_L$ and $A_{OL} \approx g_{mDP} \cdot R_{OUT}$ respectively, where C_L is the capacitive load connected to the output of the Op-Amp, R_{OUT} is the output resistance and g_{mDP} is the transconductance gain of the input differential pair. In a conventional Op-Amp $g_{Mdp} \approx g_m$, corresponds to the transconductance gain of the input stage transistors. One possible solution to tackle the problem of low intrinsic gain in modern CMOS technology is to increase g_{mDP} by replacing the differential input stage by the circuit of Fig. 12.5(c) with R=0 [9]. In this case the effective transconductance gain g_m'is given by $g_m' = g_{m1}g_{m2}r_{o2} \approx g_m^2 \cdot r_0$. Therefore, both GB and AOL are increased by the factor $A_i = g_m r_o$ achieving $GB \approx g_m^2 \cdot r_0/C_L$ and $A_{OL} \approx g_m^2 \cdot r_0 \cdot R_{OUT}$. Note that the g_m enhancement factor depends on the FVF used. For example, if the cascoded FVF is used instead of the FFVF of Fig. 12.5(c) the gm (and gain) enhancement factor is A_i^2.

12.3 Basic Single Input FVF Based Structures

For simplicity, in this section we will only refer to one implementation of the FVF, usually the basic one of Fig. 12.1(a). Note that, in every case, we can replace the FVF by the generic cell with four terminals X, Y, Z and W presented in Fig. 12.2. In this way, by FVF we can refer not only to the original FVF presented in section 2.1, but also to all improved FVF variations studied along section 2.

12.3.1 FVF Current Sensor (FVFCS)

The FVF can be used as a current sensing cell [1], where node X in Fig. 12.6(a) is the input current sensing node. This very low impedance node X can source large current variations. They are translated by the FVF into compressed voltage variations at node Y. Fig. 12.6(b) shows the DC response of the circuit in Fig. 12.6(a). The output and the input currents are related through the expression $I_{out} = I_{in} + I_b$.

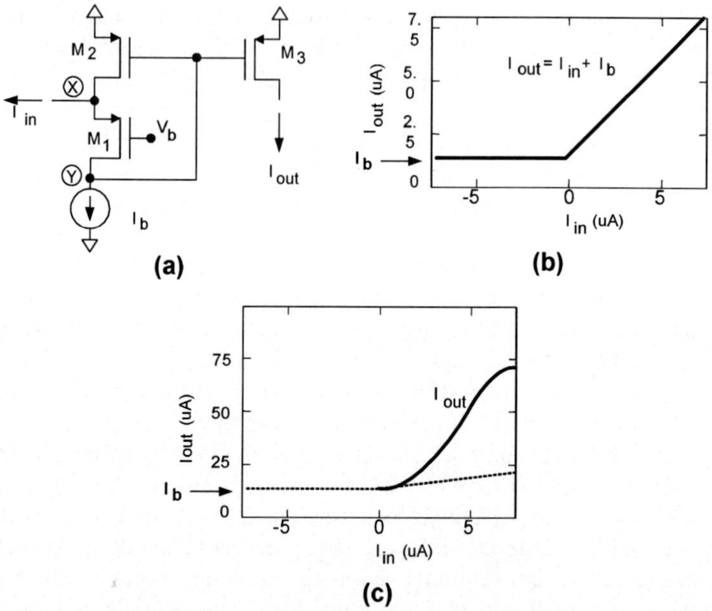

Fig. 12.6 FVF Current Sensor (FVFCS): (a) Basic implementation, (b) DC response with M_2 in saturation, (c) DC response with M_2 biased near the linear region

If transistor M2 is biased with voltage V_b near the linear region and M3 is maintained in the saturation region, the output current can increase by a very large factor compared to the input current (Fig. 12.6(c)). This is due to the fact that with input current increases M2 enters triode mode and a large source-gate voltage, V_{SG2}, is developed in M2. This can be used to achieve class AB operation as was demonstrated in [10].

12.3.1.1 FVFCS Applications

The FVFCS has been used in the past for different applications [11-13]. For example, in [12] the FVFCS was used as a part of a power amplifier. A well-known application is at the input stage of a high performance low-voltage current mirror [11,12,14], but many other applications exist, such as in digital IDDQ and Mixed-signal iDD testing [15], which is employed to detect the presence of faults in CMOS integrated circuits.

A recent useful application of the FVF has been found in Low Dropout Voltage Regulators (LDO) [16,17]. An LDO is a circuit which provides a nearly constant DC output voltage despite changes in load current or its input (supply) voltage with a small drop between its supply and output regulated voltage.

Generic LDO structures have two high impedance nodes and a tradeoff exists between accuracy and feedback stability: A high loop gain improves the line regulation and load regulation factors, but degrades closed-loop stability. Besides low drop in the pass transistor, stability is an important issue in LDOs because they are required to operate with a wide range of load currents. This causes very large variations in the output pole of the feedback loop. Therefore, different methodologies such as advanced pole–zero cancellation schemes, load-dependent reference voltage concept, pole-splitting schemes, and extremely large capacitive loads were proposed [16]. However, this increases the complexity and the power consumption and cost of the devices.

The properties of the FVF as an enhanced voltage buffer make it a simple and excellent solution to overcome these drawbacks. The FVF's low supply requirements and reduced output impedance due to shunt feedback connection [16], is the key for obtaining high regulation and achieving frequency compensation with wide load variations. The FVF has a feedback loop with only one high impedance node. In fact, the enhancement of the output resistance by means of cascoding techniques improves its characteristics in terms of line and load regulation, with negligible impact in stability. In addition, the FVF is very simple, can be biased with low currents and has very low drop, leading to high power efficiency.

12.3.2 FVF Differential Structure (FVFDS)

The FVF differential structure (FVFDS) is built by adding an extra transistor M3 connected to node X of the FVF cell, as it is shown in Fig. 12.7(a) [11]. As the impedance at node X is very low, its voltage remains approximately constant for large input currents.

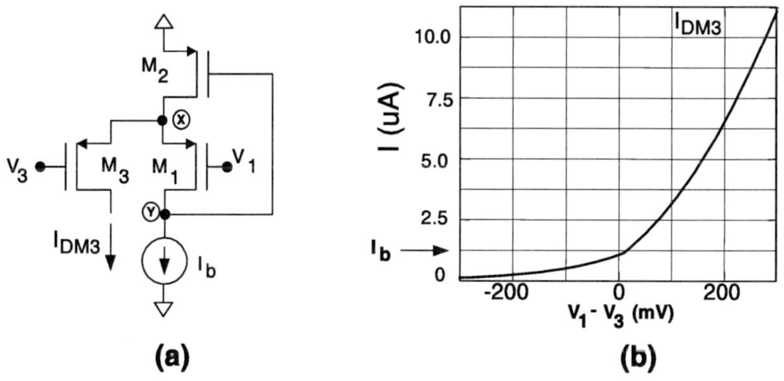

Fig. 12.7 (a) FVF Differential structure (FVFDS), (b) DC transfer characteristic

Under quiescent conditions (V_1=V_3) assuming that transistors M1 and M3 have the same size then $I_{DM1} = I_{DM3} = I_b$. A nonzero differential voltage V_1–V_3 adds to V_{SG3} and generates current variations in M3 that follow approximately the MOS square law. These current variations are supplied by M2. In this way, the FVFDS maximum output current I_{DM3} can be much larger than the quiescent current. Fig. 12.7(b) shows the DC transfer characteristic for I_{DM3} vs. V_1–V_3. The typical class AB behavior can be observed. On the other hand the FVFDS shows similar high common mode rejection (CMRR) as a conventional differential pair (DP). Since common mode input voltage variations do not lead to changes in IDM3.

12.3.2.1 FVFDS Applications

FVFDSs are mainly applied to build low-power, low-voltage, class-AB stages with a wide variety of applications: transconductance operational amplifiers [18], output stages [19] and buffers [1].

12.3.3 FVF Pseudo-Differential Pair (FVFPDP)

The FVF pseudo-differential pair (FVFPDP) is constructed from the FVF by adding an extra input transistor (M4) connected to node X, as it is shown in Fig. 12.8(a) [20].

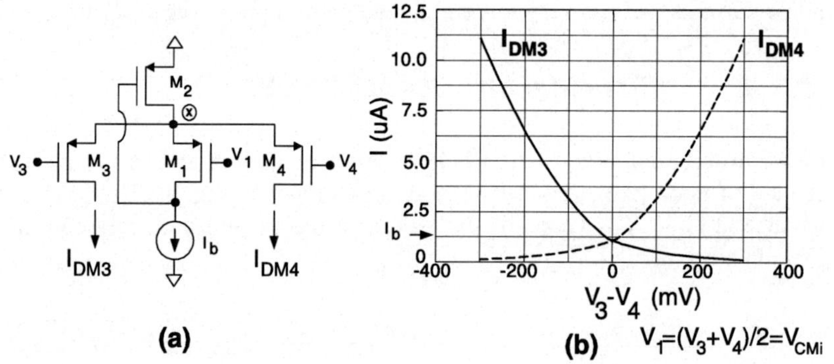

Fig. 12.8 (a) FVF Pseudo-Differential pair (FVFPDP), (b) DC transfer characteristic

Fig. 12.8(b) shows the DC output currents I_{DM3} and I_{DM4} versus the differential input voltage $V_{34} = V_3$–V_4, in a typical case. The pseudo-differential pair also exemplifies the characteristic behavior of a class-AB circuit, where the quiescent output current I_b can be much lower than the peak output current. In this case, we have considered that, under quiescent conditions, $V_1 = V_3 = V_4$. That is, the voltage at the gate of M1 corresponds to the common mode of M3 and M4: V_1=(V_3+V_4)/2=V_{CMi}. If the common mode value V_{CMi} of input voltages V_3 and V_4

is not equal to V_1 the DC output characteristic will have the same shape, but a DC level shift will appear.

The main difference between the FVFDS and the FVFPDP is that the latter has a true differential output. The output current I_{DM3} of the FVFDS can be large if V_1-V_3 is positive and zero if $V_1 - V_3$ is negative, while in the FVFPDP we can have positive or negative large differential output currents ($I_{out} = I_{DM3} - I_{DM4}$) depending on the value of the input differential voltage ($V_{in} = V_3 - V_4$). This pseudo-differential pair can be also operated with a minimum supply voltage of $V_{DDmin} = |V_{TP}| + 2\,|V_{DSsat}|$, as in the case of the FVFCS and FVFDS. The FVF formed by M1,M2 and Ib can be replaced by any of the variations discussed in section 2 in order to increase the input signal swing. The FVFPDP requires a common mode input voltage detector or complementary input signals with a well defined constant common mode voltage applied to the gate of M1. This is available in fully differential circuits where the common mode voltage of a signal is set by the output common mode feedback network of the previous stage.

12.3.3.1 FVFDP Applications

FVFPDPs can be applied as input stage in Class-AB Op-Amps [20], Class AB operational transconductance amplifiers for switched capacitor (SC) applications, Class AB operational transconductance amplifiers for continuous time operation [21] and multipliers [22].

12.3.4 Other Structures

The FVF cell can be used in many other applications such as in Translinear Circuits [1]: FVF Voltage Translinear loops and Static Nonlinear Circuits (Geometric-Mean circuit, Squarer/Divider circuit ([1], [23]), Multiplier/Divider circuit and Dynamic linear and nonlinear circuits [1].

12.4 Differential Input FVF Structures

12 4.1 Differential Voltage Follower (DVF)

Fig. 12.9(a) shows a simple buffer without DC level shift between the input and the output terminals. It consists of an active loaded differential pair with unity gain negative feedback (shown in dashed lines connecting nodes 2 and 3). This class A circuit has an output resistance given by $R_{out} = 1/g_{m1}$, gain-bandwidth product GB $= g_{m1}/(2\pi C_L)$ and symmetrical slew rate SR $= 2I_b/C_L$, which is determined by the maximum output current $I_{outMAX} = 2I_b$. The input swing is $V_{inpp} = V_{DD} - V_{SS} - 2V_{DSat} - |V_{TN}|$. The minimum supply voltage is $V_{DDMIN} = V_{GS} + V_{DSsat} + |V_{TP}| - V_{TN}$, while the output swing is $V_{outpp} = V_{DD} - V_{SS} - 2V_{DSat} - |V_{TN}|$.

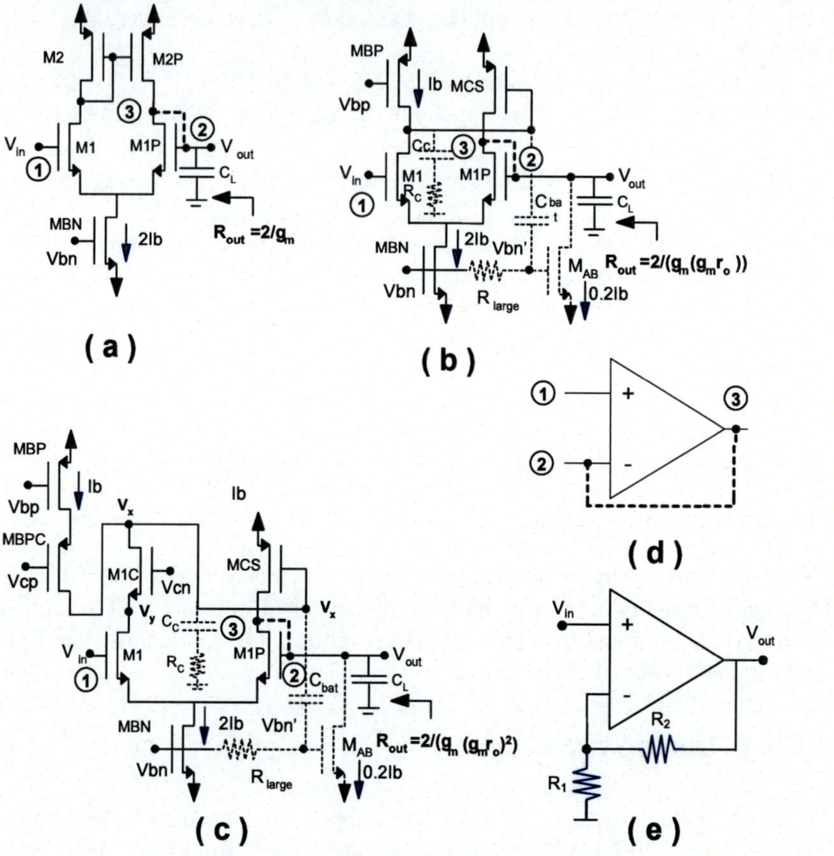

Fig. 12.9 Differential buffers: (a) Conventional DP buffer, (b) Differential FVF buffer (c), Cascoded Differential FVF buffer, (d) Generic differential amplifier representation, (e) DFVF used as op-amp with resistive feedback network

12.4.2 Differential Flipped Voltage Follower (DFVF)

The Differential Flipped Voltage Follower was reported as a part of a current conveyor in [24] and as general three terminal structure in [25]. The DFVF of Fig. 12.9(b) can be derived from the FVF of Fig. 12.1(b) by replacing the transistor M1 by a differential pair M1-M1P and a tail current source as explained in section 2.6.5 and illustrated in Fig. 12.5. This is done by diode connecting M1P and flipping the currents sensing transistor MCS from NMOS to PMOS type. Besides the fact that there is no DC level shift between the input and output, the input range V_{inpp} is higher than in the FVF, dependent on V_{DD}. It is given by $V_{inpp} = V_{DD} - V_{SS} - 3V_{DSat}$. The minimum supply voltage is $V_{DDMIN} = V_{GS} + 2V_{DSsat} - V_{TN}$, while the output swing is $V_{outpp} = V_{DD} - V_{SS} - 3V_{DSat} - V_{TN}$. Other features of this circuit are lack of body effect attenuation and low output resistance $R_{out}=1/[g_m(g_m r_o)]$ (~ 50Ω).

12.4.3 Cascoded Differential Flipped Voltage Follower (CDFVF) [24]

In Fig. 12.9(c) an additional cascoding transistor M1C is introduced in the negative feedback loop formed by M1-M1P, MCS and M1C. In this case the FVF corresponds to a folded double cascoded amplifier with unity gain negative feedback. M1C increases the open loop gain of the cascode amplifier negative feedback by the factor $g_{m1C}r_{o1C}$ and leads to an extremely low output resistance given by $R_{out}=1/[g_m(g_m r_o)^2]$ (~ tenths of Ω). The input swing is slightly reduced since the maximum input voltage now is given by $V_{inMAX} = V_{DD} - V_{SS} - 3V_{DSat} + V_{TN}$. The minimum supply voltage is $V_{DDMIN} = V_{GS} + 4V_{DSsat} - V_{TN}$, while the output swing is $V_{outpp} = V_{DD} - V_{SS} - 2V_{DSat} - V_{TN}$.

12.4.4 Other Improvements of Differential Input FVF Structures

12.4.4.1 Class AB Operation

The DFVFs of Fig 12.9(b) and Fig 12.9(c) are class A circuits characterized by nonsymmetrical slew rate. They all have large current sinking capability thanks to the current sensing transistor MCS but their current sourcing capability is limited by the bias current $2I_b$. Similar to single input FVFs, a simple modification of the basic FVF provides class AB operation to DVFVs with both large current sourcing and sinking capabilities. This modification is based on the quasi floating gate technique described in [26], and it is shown in discontinuous line in Figs. 12.9(b) and 12.9(c). It requires an additional transistor M1AB whose gate is connected to the DC biasing voltage V_{bn} through a very large valued resistive element R_{large} (~ 100 GΩ) and the connection of the gate of the current sensing transistor MCS through a capacitor C_{bat}. This transistor has a quiescent gate voltage $V_{bn}' = V_{bn}$ and small dimensions $(W/L)_{AB} = (W/nL)$ with n > 3 and for this reason it has a small quiescent drain current (~$2I_b/n$). Under dynamic conditions R_{large} prevents flow of charge into the gate of MAB. For this reason C_{bat} performs as a floating battery so that V_{bn}' follows the gate voltage variations of MCS. During negative slewing (when a negative current flows into C_L), the current of M1 and M1P remains constant and the drain current if MCS increases so it does its gate voltage. As a consequence, the gate voltage V_{bn}' of MAB increases so that its drain current can increase significantly over the quiescent value $2I_b/n$. During positive slewing the gate of MCS decreases. Due to this it can provide very large positive currents to C_L which can be much larger than $2I_b$. At the same time MAB decreases its current and turns off.

As shown in Fig. 12.9(d) the DFVF can be seen as a more general three terminal amplifier (or Op-Amp) with two differential input terminals (1 and 2) and an output terminal (3). This was reported in [25]. As such it has an open loop output resistance $R_{out}=1/g_m$ and the same open loop gain $A_{ol}= g_m r_o/2$ as the conventional circuit of Fig. 12.16(a) which has a much higher open loop output resistance $R_{out}=r_o/2$. The circuits of Fig. 12.9(b) (and 12.9(c) with even higher open loop gain and lower output resistance) can be considered transposed versions of the conventional circuit of Fig. 12.9(a). They have the same power dissipation and complexity but given it has lower open loop output resistance it can be used as a compact buffered Op-Amp and operate with resistive feedback as shown in Fig. 12.9(e). This is in general only possible using more complex two stage (or three stage) Miller op-amps.

Table 12.2 shows a comparison summarizing the characteristics of the differential input FVFs discussed and those of the conventional differential input voltage follower.

Table 12.2 Comparison of buffer performance characteristics

Differential Input Buffer comparison					
Circuit	R_{out}	V_{DD}^{min}	V_{outPP}	V_{inpp}	I_{DD}
DVF	$1/g_m$	$V_{GS}+V_{DSsat}+\lvert V_{TP}\rvert-V_{TN}$	$V_{DD}-V_{SS}-3V_{DSat}$	$V_{DD}-V_{SS}-2V_{DSat}-\lvert V_{TP}\rvert$	2Ib
DFVF	$1/(g_m\,(g_m r_o))$	$V_{GS}+2V_{DSsat}-V_{TN}$	$V_{DD}-V_{SS}-3V_{DSat}-V_{TN}$	$V_{DD}-V_{SS}-3V_{DSat}$	2Ib
CDFVF	$1/(g_m\,(g_m r_o)^2)$	$V_{GS}+4V_{DSsat}-V_{TN}$	$V_{DD}-V_{SS}-5V_{DSat}-V_{TN}$	$V_{DD}-V_{SS}-5V_{DSat}$	2Ib

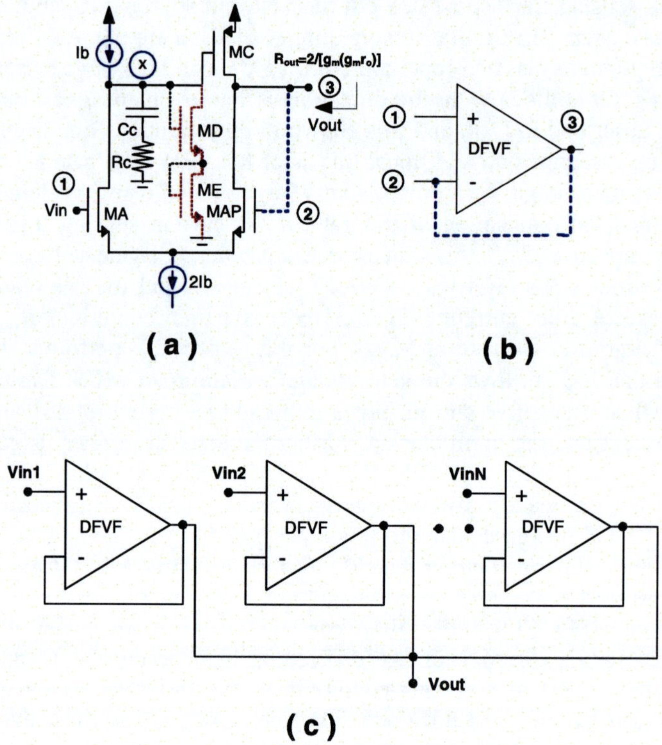

Fig. 12.10 (a) Differential flipped voltage follower cell, (b) Symbol, (c) WTA MAX Circuit based on DFVF cells with their output connected in parallel

12.4.5 Applications of the Differential Input FVF Structures

Due to its characteristics and compactness, the main application of differential input FVF structures are in high performance voltage buffers [25], but recently they have also been employed in non-linear circuits such as winner-take-all circuits (WTA) [27,28], programmable gain voltage amplifiers [29] and current conveyors [30].

WTA circuits are widely employed in non-linear systems. They are typically based on the parallel connection of identical buffer cells (driven by the input voltage or current signals) to a common low-impedance output node. Input signals "compete" to set the voltage at the output branch, so that, in a WTA, only the branch with the highest or lowest voltage or current input signal remains active and sets the output voltage (current), depending if the circuit performs the maximum (MAX) or minimum (MIN) function, respectively.

Fig. 12.10 shows an implementation of a high-performance WTA circuit based on the differential FVF (DFVF) [28]. The circuit features high precision, high speed, very low supply voltage requirements and high input/output signal swing which is dependent on V_{DD}. The use of DFVF versus the single input FVF presents several advantages: with the same low output resistance and supply requirements as the FVF, it does not introduce a DC level shift between the input and the output terminals; it has wide input signal swing and it is not subject to the body effect attenuation, featuring close to unity gain; it has wide input/ output swing which is dependent on V_{DD}. As opposed to most other WTA circuits reported in literature this circuit can operate with a single supply close to a transistor's threshold voltage and with high speed and accuracy, finding application in the sub-volt supply environment required by modern deep sub-micrometer CMOS technologies.

Differential input FVFs can be also employed in the implementation of high performance, second generation current conveyors (CCII) and current feedback operational amplifiers (CFOAs) [30]. They are used in many current mode analog circuits as basic building blocks due to the suitability of current mode signal processing for the implementation of low voltage, low power, and high bandwidth analog circuits.

12.5 Conclusion

In this work a cell denoted "Flipped Voltage Follower" has been revisited. In this revision, several improved FVF cells (with single and differential input) and structures derived from the basic cell have been discussed, comparing their performances and characteristics. It has also been shown to be a useful cell with many applications in low-power, low-voltage analog design. The large number of linear and nonlinear structures that can use and be derived from this cell demonstrate its usefulness for low-power, low-voltage analog circuit design. However, FVF applications do not finish here. New fields and new possibilities exist. An area where the applicability of FVF has not been studied, is in line drivers and in digital circuits. But there are many additional possibilities essentially any application where a very high performance buffer is required.

References

[1] Carvajal, R.G., Ramírez-Angulo, J., López-Martin, A., Torralba, A., Galán, J., Carlosena, A., Muñoz, F.: The Flipped voltage follower: A useful cell for low-voltage low-power circuit design. IEEE Transactions on Circuits and Systems I 52, 1276–1291 (2005)

[2] Gupta, S., Padilla, I., Ramirez-Angulo, J., Torralba, A., Jimenez, M.: Comparison of conventional and new flipped voltage follower structures. In: IEEE Midwest Symposium on Circuits and Systems, pp. 1151–1154 (2005)

[3] Gray, P.R., Hurst, P.J., Lewis, H., Meyer, R.G.: Design of analog integrated circuits, 4th edn. John Wiley and Sons, New York (2001)

[4] Ramirez-Angulo, J., Lopez, A.: MITE Circuits: The continuous time counterpart to switched capacitor circuits. IEEE Transactions on Circuits and Systems II 48, 45–55 (2001)

[5] Torralba, A., Carvajal, R.G., Jiménez, M., Muñoz, F., Ramírez-Angulo, J.: A new low-voltage CMOS unity-gain buffer. In: Proc. of the IEEE International Symposium on Circuits and Systems, pp. 919–922 (2006)

[6] Torralba, A., Carvajal, R.G., Jiménez, M., Muñoz, F., Ramírez-Angulo, J.: A new, true low voltage, Class-AB current mirror. IEICE Electronics Express ELEX 2, 103–107 (2005)

[7] Acosta, L., Jimenez, M., Lopez-Martin, A., Ramirez-Angulo, J.: Highly linear tunable CMOS Gm-C low-pass filter. IEEE Transactions on Circuits and Systems 56, 2145–2158 (2009)

[8] Acosta, L., Carvajal, R.G., Jimenez, M., Ramirez-Angulo, J., Lopez-Martin, A.: A CMOS transconductor with 90 dB SFDR and low sensitivity to mismatch. In: Proc. of the IEEE International Symposium on Circuits and Systems (2006)

[9] Sanchez-Rodriguez, T., Ramirez-Angulo, J., Lopez-Martin, A., Carvajal, R.G., Patil, C.: Gain enhancement and low voltage techniques for analogue circuits in deep submicrometer CMOS Technology. In: Krysztof, I. (ed.) CMOS Nanoelectronics: Analog and RF VLSI Circuits. McGraw Hill, New York (2009)

[10] You, F., Embabi, S.H.K., Sánchez-Sinencio, E.: Low-voltage class AB buffers with quiescent current control. IEEE Journal of Solid-State Circuits 33, 915–920 (1998)

[11] Peluso, V., Steyaert, M., Sansen, W.: Design of low-voltage low-power Sigma-delta A/D converters. Kluwer Academics Publishers, Boston (1999)

[12] Pardoen, M.D., Degrauwe, M.G.: A rail-to-rail CMOS input/output power amplifier. IEEE Journal of Solid-State Circuits 25, 501–504 (1990)

[13] Ramírez-Angulo, J., Carvajal, R.G., Torralba, A.: Low supply voltage high-performance CMOS current mirror with low input and output voltage requirements. In: Proc. of the IEEE Midwest Symposium on Circuits and Systems, pp. 510–513 (2000)

[14] Torralba, A., Carvajal, R.G., Ramirez-Angulo, J., Munoz, F.: Output stage for low supply voltage, high-performance CMOS current mirrors. IEE Electronics Letters 38, 1528–1529 (2002)

[15] Ducoudray, G.O., Carvajal, R.G., Ramírez-Angulo, J.: A High-speed dynamic current sensor scheme for iDD test using a flipped voltage follower. In: Proc. of the Southwest Symposium on Mixed-Signal Design, pp. 50–53 (2003)

[16] Man, T.Y., Leung, K.N., Leung, C.Y., Mok, P.K.T., Chan, M.: Development of single transistor control LDO based on flipped voltage follower for SoC. IEEE Transactions on Circuits and Systems I 55, 1392–1401 (2008)
[17] Chen, H., Leung, K.N.: A fast transient LDO based on buffered flipped voltage follower. In: Proc. of the IEEE Electron. Devices and Solid State Circuits (2010)
[18] Peluso, V., Coreland, P.V., Steyaert, M., Sansen, W.: A 900 mV differential class AB OTA for switched op-amp applications. IEE Electronics Letters 33, 1455–1456 (1997)
[19] Torralba, A., Carvajal, R.G., Martínez-Heredia, J., Ramírez-Angulo, J.: Class AB output stage for low voltage CMOS Op-Amps with accurate quiescent current control. IEE Electronics Letters 36, 1753–1754 (2000)
[20] Carvajal, R.G., Galán, J., Ramírez-Angulo, J., Torralba, A.: Low-power low-voltage differential class-AB OTAs for SC circuits. IEE Electronics Letters 38, 1304–1305 (2002)
[21] Ramírez-Angulo, J., Carvajal, R.G., Torralba, A., Nieva, C.: A new class AB differential input stage for implementation of low-voltage high slew rate op amps and linear transconductors. In: Proc. of the IEEE International Symposium on Circuits and Systems, pp. 671–674 (2001)
[22] Ramirez-Angulo, J., Thoutam, S., Lopez-Martin, A., Carvajal, R.G.: Low voltage CMOS analog four quadrant multiplier based on flipped voltage followers. IEE Electronic Letters Proc. 39, 1771–1772 (2003)
[23] Lopez-Martin, A., Ramirez-Angulo, J., Chintham, R., Carvajal, R.G.: Class AB CMOS analogue squarer circuit. IEE Electronics Letters 43, 1059–1060 (2007)
[24] Palmisano, G., Palumbo, G.: A simple CMOS CCII+. International Journal of Circuit Theory and Applications 6, 599–603 (1995)
[25] Ramirez-Angulo, J., Gupta, S., Carvajal, R.G., Lopez-Martin, A.: New improved class AB buffers based on differential flipped voltage followers. In: Proc. of the IEEE International Symposium on Circuits and Systems, pp. 3914–3917 (2006)
[26] Ramirez-Angulo, J., Lopez-Martin, A., Carvajal, R.G., Chavero, F.M.: Very low-voltage analog signal processing based on quasi-floating gate transistors. IEEE Journal of Solid State Circuits 39, 434–442 (2004)
[27] Ramírez-Angulo, J., Ducoudray-Acevedo, G., Carvajal, R.G., López-Martín, A.: Low voltage high performance voltage mode and current mode WTA circuits based on flipped voltage follower. IEEE Transactions on Circuits and Systems II 52, 420–423 (2005)
[28] Ramirez-Angulo, J., Molinar-Solis, J.E., Gupta, S., Carvajal, R.G., Lopez-Martin, A.: A high swing, high speed CMOS WTA using differential flipped voltage follower. IEEE Transactions on Circuits and Systems II 54, 668–672 (2007)
[29] Bikumandla, M., Ramírez-Angulo, J., Urquidi, C., Carvajal, R.G., Lopez-Martin, A.: Biasing CMOS amplifiers using MOS transistors in subthreshold region. IEICE Electronics Express 1, 339–345 (2004)
[30] Marri, S., Ramirez-Angulo, J., Lopez-Martin, A., Carvajal, R.G.: New high performance second generation CMOS current conveyor. In: Proc. of the XXIV Conference on Design Circuits and Integrated Systems, pp. 388–391 (2009)

Chapter 13
Synthesis of Analog Circuits Using Only Voltage and Current Followers as Active Elements

Raj Senani, D.R. Bhaskar, A.K. Singh, and R.K. Sharma

Abstract. There has been considerable interest in recent technical literature on the use of only unity gain cells (i.e. unity gain voltage followers (VF) and unity gain current followers (CF)) as alternatives to other more complex building blocks in the design of analog circuits. This has been possible due to the advances made in IC technology. It is possible to design VFs and CFs providing wider bandwidth, lower power consumption and simpler architectures as compared to other more complex building blocks. Moreover, using CFs and VFs it becomes possible to design all the four controlled sources in an open loop arrangement without encountering the gain-bandwidth-conflicts. These advantages have stimulated considerable interest in the use of VFs and CFs in a variety of linear as well as non-linear applications over the past two decades. Simultaneously, significant work has also been done on improving the design of VFs and CFs themselves for both bipolar and CMOS technologies. This chapter presents a brief account of some prominent works done on the analog circuit design using VFs and CFs together with the design of VFs and CFs themselves.

Raj Senani
Division of Electronics and Communication Engineering, Netaji Subhas Institute of Technology, Sector-3, Dwarka, New Delhi-110078, India
e-mail: senani@nsit.ac.in

D.R. Bhaskar
Department of Electronics and Communication Engineering, Faculty of Engineering and Technology, Jamia Millia Islamia, Jamia Nagar, New Delhi-110025, India
e-mail: dbhaskar@jmi.ac.in

A.K. Singh
Department of Electronics and Communication Engineering,
HRCT Group of Institutions, Faculty of Engineering and Technology,
7thKm, Delhi-Meerut Road, Ghaziabad-201003, U.P. India
e-mail: abdheshks@yahoo.com

R.K. Sharma
Department of Electronics and Communication Engineering, Ambedkar Institute of Advanced Communication Technologies and Research, Geeta Colony, Delhi-110031, India
e-mail: 21.ravindra@gmail.com

M. Fakhfakh et al. (Eds.): *Analog/RF & Mixed-Signal Circuit Sys. Design*, LNEE 233, pp. 289–315.
DOI: 10.1007/978-3-642-36329-0_13

13.1 Introduction

Due to their relatively wider bandwidth, simpler architectures as compared to other more complex building blocks and low power consumption, unity gain voltage followers (VF) and unity gain current followers (CF; henceforth, both referred to as *followers*) have attracted considerable attention in recent technical literature on analog circuit design and analog signal processing. This chapter gives a brief account of various applications of the *followers* in the synthesis of universal biquad filters, sinusoidal oscillators, impedance converters/inverters, precision rectifiers and chaotic oscillators. The realization of VFs and CFs in both bipolar and CMOS technologies has also been dealt with. Other than their direct applications, VFs and CFs are also recognized to be basic building blocks in the synthesis of CCIIs and CFOAs. Finally, recent directions of research in this area and some suggestions for future work have been made.

13.2 Realization of Controlled Sources Using VFs and CFs with Gain-Bandwidth-Decoupling

The traditional IC-op-amp is generally regarded to be the work horse of analog circuit design and it is said that almost anything can be done with IC op-amps in the area of analog circuits. However, it is well known that in several cases, op-amp circuits impose a number of limitations such as

(i) There is a well-known *gain-bandwidth-conflict* in the design of controlled sources. For instance, it can be easily worked out that the non-inverting amplifier circuit of Fig. 13.1(a) has the maximum gain = K and 3–dB bandwidth $=\frac{\omega_t}{\kappa}$ whereas the inverting amplifier of Fig. 13.1(b) has maximum gain = –K and 3–dB bandwidth= $\frac{\omega_t}{\kappa+1}$ .where $\omega_t = A_o\omega_p$ is the gain bandwidth product of the op-amp used. Thus, changing gain 'k' also changes the bandwidth.

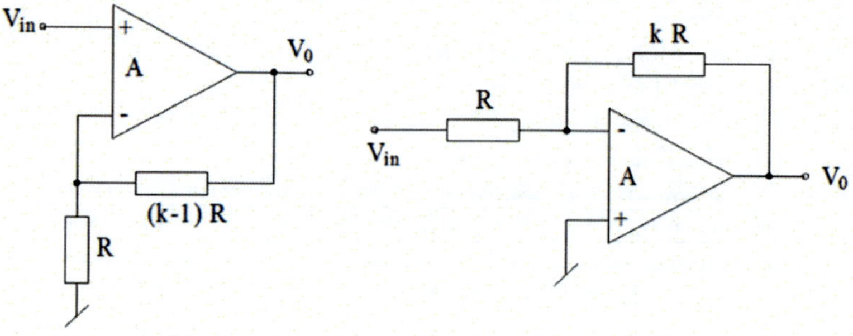

Fig. 13.1 Op-amp based amplifiers: (a) Non-inverting amplifier, (b) Inverting amplifier

(ii) In a number of applications one has to use more number of passive components to realize the given function than minimum number necessary. In the realization of VCCS, single op-amp based non-inverting integrators and differentiators and instrumentation amplifiers etc., the traditional op-amp circuits compulsorily require component-matching conditions or realization constraints, the non-fulfillment of which may either lead to non-realization of the intended function and/or instability problem as encountered in the design of single op-amp based non-inverting integrators, single op-amp based VCCS circuits and three or more op-amps based instrumentation amplifiers.

(iii) The finite slew rate of the general purpose commonly employed op-amps such as uA741 results in slew-induced distortion or limitations in operational frequency range.

In the following, we show that use of CFs and VFs overcomes many of these limitations. However, before proceeding to discuss the various analog circuits realizable using VFs and CFs and advantages gained in doing so, it is useful to introduce the symbolic notations of VFs and CFs which are shown in Fig.13.2.

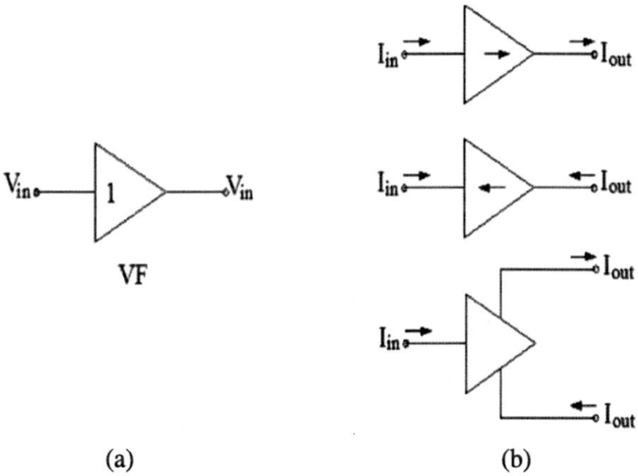

Fig. 13.2 Symbolic notations of VFs and CFs (a) VF (b) Non-inverting CF, Inverting CF and dual complimentary output CF

We now show that the use of VFs and CFs leads to realizations of controlled sources with a bare minimum of passive components. These circuits, using the symbolic notation of VFs and CFs are shown in Fig.13.3 from which it is clear that using followers, all the four controlled sources can be realized with a minimum possible number of resistors without requiring any component-matching [1].

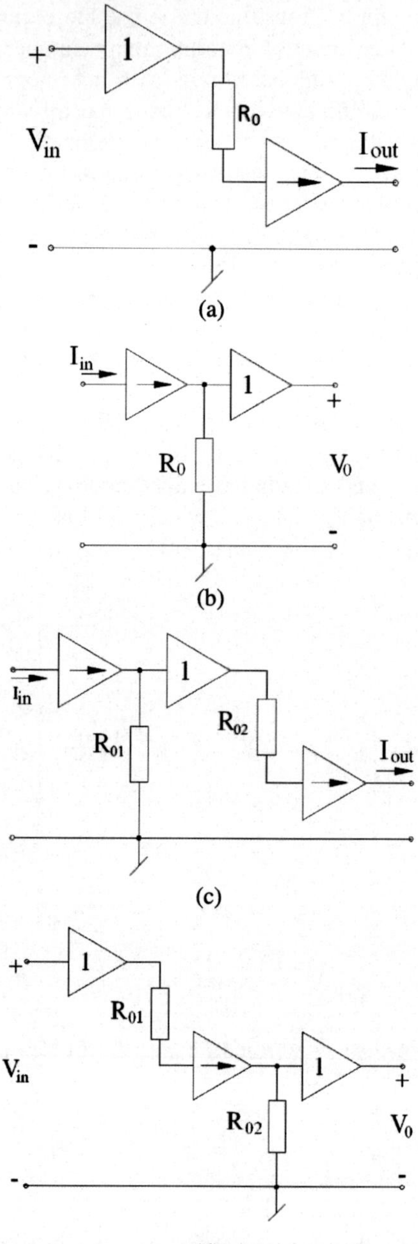

Fig. 13.3 Constant-bandwidth variable-gain realizations of the four Controlled sources using only followers as active elements

The VCCS of Figure 13.3 (a) is characterized by $I_{out} = \frac{V_{in}}{R_o}$ whereas the CCVS of Figure 13.3 (b) has output voltage given by $V_o = R_0 I_{in}$. On the other hand, the VCVS of Figure 13.3(d) has output voltage $V_o = \frac{R_{02}}{R_{01}} V_{in}$ and the CCCS of Figure 13.3 (c) has output current given $I_{out} = \frac{R_{01}}{R_{o2}} I_{in}$.

An important property of all the configurations is that they exhibit constant bandwidth and variable gains and thus, have no gain bandwidth conflict [1].

13.3 Impedance Synthesis Using VFs and CFs

It has been known for the past four decades that although one usually requires two op-amps, one capacitor and four or more resistors to simulate a lossless grounded inductance but nearly double the number of active and passive components are needed to simulate its floating counterpart using op-amps. In the following, we show how using only VFs and/or only CFs or a combination thereof, a variety of interesting inductance simulation circuits can be devised.

It has been known that using a single unity gain VF one can realize a grounded lossy inductor and using only two VFs and a minimum of only five passive components one can realize a floating lossy inductance. A well-known circuit for realizing grounded lossy (series RL) inductor is shown in Figure 13.4 (a) whereas two circuits for simulating the same kind of lossy inductor in floating form are shown in Figure 13.4 (b) with their equivalent circuit shown in Figure 13.4(c).

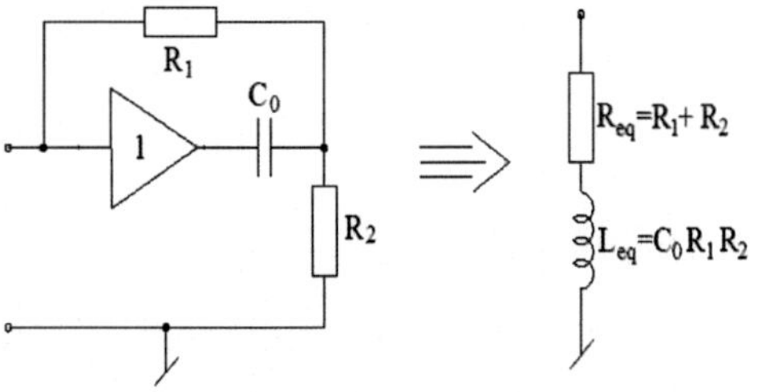

(a)

Fig. 13.4 Simulation of lossy grounded and floating inductors using VFs (a) Grounded lossy inductor (adapted from [2] ©1966 IEE) (b) Floating lossy inductors (adapted from [3] ©1974 IEE [4] ©1974 IEEE) and an alternative[1] version (adapted from [5] © 1985 IEEE) (c) An equivalent of the circuits of Fig. 13.4(b)

[1] This topology was proposed earlier in "A study of Active-RC Floating-Inductance-Simulators employing operational amplifiers", M.E Thesis by M.K. Jain (Supervisor: Raj Senani) 1980, University of Allahabad, India.

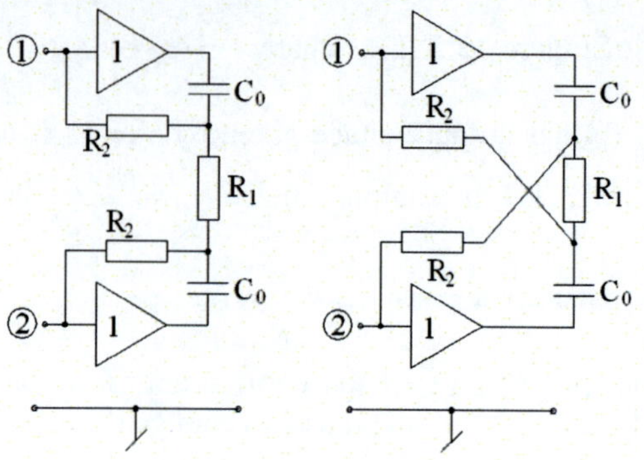

(b)

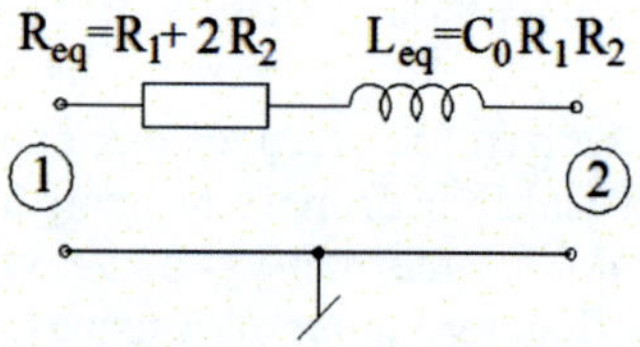

(c)

Fig. 13.4 (*continued*)

It is interesting to note, however that the fact that exactly same topologies can also be realized by using CFs rather than VFs does not appear to have been recognized explicitly in the literature so far. The circuits simulating the same type of grounded and floating inductors using CFs are shown in Fig.13.5 and surprisingly have never been perceived explicitly in the technical literature earlier.

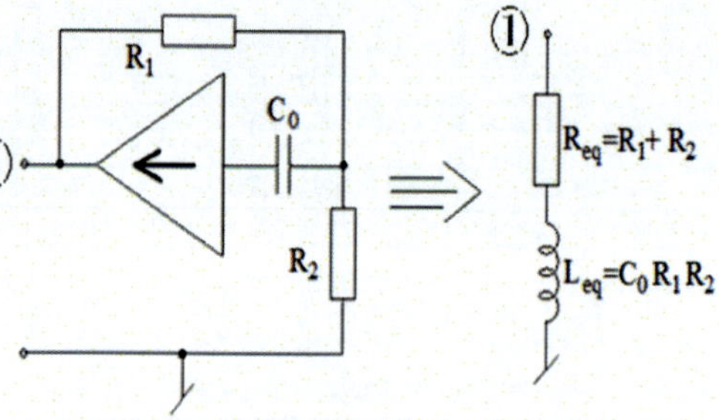

(a)

Fig. 13.5 Simulation of lossy grounded and floating inductors using CFs: (a) Grounded lossy inductor, (b) Floating lossy inductors, (c) An equivalent circuit of the circuit of Figure 13.5(b)

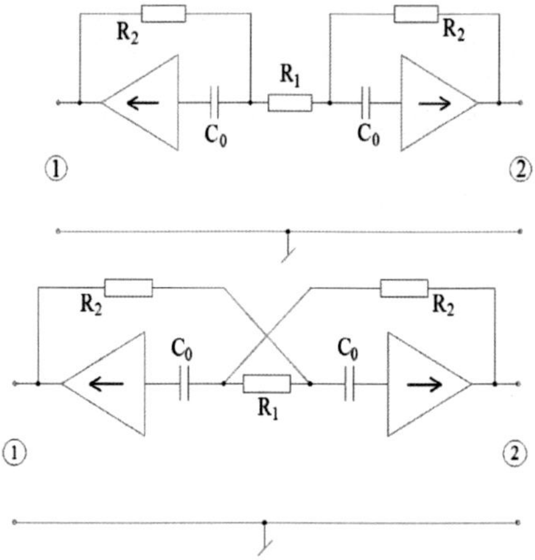

(b)

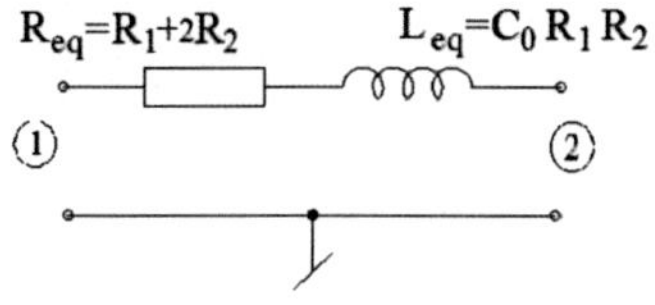

(c)

Fig. 13.5 (*continued*)

We now show that if we combine VFs and CFs then a number of interesting inductor simulation circuits can be devised using no more than a bare minimum of only two resistors alongwith a single grounded capacitor as preferred for IC implementation. A circuit for simulating a lossless grounded inductance is shown in Figure 13.6. Note that the inductance value can be adjusted/tuned by either or both of R_1 and R_2.

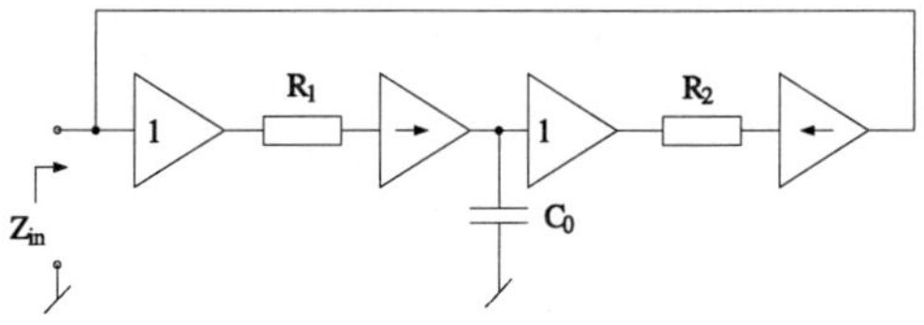

Fig. 13.6 Simulation of a lossless grounded inductor using VFs and CFs: $Z_{in}(s) = sC_0R_1R_2$

The total number of VFs and CFs can be reduced to only three with the use of grounded capacitor as well as tunability of the inductance value through a single resistance, both remaining intact, if an attempt is made to simulate lossy inductors instead of lossless one. Two such circuits are shown here in Figure 13.7. It may be observed that in these circuits, the inductance value is single resistance controllable through R_1.

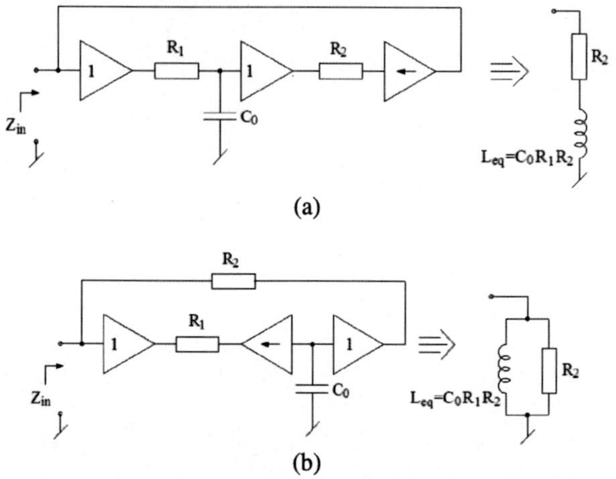

Fig. 13.7 Simulation of lossy grounded inductors: (a) series RL type, (b) parallel RL type

In a more generalized context, one can also think of formulating follower-based impedance inverter and impedance converters. In principle, an impedance inverter can simply be obtained by a parallel back-to-back connection of two VCCSs having opposite polarities. The circuit of Figure 13.8 is an embodiment of this idea. Interestingly, this circuit can also be obtained by deleting capacitor C_0 from the circuit of Figure 13.6 and treating the port thus created as port 2. From this circuit, a negative impedance inverter would be obtainable by having both the CFs with identical polarities (either both positive and both negative). When port 2 of this circuit is terminated into impedance Z_3, the input impedance looking into port 1 is given by:

$$Z(s) = \frac{Z_1 Z_2}{Z_3} \tag{13.1}$$

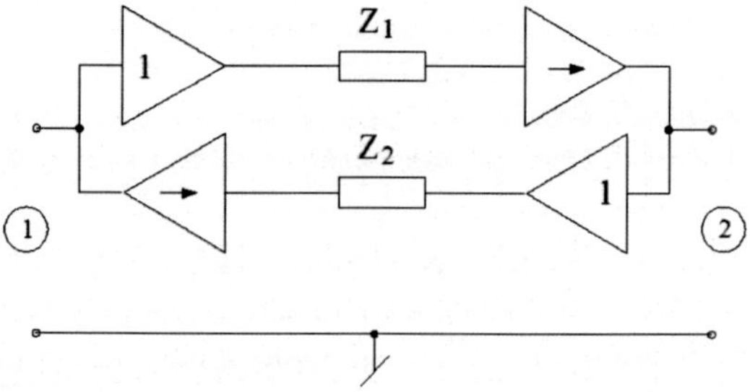

Fig. 13.8 Generalized positive impedance inverter

Thus, circuit can be looked upon as a Generalized Positive Impedance Inverter (GPII).

Lastly, it is worth pointing out that a more generalized positive impedance simulator is easily obtainable by a cascade of two such GPIIs terminated into load impedance Z_5 thereby yielding input impedance

$$Z_{in} = \frac{Z_1 Z_3 Z_5}{Z_2 Z_4} \tag{13.2}$$

which is analogous to that obtained from op-amp-based Antoniou's GIC [6].

If multiple output current followers (MOCF) are considered, positive impedance Inverter (PII) can be realized by only two MOCFs. One such circuit, equivalent to that shown in Figure 13.8, is shown in Figure 13.9.

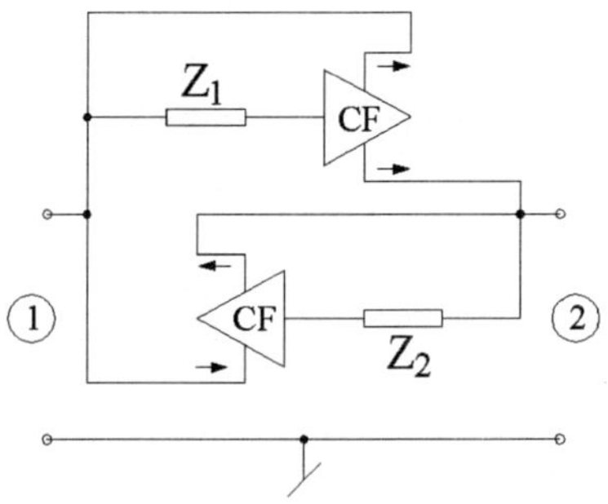

Fig. 13.9 An entirely CF-based PII

13.4 Sinusoidal Oscillator Realization Using VFs and CFs

In this section, we present some important contributions made during the last 25 years in the synthesis of sinusoidal oscillators using only unity gain VFs and CFs.

13.4.1 Oscillator Realization Using Only VFs as Active Elements

It was demonstrated, for the first time, by Senani in 1985 [7] that it is possible to realize single resistance controlled oscillator (SRCO) (i.e. an oscillator circuit in which condition of oscillation (CO) and frequency of oscillation (FO) both can be independently varied/adjusted through single variable resistances without affecting each other) using only two VFs as active elements. In [7], a configuration was presented which employed two unity gain amplifiers (UGA) made from op-amps, only three resistors and only three capacitors; see Fig. 13.10.

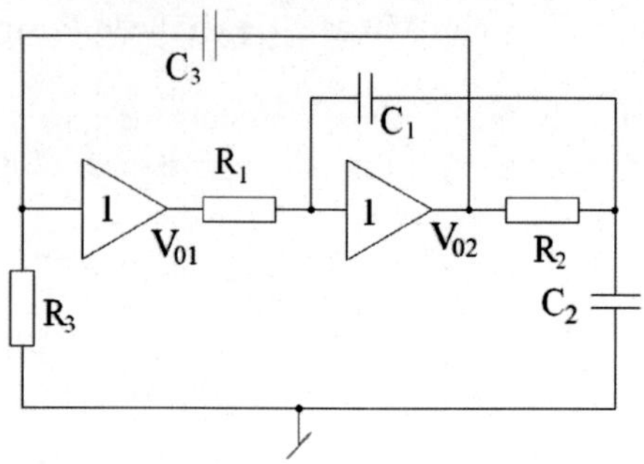

Fig. 13.10 SRCO using two VFs (adapted from [7] ©1985 IEE)

By a straight forward analysis, the CO and the FO of this circuit are found to be

$$C_3R_3 = R_2(C_1+C_2) \tag{13.3}$$

whereas the frequency of oscillation is given by

$$f_o = \frac{1}{2\pi}\sqrt{\frac{1}{C_1C_2R_1R_2}} \tag{13.4}$$

Thus, f_0 can be independently adjusted through a single variable resistor R_1 without affecting the CO which is also independently adjustable through a grounded resistance R_3 (or through C_3).

Apart from the novelty of independent control this circuit was superior to all previously known oscillators/ SRCOs as it achieved, for the first time ever, the highest possible frequency stability factor $S_F = \sqrt{n}$ (for $C_1 = C_2 = C$, $C_3 = 2C$, $R_2 = R_3 = R$ and $R_1 = n\,R$) whereby large values of S_F could be obtained by keeping n large.

Another advantage of this circuit was that it could achieve highest possible oscillation frequency, typically around 400-500 KHz, among all op-amp based SRCOs known till then because both the op-amps were employed in unity gain connection exhibiting the highest possible closed loop bandwidth (equal to the unity gain frequency of the op-amps employed).

Soon after, Abuelma'atti in [8] demonstrated that, starting from a general scheme; ten such two-UGA-based oscillators can be derived. About the same time, Boutin [9] demonstrated that an oscillator using only a single UGA is also possible using only three resistors and three capacitors but in such a circuit, the control of FO is possible only through a potentiometer realizing simultaneously two specific resistances in the circuit which share a common node; see Fig.13.11.

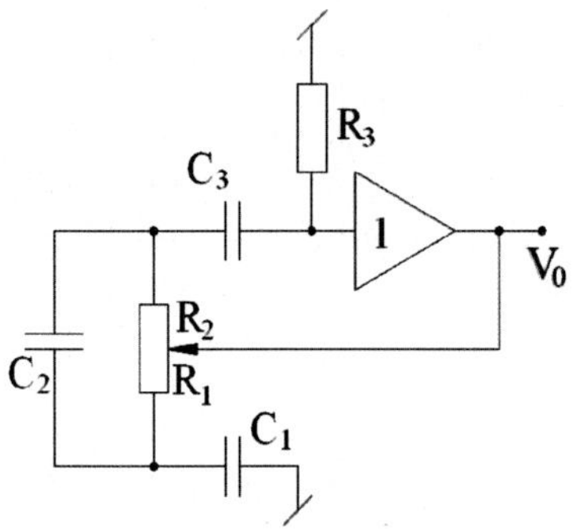

Fig. 13.11 Single-UGA Oscillator (adapted from [9] (©1986 IEE)

The CO and FO for circuit of Fig.13.11 are given by

$$R_3 = 6(R_1 + R_2): \text{ assuming } C_1 = C_2 = C_3 = C \tag{13.5}$$

$$f_0 = \frac{1}{2\pi C\sqrt{3R_1R_2}} \tag{13.6}$$

Thus, by adjusting the potentiometer the product R_1R_2 changes and hence, FO changes whereas sum (R_1+R_2) and hence, CO remains intact.

A limitation of the circuits quoted above has been that all of them are necessarily third order circuits therefore, a natural question arises whether or not a second order SRCO using only VFs is possible? From the basic considerations it was subsequently found[2] that a second order *canonic* SRCO *using only VFs* is not possible.

It is interesting to point out that the circuits of Figure 13.10 and Figure 13.11 both can be realized by CFs also. For this, the VF(s) are to be replaced by CF(s) with input output port reversed (as done in the circuits of Figure 13.5 obtained from those of Figure 13.4).

However, with the advent of current-mode circuits later on, interest in current-follower-based circuits picked up again around 1990 onwards. During the course of investigations concerning the realization of analog signal processing circuits using VFs and CFs as building blocks [10], [11], it was discovered that if one combines VFs together with CFs then it does become possible to realize second order SRCOs using only two capacitors (both of which can be grounded as preferred for IC implementation) and no more than three resistors such that CO and FO both can be independently controlled through separate variable resistors [10]. In the next section, we present the details of systematic method of synthesizing a variety of SRCOs using VFs and/or CFs as active elements.

---

[2] R. Senani, "On the realization of RC-active oscillators using unity gain amplifiers", November 1985, *unpublished.*

13.4.2 State Variable Synthesis of VF and CF Based SRCOs

In this section we demonstrate a systematic method of synthesizing SRCOs using VFs and CFs through a state variable methodology which is briefly outlined here. In general, a second order oscillator employing only two capacitors can be characterized by the following autonomous state equations.

$$\begin{bmatrix} \frac{dx_1}{dt} \\ \frac{dx_2}{dt} \end{bmatrix} = \begin{bmatrix} a_{11} & a_{12} \\ a_{21} & a_{22} \end{bmatrix} \begin{bmatrix} x_1 \\ x_2 \end{bmatrix} \tag{13.7}$$

where x_1 and x_2 are the voltages across the capacitors in the circuit to be synthesized.

Consider now the following [A] matrix

$$[A] = \begin{bmatrix} 0 & -\frac{1}{C_1 R_2} \\ \frac{1}{C_2 R_3} & \frac{1}{C_2}\left(\frac{1}{R_1} - \frac{1}{R_3}\right) \end{bmatrix} \tag{13.8}$$

The corresponding CO and FO of the SRCOs (to be synthesized from the above matrix), would be:

$$R_1 = R_3 \tag{13.9}$$

$$f_O = \frac{1}{2\pi\sqrt{C_1 C_2 R_2 R_3}} \tag{13.10}$$

from which it is seen that both CO and FO would be non-interactively controllable by R_1 and R_2, respectively.

Eq. (13.7) in conjunction with (13.8) leads to the following node equations (NEs):

$$C_1 \frac{dx_1}{dt} = -\frac{x_2}{R_2} \tag{13.11}$$

$$C_2 \frac{dx_2}{dt} = \frac{x_1 - x_2}{R_3} + \frac{x_2}{R_1} \tag{13.12}$$

Let us choose the state variables x_1 and x_2 as voltages across the grounded capacitors C_1 and C_2, respectively. Now the terms on the left hand side of the

Eqs. (13.11) and (13.12) can easily be seen as currents flowing into grounded capacitors C_1 and C_2. The currents flowing through R_1, R_2 and R_3 can be generated with the use of two VFs, one CF+ and one CF– such that the Eqns. (13.11) and (13.12) can be satisfied. The SRCO, thus generated, is shown in Figure 13.12. The various currents have been marked in the circuit to make the synthesis clear.

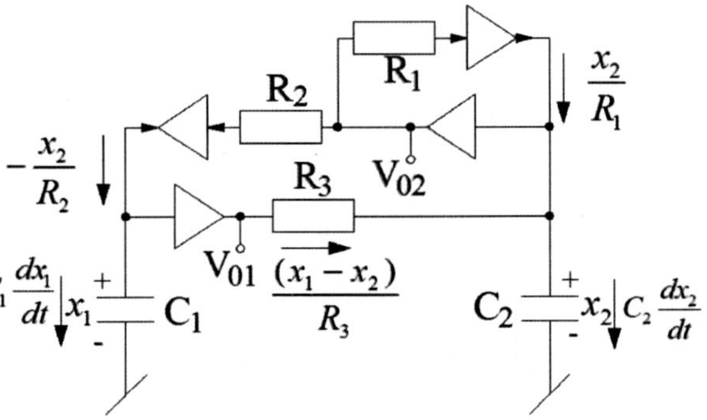

Fig. 13.12 SRCO with explicit voltage mode output (adapted from [10] © 2006 Springer)

It may be noted that although this SRCO does provide an explicit voltage-mode output however, it does not provide explicit current-mode output (i.e. from a high output impedance terminal). Here we show how to accomplish this.

For the synthesis of SRCO with the above desired features, the following matrix [A] is a suitable choice:

$$[A] = \begin{bmatrix} \frac{1}{C_1}\left(\frac{1}{R_1} - \frac{1}{R_3}\right) & -\frac{1}{C_1 R_2} \\ \frac{1}{C_2 R_3} & 0 \end{bmatrix} \tag{13.13}$$

which leads to the following node equations:

$$C_1 \frac{dx_1}{dt} = \frac{x_1}{R_1} - \frac{x_1}{R_3} - \frac{x_2}{R_2} \tag{13.14}$$

$$C_2 \frac{dx_2}{dt} = \frac{x_1}{R_3} \tag{13.15}$$

Implementation of the above equations into physical circuit results into SRCO shown in Figure 13.13 which employs one voltage-follower and two dual-output CFs.

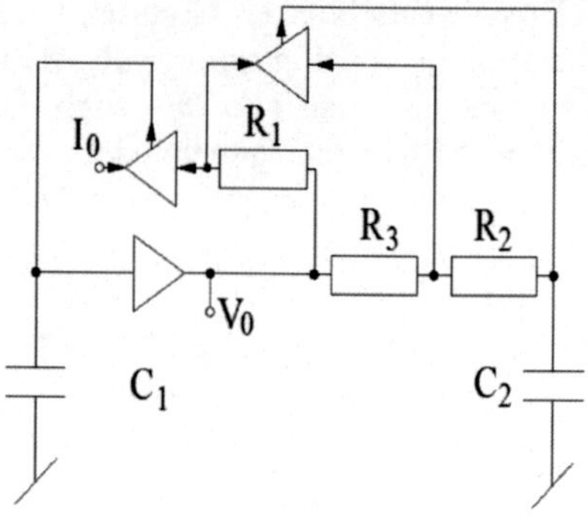

Fig. 13.13 SRCO providing voltage output and explicit Current output both (adapted from [10] © 2006 Springer)

13.4.3 SRCO Using Only Two Unity-Gain Cells

Finally, we show a grounded-capacitor SRCO which employs only one voltage-follower and one dual-output current follower which is shown in Figure 13.14. This circuit does not provide an explicit current-mode output but has the advantages of employing only one VF, only one dual-output current-follower and both grounded-capacitors as desirable for IC implementation.

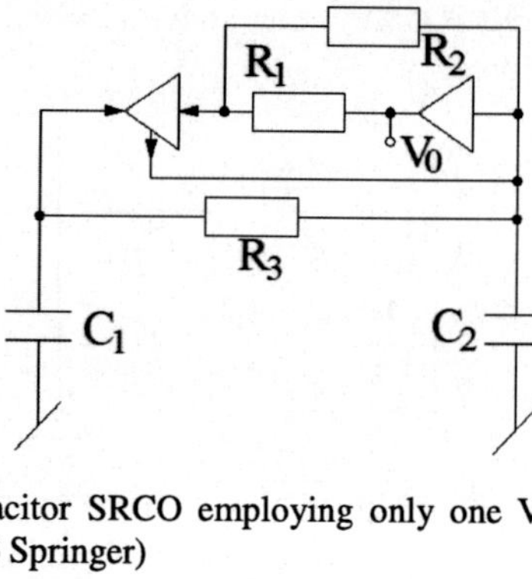

Fig. 13.14 Grounded-capacitor SRCO employing only one VF and one dual output CF (adapted from [10] © 2006 Springer)

The circuit has CO: $R_1 = \left(\dfrac{C_1}{C_1 + C_2}\right) R_3$ and FO: $f_0 = \dfrac{1}{2\pi\sqrt{C_1 C_2 R_2 R_3}}$

13.5 Synthesis of Universal Current-Mode Biquads Using VFs and CFs

Although the idea of using only single VF to realize active filters was first proposed by Sedra as early as in 1972 [12], however, all the nine circuits considered therein were non-canonic since they required three to four capacitors to realize various second order filters. With the non-obvious idea of using VFs and CFs both, to implement analog processing circuits, it subsequently became known only after nineties that by doing so, canonic filters using no more than two capacitors can be derived. As a consequence, several researchers investigated the realization of all the five standard filter functions using only CF or a combination of both VFs and CFs while using no more than two capacitors (see [14] and [15] and references cited therein). In this section, we omit the discussion about circuits which realize only one function at a time and mainly concentrate on circuits which can realize a universal biquad configuration capable of realizing all the five basic functions from the same structure.

For the generation of universal biquad employing only CFs, we have used Senani's method of [13] to show how a number of such circuits can be systematically derived. This method is based upon the RLC parallel resonator shown in Figure 13.15, according to which if we denote the currents in the passive elements C, L and R as I_{HP}, I_{LP} and I_{BP}, respectively, the three transfer functions realized are given by:

$$\frac{I_{LP}}{I_{in}} = \frac{\frac{1}{LC}}{s^2 + \frac{1}{CR}s + \frac{1}{LC}} \tag{13.16}$$

$$\frac{I_{BP}}{I_{in}} = \frac{\frac{1}{CR}s}{s^2 + \frac{1}{CR}s + \frac{1}{LC}} \tag{13.17}$$

$$\frac{I_{HP}}{I_{in}} = \frac{s^2}{s^2 + \frac{1}{CR}s + \frac{1}{LC}} \tag{13.18}$$

Thus, if one can sense the currents I_{LP} and I_{HP} as it is and the current I_{BP} with a negative sign then the circuit of Figure 13.15(a) enables the realization of a universal filter which will provide all the five standard responses explicitly in current-mode without any constraints/ condition (as the notch can be obtained just by adding I_{LP} and I_{HP} and all pass response by adding all the three currents).

One possible way to accomplish this by employing VFs and various CFs is shown in Figure 13.15 (b) where VF is characterized by the equations $i_y = 0, v_w = v_y$ with i_w being arbitrary and various CFs are characterized by the equations $v_x = 0, i_z^+ = -i_z^- = i_x$, with v_z being arbitrary. This realization uses two VFs and four CFs. Note that if the current through R_2 could be returned to node A, VF_1 can be eliminated and similarly, VF_2 can be eliminated by returning the current through R_3 to node B. The circuit obtained by this process is shown in

Figure 13.15 (c) which employs only four CFs. Further reduction in the number of CFs can be obtained by simulating parallel RL and sensing the currents I_{BP} and I_{LP}. This has been done in Figure 13.15(d) which is a minimal realization of universal current-mode biquad employing only three CFs.

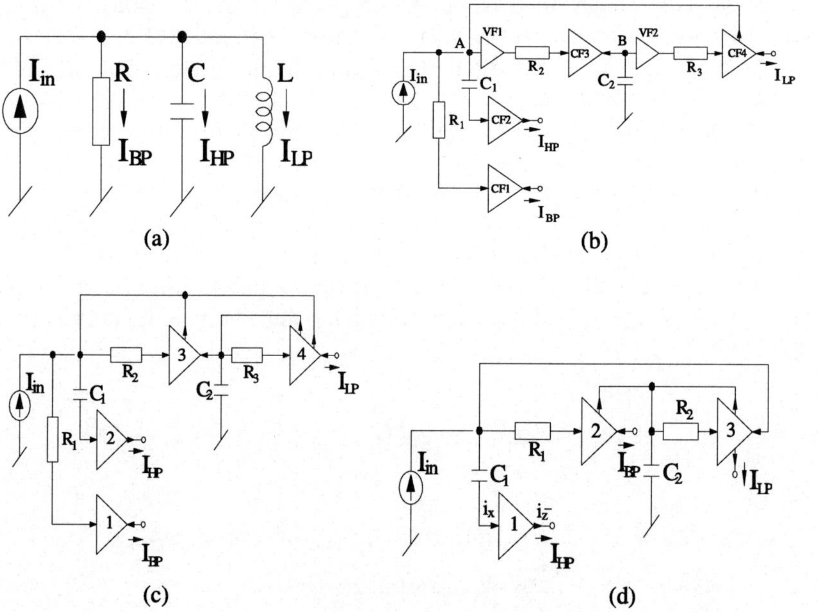

Fig. 13.15 (a) LCR resonator, (b) Universal biquad employing two VFs and four CFs, (c) Universal biquad employing four CFs, (d) The minimal universal biquad using only three CFs (adapted from [14] © 2011 Brno University of Technology Czech Republic)

A routine analysis of the circuit of Figure 13.15(d)[3] reveals the following three transfer functions:

$$\frac{I_{HP}}{I_{in}} = s^2 / D(s) = H_0 s^2 / D(s) \tag{13.19}$$

$$\frac{I_{BP}}{I_{in}} = -\left(\frac{1}{C_1 R_1}\right) s / D(s) = -H_0\left(\frac{\omega_0}{Q_0}\right) s / D(s) \tag{13.20}$$

[3] For brevity, we have omitted the details of the functions realized by the circuits of Fig.13.15 (b) and (c); the interested reader is referred to Senani R. and Gupta S. S. (2006) New universal filter using only current followers as active elements. Int. J. Electron Commun. (AEU), 60, pp.251-256.

$$\frac{I_{LP}}{I_{in}} = \left(\frac{1}{C_1C_2R_1R_2}\right) / D(s) = H_0\omega_0^2 / D(s) \tag{13.21}$$

$$\text{where} \quad D(s) = s^2 + \left(\frac{1}{C_1R_1}\right)s + \frac{1}{C_1C_2R_1R_2} \tag{13.22}$$

where the parameters ω_0, H_0 and Q_0 have their usual meanings.

It is seen that the current-mode outputs I_{HP} and I_{LP} can be tied together to obtain current-mode notch response directly without any condition and I_{HP}, I_{LP} and I_{BP} can be added together to obtain current-mode all pass response also without any condition. These current-mode transfer functions are given by:

$$\frac{I_{Notch}}{I_{in}} = \left(s^2 + \frac{1}{C_1C_2R_1R_2}\right) / D(s) \tag{13.23}$$

$$\frac{I_{AP}}{I_{in}} = \left(s^2 - \frac{s}{C_1R_1} + \frac{1}{C_1C_2R_1R_2}\right) / D(s) \tag{13.24}$$

where $D(s)$ is same as in Eq. (13.22).

It may be observed that all the five standard filter functions are available from the same structure, no design constraints/ cancellation condition is required for any response, explicit current-mode outputs are available from high output impedance terminals, f_0 can be sequentially tuned by R_2 after adjusting the bandwidth by R_1 and out of two capacitors used, one is virtually grounded and the other is grounded.

13.6 Building Blocks for the Design of Digitally-Controllable Biquad Filters

Although we have described above a number of current mode filters based on CFs and/or VFs which offer the inherent advantages of larger bandwidth, higher linearity and lower power consumption as compared to filters realizations using other building blocks. However, a major limitation associated with the above described filters is the absence of programmability feature.

In the recent literature two interesting techniques have been advanced to realize digitally programmable active filters. In the first approach advanced by Alzaher [16] digitally controlled current amplifier (DCCA) is used which provide precise frequency and or gain characteristic that can be digitally tuned over a wide range. A typical DCCA structure which makes use of a current division network is shown in Figure 13.16.

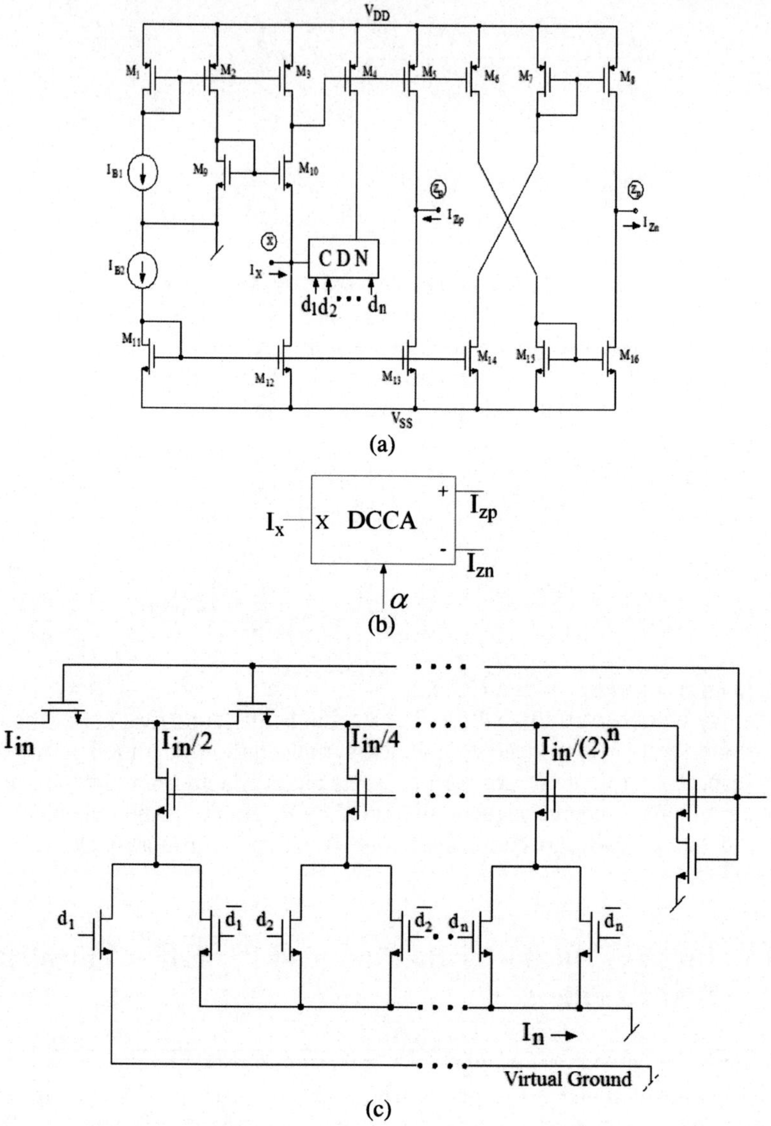

Fig. 13.16 Digitally controlled current amplifier (DCCA): (a) CMOS implementation, (b) Symbolic notation, (c) current division network (adapted from [16] © 2008 Springer)

The transfer current characteristic of the DCCA is given by

$$I_Z = \alpha I_X \text{ with } \alpha = 1/\sum_{i=1}^{n} d_i 2^{-i} \tag{13.25}$$

where d_i is the ith digital bit and n is the size of control word. It has been shown in [16] that using DCCAs, a number of interesting circuits with programmability features can be formulated. For example, Figure 13.17 shows the realization of four different types of amplifiers whereas realization of ideal integrator and lossy integrator are shown in Figure 13.18. On the other hand, Figure 13.19 shows a negative impedance converter (NIC) and Figure 13.20 shows a programmable lossless inductor, a series RL and a parallel RL.

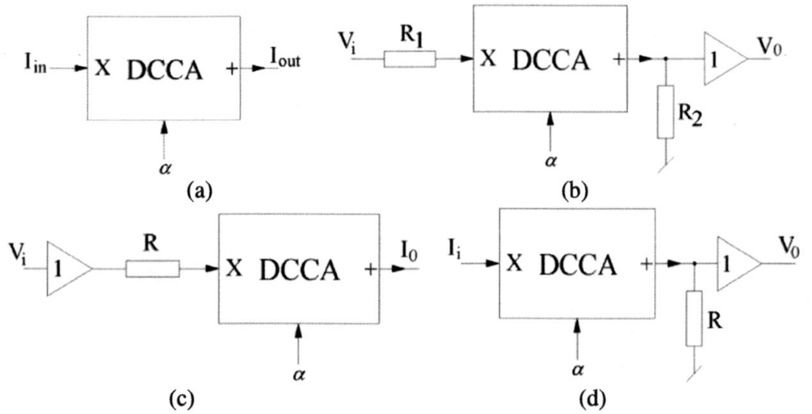

Fig. 13.17 The four different types of amplifiers: (a) Current, (b) Voltage, (c) Transconductance, (d) Transresistance (adapted from [16] © 2008 Springer)

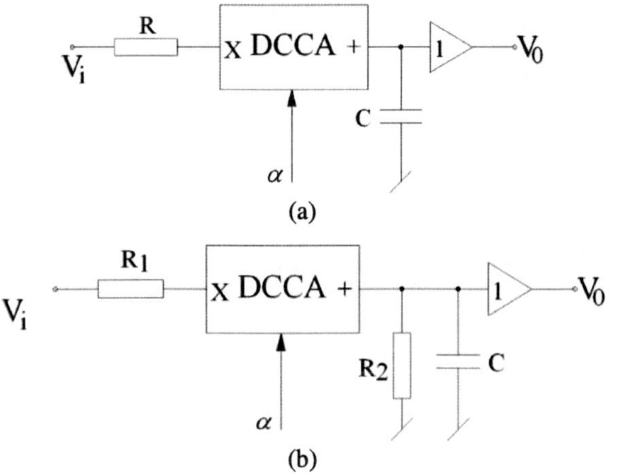

Fig. 13.18 Integrators: (a) Ideal, (b) Lossy with grounded elements (adapted from [16] © 2008 Springer)

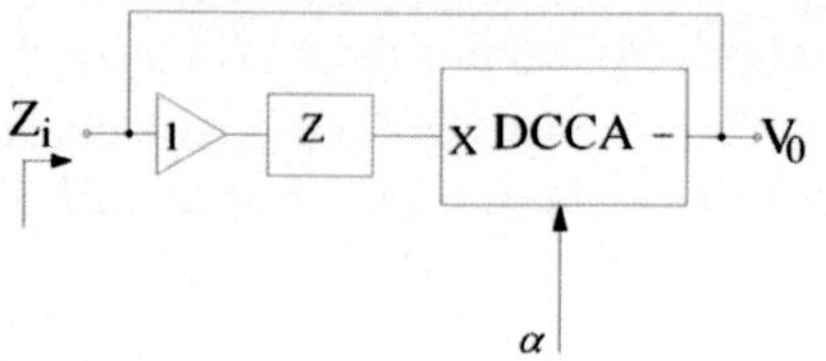

Fig. 13.19 Negative impedance converter (NIC)

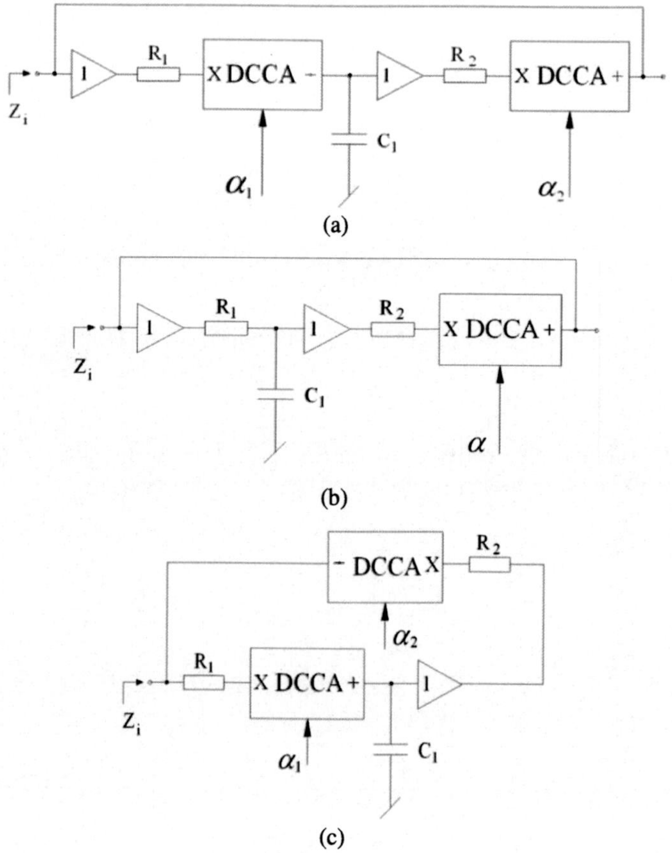

Fig. 13.20 Simulated inductors: (a) another lossless inductor, (b) series R-L, (c) parallel R-L (adapted from [16] © 2008 Springer)

These basic circuits can be employed to devise a number of interesting universal biquad filters, the details of which can be seen in [16].

In reference [17], on the other hand, a digitally-programmable current follower has been proposed which is employed to realize digitally programmable current mode universal filter and a digitally programmable current mode full wave rectifier which is essentially similar to the circuit described in Figure 13.24 of this chapter in section 13.8.

13.7 Realization of Chua's Chaotic Oscillator

Due to the importance of chaotic phenomena, in the understanding and design of nonlinear electronic circuits, in general and that of Chua's oscillator in particular, there have been many studies in devising Chua's Chaotic Oscillator circuit using a variety of devices such as op-amps, current conveyors, current feedback op-amps and even log-domain circuits, for instance, see references [18] to [22]. In this section we show how Chua's Chaotic Oscillator can be designed using unity gain cells.

In this context we know that the required grounded inductance simulator (Figure 13.21)[4] with unity gain cells can be readily realized with the PII circuit presented in the earlier section by simply loading the port-2 into a grounded capacitance as shown in Figure 13.6.

Together with this, the non-linear resistor characteristic (so-called *Chua diode*) can also be realized by a specially devised negative resistance (NR) circuit using unity gain cells as shown in Fig. 13.22. Using these building blocks, the complete Chua's oscillator using unity gain cells turns out to be as shown in Figure. 13.23.

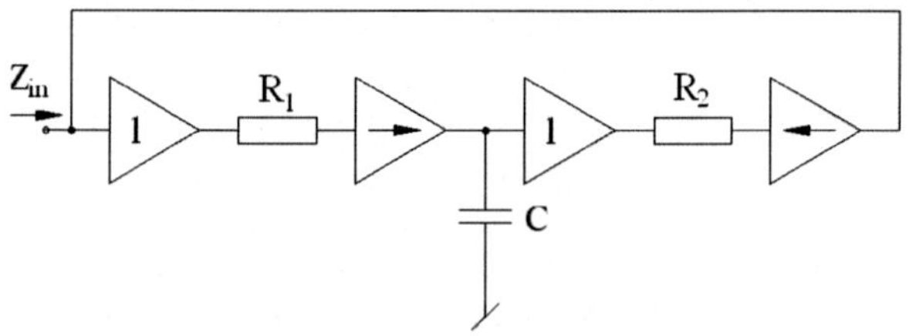

Fig. 13.21 Grounded inductance simulation using only unity-gain cells (adapted from [22] © 2008 IEEE)

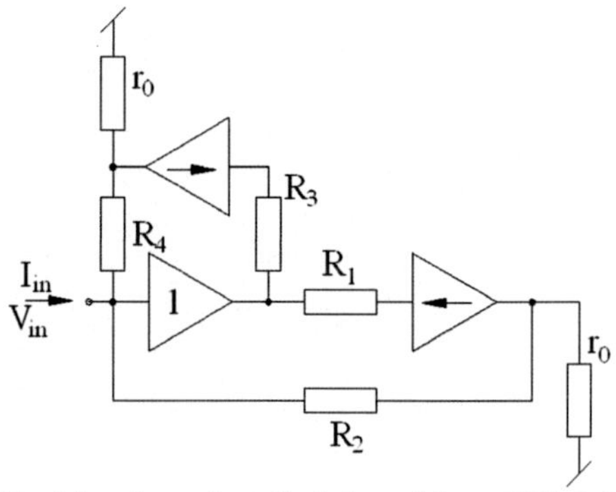

Fig. 13.22 Design of the NR with unity-gain cells (adapted from [22] © 2008 IEEE)

[4] This *specific* grounded inductor was earlier described in Ch. 9 titled ` Some Contributions to the Realization of Universal Biquad Filters' in the Ph.D. thesis of S. S. Gupta (Supervisor R. Senani) entitled `Realizations of Some Classes of Linear/Nonlinear Analog Electronic Circuits, University of Delhi, 2001-2005.

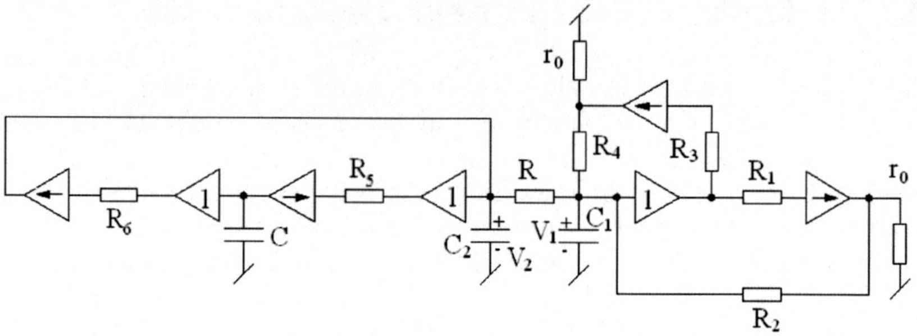

Fig. 13.23 The complete Chua's oscillator using unity gain cells (adapted from [22] © 2008 IEEE)

13.8 Full Wave Current Mode Precision Rectifier Using Unity Gain Cells

A precision rectifier can be made from a single dual output current follower and four diodes capable of rectifying either a current mode sinusoidal signal or a voltage mode signal using schematic of Figure 13.24. Assuming identical diodes, the operation of these circuits can be explained as follows. When I_{in} or V_{in} is positive, diodes D_1 and D_3 are forward biased and current is conducted to the load through them. When the polarity of the input signal changes, D_1 and D_3 are off and D_2 and D_4 are on. Thus, positive full wave rectified current flows into R_1 since negative current flows in R_2. It may be noted that if current outputs are required then resistors R_1 and R_2 are not needed and the output of the diode bridge can be directly connected to the load. If the input signal is a voltage, it can be applied through a resistor R_i connected between the input voltage source V_{in} and the input terminal of the CF which has virtual ground potential thereby making $I_{in} = \frac{V_{in}}{R_i}$ (Figure 13.25). Furthermore, if input impedance Z_{in} is desired to be ideally infinite, then the input resistor R_i can be preceded by a unity gain VF as shown in Figure 13.25.

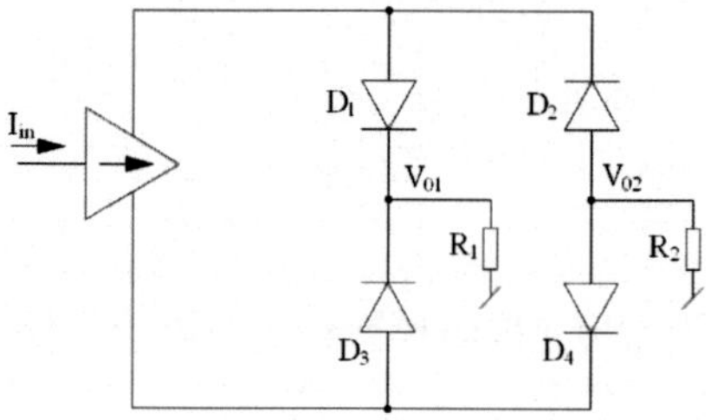

Fig. 13.24 First full-wave precision rectifier scheme (adapted from [23] © 2003 Pateikta Spaudai)

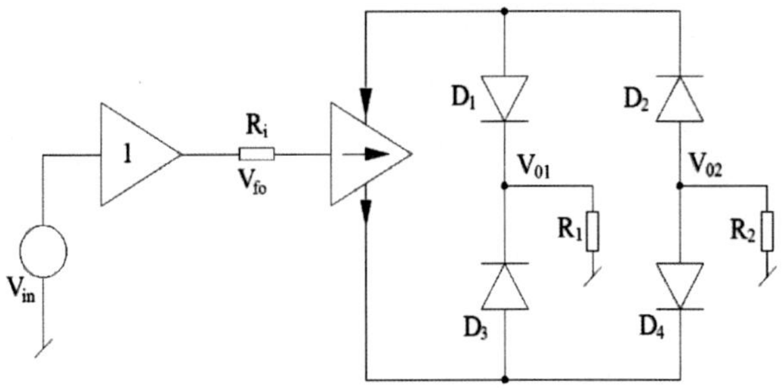

Fig. 13.25 High input impedance current-mode precision rectifier (adapted from [23] © 2003 Pateikta Spaudai)

13.9 Hardware Implementation of VFs and CFs

In view of numerous applications of VFs and CFs in both linear and non-linear circuit designs as outlined in this chapter, there has been quite a lot of interest in designing the internal hardware of VFs and CFs for both bipolar and CMOS technologies. Quite often, a basic CCII based current conveyor structure has been employed by several researchers to implement CFs with a single or multiple outputs which can be created easily by grounding terminal Y and duplicating the Z output terminals and creating a complimentary output merely by using additional pairs of current mirrors. A typical circuit design of dual output CFs, which has been employed by several authors, is shown here in Figure 13.26.

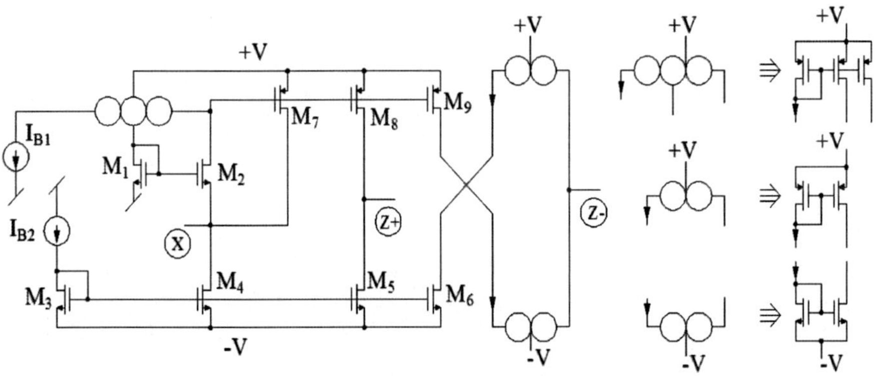

Fig. 13.26 Current Follower (adapted from [15] © 1999 IEE)

It is well known that both VF and CF can be represented as a pair of a nullator and norator, as shown in Figure 13.27.

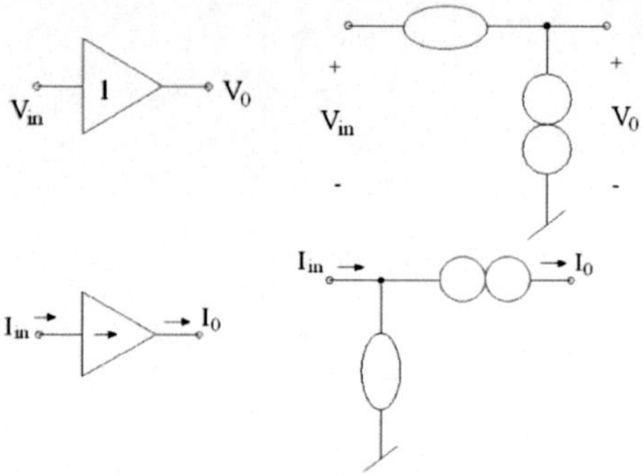

Fig. 13.27 Nullor models of VF and CF

In the recent years, two new pathological elements namely the current mirror and voltage mirror were defined in [24]. Motivated by these, there have been a number of attempts on systematically synthesizing VFs and CFs by manipulation of generic cells as well as by applying genetic algorithms, for instance, see [25], [26]. An exemplary, genetically-derived CMOS VF from [26] is shown in Figure 13.28.

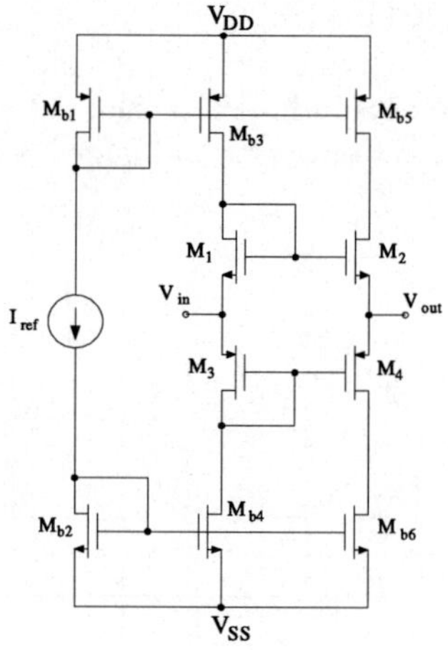

Fig. 13.28 A genetically derived voltage follower (adapted from [26] © 2008 Birkhauser)

These developments have also lead to the computational synthesis of CMOS VFs [27]. On the other hand, the genetic algorithm based synthesis has been employed to the synthesis of CCIIs by superimposing VFs and CFs in [28]. There have been many attempts on designing VFs using CMOS transistors such as [29], [30] for improving various performance criterions.

Attention has also been given in recent literature in evolving low power fully differential CFs. One such circuit, proposed by Azhari and Safari, is shown here in Figure 13.29 [31].

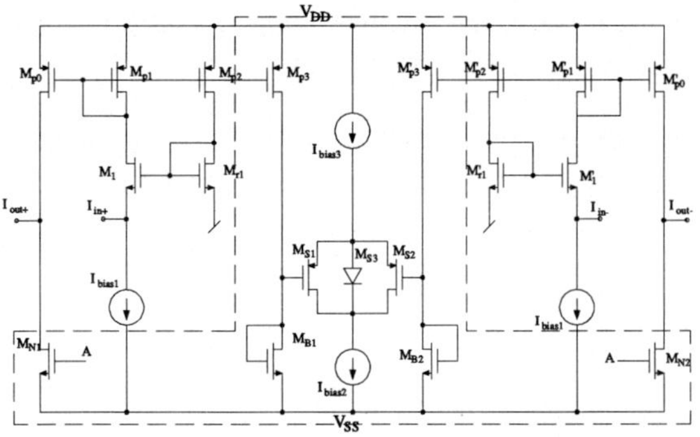

Fig. 13.29 Fully-differential Current buffer (adapted from [31] © 2010 IEICE)

13.10 Use of CFs and VFs in the Synthesis of Other Building Blocks

Interesting and useful work has been done by Tlelo-Cuautle and his co-workers in using CFs and VFs systematically for synthesizing CCIIs and CFOAs [28], [32]. In [28], a genetic algorithm based synthesis has been proposed wherein a number of new topologies have been evolved for implementation of CCIIs. On the other hand, in [32], a systematic synthesis has been evolved for generating the hardware for CCs and CFOAs by manipulation VFs and CFs by executing four basic steps (i) selection of generic cells using nullators and norators (ii) addition of norators and nullators to form joined nullator-norator pairs (iii) addition of DC biasing circuitry and (iv) synthesis of the joint-pairs by transistors which could be either BJTs or MOSFETs. This methodology has been shown to be suitable for generating new transistor implementation of CCs or CFOAs and appear well suited for automatic design of these building blocks.

13.11 Concluding Remarks and Directions of Future Research

This chapter has presented, in a tutorial-review format, various analog signal processing and signal generation circuits using only voltage followers and current followers as active elements which have attracted considerable interest in recent analog circuit design owing to the advantages of wider bandwidth, higher linearity and lower power consumption as well as possessing simpler circuit architectures as compared to other more complex active building blocks. Apart from enumerating a number of circuits which are known in literature, a few alternative circuits have also been suggested which have not been explicitly disclosed in the earlier literature so far, such as the circuits based upon CFs derived from those employing VFs as exemplified in sections 13.3 and 13.4. Apart from their numerous applications such as impedance simulation, oscillator synthesis, universal filter realization, full-wave rectification, chaotic oscillators and digitally programmable filters, other applications still await to be explored and this constitutes an important area of research. On the other hand, although numerous techniques are currently being investigated for synthesizing CMOS VFs and CFs, including those possessing digital programmability of the gain, the work reported so far is by no means complete. The development of optimum CMOS VFs and CFs structures is still open to investigation. Continued work on the use of CFs and VFs in systematically evolving CCII and CFOA architectures as in [28], [32] may likely to result in optimum topologies. Furthermore, this work may also be carried over for the synthesis of more complex building blocks like current differencing buffered amplifiers (CDBAs), Differential difference buffered amplifiers (DDBAs) and others.

Acknowledgement. The authors gratefully acknowledge the dedicated and meticulous help of Ms. Shashi Rawat in the preparation of this manuscript.

References

[1] Toumazou, C., Lidgey, F.J., Haigh, D.G.: Analogue IC design: the current-mode approach. Peter Peregrinus Ltd. (1990)
[2] Prescott, A.J.: Loss-compensated active gyrator using differential-input operational amplifiers. Electron. Lett. 2, 283–284 (1966)
[3] Wise, D.R.: Active simulation of floating lossy inductances. Proc. IEE 121, 85–87 (1974)
[4] Dutta Roy, S.C.: A circuit for floating inductance simulation. Proc. IEEE 62, 521–523 (1974)
[5] Takaji, S., Fujii, N.: High-frequency active RC simulation of impedance-scaled LC filters using voltage followers. In: Proc. ISCAS, pp. 295–298 (1985)
[6] Antoniou, A.: Realization of gyrators using operational amplifiers, and their use in RC-active-network synthesis. Proc. IEE 116, 1838–1850 (1969)
[7] Senani, R.: New RC-active oscillator configuration employing unity-gain amplifiers. Electron. Lett. 21, 889–891 (1985)
[8] Abuelma'atti, M.T., Humood, N.A.: Ten faster op-amp oscillators. Electron. and Wireless World 93, 1053–1054 (1987)
[9] Boutin, N.: Synthesis of oscillator circuits employing only one unity-gain amplifier. Electron. Lett. 22, 22–23 (1986)
[10] Gupta, S.S., Senani, R.: New single-resistance-controlled oscillator configurations using unity-gain cells. Analog Integr. Circuits Signal Process. 46, 111–119 (2006)
[11] Gupta, S.S., Senani, R.: New single resistance controlled oscillators employing a reduced number of unity-gain cells. IEICE Electronics Express 1, 507–512 (2004)
[12] Sedra, A.S.: A class of stable active filters using unity-gain voltage followers. IEEE J. Solid State Circuits 7, 311–315 (1972)
[13] Senani, R.: A simple approach of deriving single-input-multiple-output current-mode biquad filters. Frequenz 50, 124–127 (1996)
[14] Senani, R., Gupta, S.S.: Current-mode universal biquad using current followers: a minimal realization. Radioengineering 20, 898–904 (2011)
[15] Alzaher, H.A., Ismail, M.: Current-mode universal filter using unity gain cells. Electron. Lett. 35, 2198–2200 (1999)
[16] Alzaher, H.A.: A CMOS digitally programmable filter technique for VLSI applications. Analog Intgr. Circ. Sig. Process. 55, 177–187 (2008)
[17] Tangsrirat, W., Pukkalanun, T.: Digitally programmable current follower and its applications. Int. J. Electron. Commun. (AEU) 63, 416–422 (2009)

[18] Kennedy, M.P.: Robust Op Amp realization of Chua's circuit. Frequenz 46, 66–80 (1992)
[19] Elwakil, A.S., Soliman, A.M.: Current conveyor chaos generators. IEEE Trans. CAS-I Fundam. Theory Appl. 46, 393–398 (1999)
[20] Senani, R., Gupta, S.S.: Implementation of Chua's chaotic circuit using current feedback op-amps. Electron. Lett. 34, 829–830 (1998)
[21] Ozoguz, S.: Simple log-domain chaotic oscillator. Electron. Lett. 37, 1378–1379 (2001)
[22] Sanchez-Lopez, C., Trejo-Guerra, R., Tlelo-Cuautle, E.: Simulation of Chua's chaotic oscillator using unity-gain cells. In: Proc. of the 7th Int. Caribbean Conference on Devices, Circuits and Systems, pp. 28–30 (2008)
[23] Tiliute, D.E.: Full-wave current-mode precision rectifiers using unity-gain cells. Elektronika IR Elektrotechnika 7, 26–29 (2003)
[24] Awad, I.A., Soliman, A.M.: On the voltage mirrors and the current mirrors. Analog Intgr. Circ. Sig. Process. 32, 79–81 (2002)
[25] Torres-Papaqui, L., Torres-Munoz, D., Tlelo-Cuautle, E.: Synthesis of VFs and CFs by manipulation of generic cells. Analog Intgr. Circ. Sig. Process. 46, 99–102 (2006)
[26] Tlelo-Cuautle, E., Duarte-Villasenor, M.A., Guerra-Gomez, I.: Automatic synthesis of VFs and VMs by applying genetic algorithms. Circuits Syst. Signal Process. 27, 391–403 (2008)
[27] Tlelo-Cuautle, E., Torres-Munoz, D., Torres-Papaqui, L.: On the computational synthesis of CMOS voltage followers. IEICE Trans. Fundamentals E88, 3479–3484 (2005)
[28] Tlelo-Cuautle, E., Moro-Frias, D., Sanchez-Lopez, C., Duarte-Villasenor, M.A.: Synthesis of CCII-s by superimposing VFs and CFs through genetic operations. IEICE Electron. Expr. 5, 411–417 (2008)
[29] Ramirez-Angulo, J., Lopez-Martin, A.J., Carvajal, R.G., Torralba, A., Jimenez, M.: Simple class-AB voltage follower with slew rate and bandwidth enhancement and no extra static power or supply requirements. Electron. Lett. 42, 784–785 (2006)
[30] Lopez-Martin, A.J., Algueta Miguel, J.M., Acosta, L., Ramirez-Angulo, J., Carvajal, R.G.: Design of two-stage class AB CMOS buffers: A systematic approach. ETRI J. 33, 393–400 (2011)
[31] Azhari, S.J., Safari, L.: A high CMRR low power fully differential current buffer. IEICE Electron. Expr. 7, 765–771 (2010)
[32] Torres-Papqui, L., Tlelo-Cuautle, E.: Synthesis of CCs and CFOAs by manipulation of VFs and CFs. In: Proc. of IEEE International Conf. on Behavioral Modeling and Simulation, pp. 92–96 (2004)

Chapter 14
Design of Setable Active Lossy Inductors

Marian Pierzchała and Mourad Fakhfakh

Abstract. The general two-graph framework for designing series and parallel R-L immittances is presented. The transformation of passive LC-filters into active RC-circuits using signal-flow graphs in the two-graph version, is the basis of this method. The idea consists of using exclusively RC-elements and the newly introduced 'active switches'. A second novel idea is presented; it consists of the reduction of the complexity of the equivalent active circuits, i.e. reducing the number of active elements to reproduce the 'same' circuit's behavior, but at the cost of a reduction of the quality factor. SPICE simulation results are presented to show the viability of the proposed approaches.

14.1 Introduction

Active inductors play an important role in analog circuit design. They are frequently used in the design of active oscillators, active filters, and in cancellation of parasitic elements. In many cases, values of the resistance and the inductance of these inductor simulators should be set individually in order to obtain the optimal construction of such active components. Actually, the literature offers a large number of publications dealing with the realization of inductor simulators using high performance active blocks, such as current conveyors [1]-[7], trans-resistance amplifiers [8]-[9], and nullors [10]-[12]. A systematic synthesis framework for linear active-RC circuits has been also proposed [13], [14]. In this method the final node admittance matrix (NAM) is obtained by applying pivotal expansion. Next, the nullors can be introduced to change the position of the matrix elements to obtain required relationships between the input and output signals. This systematic

Marian Pierzchała
Wroclaw University of Technology, Poland
e-mail: marian.pierzchala@pwr.wroc.pl

Mourad Fakhfakh
University of Sfax, Tunisia
e-mail: mourad.fakhfakh@ieee.org, fakhfakhmourad@gmail.com

M. Fakhfakh et al. (Eds.): *Analog/RF & Mixed-Signal Circuit Sys. Design*, LNEE 233, pp. 317–336.
DOI: 10.1007/978-3-642-36329-0_14

synthesis framework has been extended in [15] to use NAM representation for mirror elements, also. This generalized approach facilitates the use of all type of pathological elements (nullor and mirror elements). However, there is a lack of publications about a general systematic method which enables designing setable active inductors, i.e. the active inductors which equivalent values of the resistance and the inductance can be set individually.

Thus, in this Chapter we propose a systematic approach which enables generating settable active inductors using manifold active elements, which are equivalent to an ideal nullor, for example current conveyors, four-terminal floating nullor, etc.

The method is based on the use of signal-flow graphs in the two-graph version. Application to the design of a series and a parallel settable inductor simulators, and a ladder filter is detailed in the following.

14.2 The Proposed Approach

Recently, a method dealing with the synthesis of lossless active inductors was proposed by the authors in [16]. It uses *RC*-elements and 'active switches' for the design of the inductor simulators. However, this method does not give the possibility to synthesize settable active inductors with losses, i.e. an active inductor with individually setting the values of the resistance and the inductance. In this Chapter we propose a modification of this method which not only enables setting the values of the resistance and the inductance of the active inductor, but also significantly simplifies the structure of these circuits.

A- The serial RLC-filter

Let's consider the lossy serial *LC*-filter shown in Figure 14.1. The circuit encompasses an independent voltage source. The inductor and the resistor are chosen to be in the tree branches, and the capacitor is, thus, placed in the cotree branch. Figure 14.2 gives the corresponding signal-flow graph associated with this circuit.

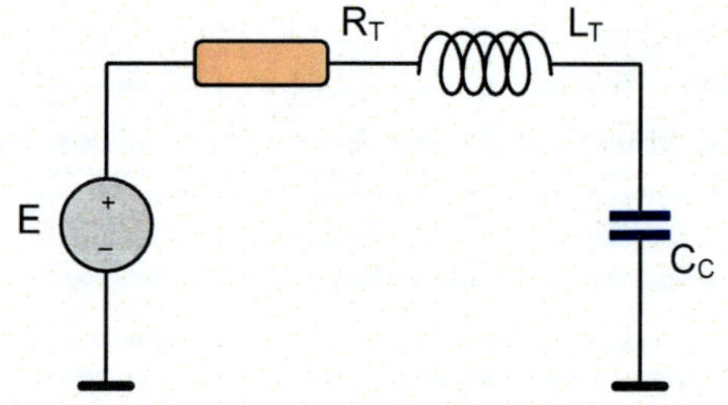

Fig. 14.1 A serial lossy *LC*-filter

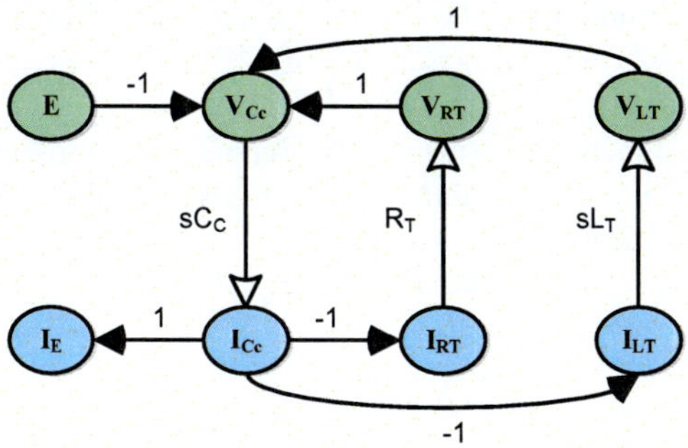

Fig. 14.2 The signal-flow graph associated with the circuit presented in Figure 14.1

The fundamental circuit matrix $(B_T^V)_{LC}$ and the *T*-matrix $(T_C^I)_{LC} = ((-Q_C^I)^t)_{LC}$ for this *LC*-filter can be presented as follows (see Table 14.1), where Q_C^I is the cutest matrix. Subscripts *T* and *C* refer to the tree position and the co-tree position, respectively. Superscripts *V* and *I* refer to the voltage graph and current graph, respectively.

Table 14.1 $(B_T^V)_{LC}$ and $(T_C^I)_{LC}$ for the lossy serial *LC*-filter

	E	V_{RT}	V_{LT}
V_{Cc}	-1	1	1

(a): $(B_T^V)_{LC}$

	I_E	I_{RT}	I_{LT}
I_{Cc}	-1	1	1

(b): $(T_C^I)_{LC}$

Table 14.2 $(B_T^V)_{RC}$ and $(T_C^I)_{RC}$ matrices for the equivalent *RC*-filter

	E	V_{RT}	V_{Rd}
V_{Cc}	-1	1	1
V_{Cd}	0	1	0

(a): $(B_T^V)_{RC}$

	I_E	I_{RT}	I_{Rd}
I_{Cc}	-1	1	0
I_{Cd}	0	0	-1

(b): $(T_C^I)_{RC}$

From Figure 14.2, and using the Masons' formula, we have:

$$Z_L = R_T + sL_T \tag{14.1}$$

The proposed idea consists of replacing the inductor with losses by a combination of suitable passive elements. Table 14.2 presents a modification of the *B* and *T* matrices given in Table 14.1.

The signal-flow graph associated with the matrices shown in Table 14.2 is given in Figure 14.3.

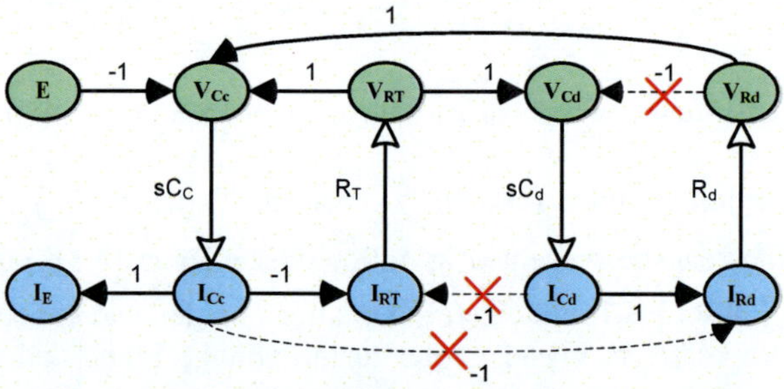

Fig. 14.3 The signal-flow graph of the lossy *RC*-active filter

According to the signal-flow graph given in Figure 14.3, the equivalent impedance (Z_{Leq}) expression of the designed active inductor is

$$Z_{Leq} = R_T + sR_T R_d C_d \tag{14.2}$$

Accordingly, the equivalent inductance L_{eq} is equal to $R_T R_d C_d$ and the equivalent serial resistor R_{eq} is equal to R_T.

Hence, the values of the serial resistance (R_T) and the inductor's inductance ($R_T R_d C_d$), can be set orthogonally, and the *Q* factor is equal to:

$$Q = \omega R_d C_d \tag{14.3}$$

In order to design the equivalent circuit that can reproduce the current and the voltage relationships shown in the signal-flow graph of Figure 14.3, one should use the 'active switches' (see [16]-[18]).

A simple procedure which enables to construct the circuits with "active switches" consists of the following:

i. Connect the elements forming the tree in the $(B_T^V)_{RC}$ and $(T_C^I)_{RC}$ matrices,

ii. For the cotree elements which have different rows in the $(B_T^V)_{RC}$ and $(T_C^I)_{RC}$ matrices, add a pair of 'active switches' (i.e. one switch for the voltage graph and the another for the current graph) at each terminal.

iii. Connect the cotree elements to the tree elements according to $(B_T^V)_{RC}$ and $(T_C^I)_{RC}$.

iv. Remove the extra switches.

As an example, let us construct the circuit with 'active switches' for the matrices from Table 14.2.

Step *i.*: In these matrices the tree encompasses the following elements: E, R_T and R_d, as shown in Figure 14.4.

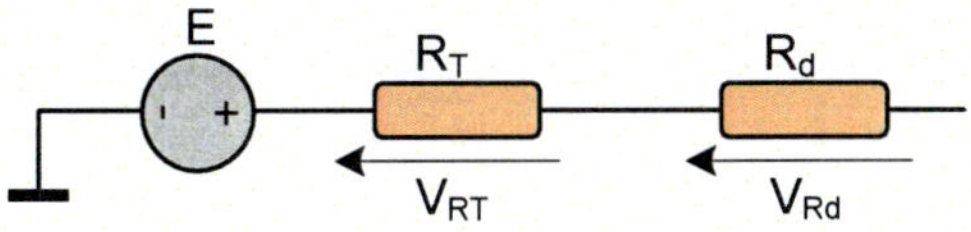

Fig. 14.4 The tree of the lossy *RC*-active filter

Step *ii.*: In the above matrices $(B_T^V)_{RC}$ and $(T_C^I)_{RC}$ both elements (C_C and C_d) have different rows, thus we draw them with the "active switches" (see Figure 14.5).

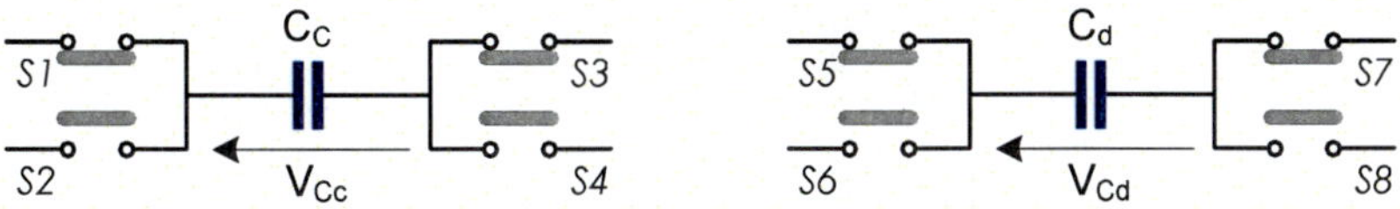

Fig. 14.5 The cotree elements with the 'active switches'

Step *iii.*: After realizing the necessary connections, the circuit will have the configuration shown in Figure 14.6.

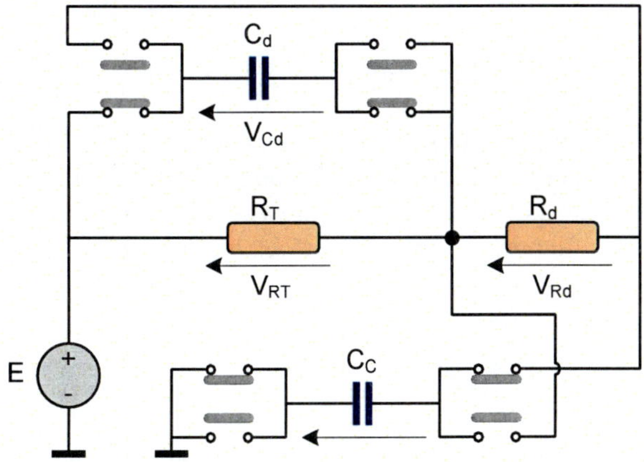

Fig. 14.6 The circuit with the 'active switches'

Step *iv.*: We can replace the parallel connection of the voltage and current switches by a short circuit and thus simplifying the circuit. Figures 14.7 and 14.8 show the 'switched' circuits that reproduce the matrices given in Table 14.2. The switches are presented in two positions, the first, i.e. Figure 14.7, for the voltage graph and the second, i.e. Figure 14.8, for the current graph.

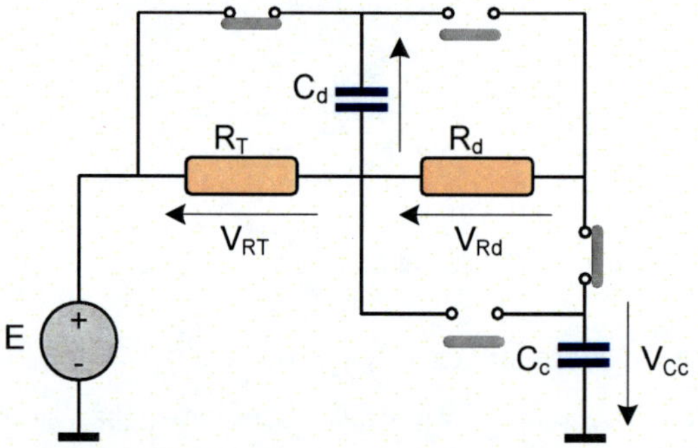

Fig. 14.7 The equivalent 'switched' circuit for the voltage graph

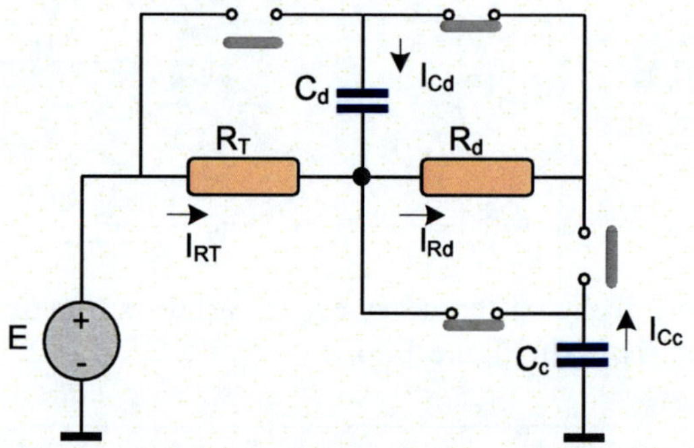

Fig. 14.8 The equivalent 'switched' circuit for the current graph

It was shown in [16] and in [18] that the behavior of such 'active switches' can be emulated by using norators and nullators, see Table 14.3. Thus, these ideal elements can be used to construct a circuit which is equivalent to the original *LC*-filter with losses. Figure 14.9 presents the obtained nullator/norator based circuit.

B- The "simplified" serial RLC-filter

Each cotree element which has different rows in the $(B_T^V)_{RC}$ and $(T_C^I)_{RC}$ needs at most two pairs of switches, as shown in Figure 14.5. It is to be noted that if we wish to have a simpler circuit, than we have to reduce the number of these rows, but at the expense of reducing the Q value.

Table 14.3 Two-graph equivalents of nullators and norators

	Symbol	Position	
		Current graph	Voltage graph
Nullator		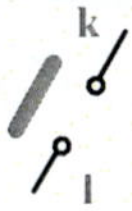	
Norator			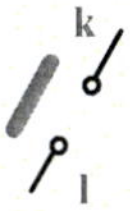

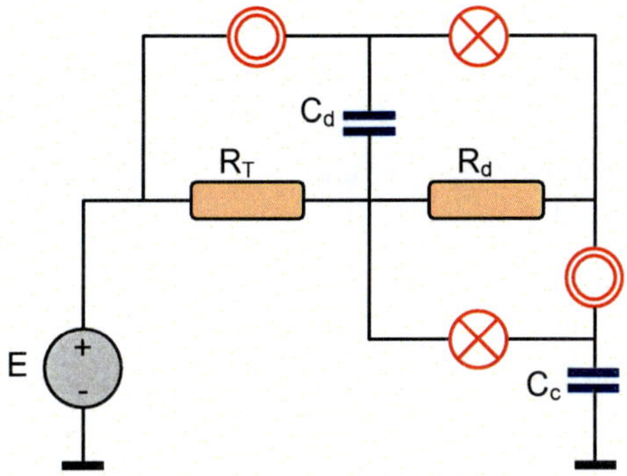

Fig. 14.9 The equivalent circuit using nullators and norators

Table 14.4 shows a modification that will result in the elimination of one pair of switches.

Table 14.4 $(B_T^V)_{RC}$ and $(T_C^I)_{RC}$ matrices for the simplified version of the lossy serial *RC*-filter

	E	$\mathbf{V_{RT}}$	$\mathbf{V_{Rd}}$
$\mathbf{V_{Cc}}$	-1	1	1
$\mathbf{V_{Cd}}$	0	1	0

(a): $(B_T^V)_{RC}$

	$\mathbf{I_E}$	$\mathbf{I_{RT}}$	$\mathbf{I_{Rd}}$
$\mathbf{I_{Cc}}$	-1	1	1
$\mathbf{I_{Cd}}$	0	0	-1

(b): $(T_C^I)_{RC}$

The signal-flow graph associated with the matrices shown above is given in Figure 14.10.

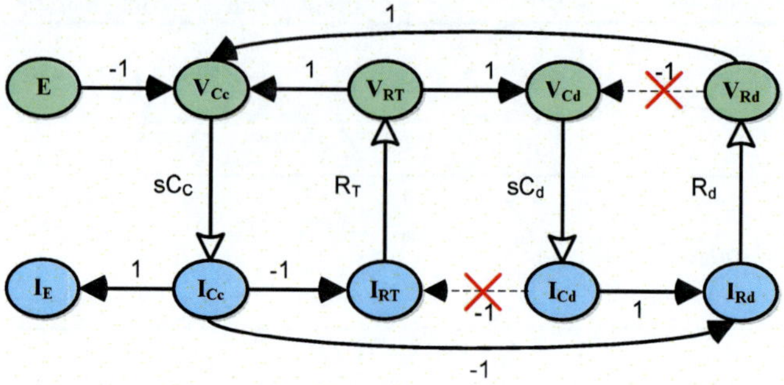

Fig. 14.10 The signal-flow graph of the simplified lossy serial *RC*-active filter

In this case, the equivalent impedance (Z_{Leq}) of the designed active inductor has the following expression:

$$Z_{Leq} = R_T + R_d + sR_T R_d C_d \tag{14.4}$$

Obviously, the Q factor of this circuit, which expression is given by (14.5), is smaller than the one of the first circuit, but the circuit offers a simpler configuration (see Figures 14.8 and 14.9).

$$Q = \omega R_d C_d \frac{R_T}{R_T + R_d} \tag{14.5}$$

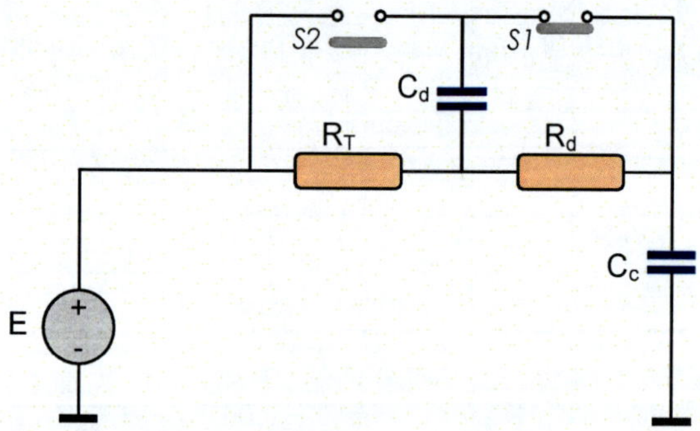

Fig. 14.11 The equivalent 'switched' (simplified) circuit. (The switches are in the position for the current graph)

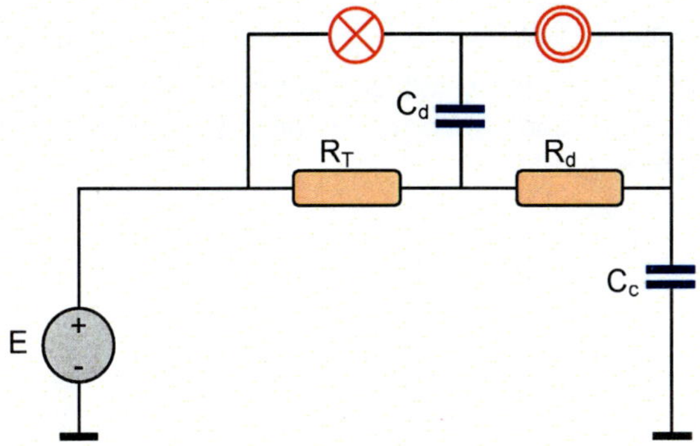

Fig. 14.12 The equivalent (simplified) circuit designed using a nullor

C- The parallel RLC-filter

In the same way, let's consider now the parallel *RLC*-filter shown in Figure 14.13.

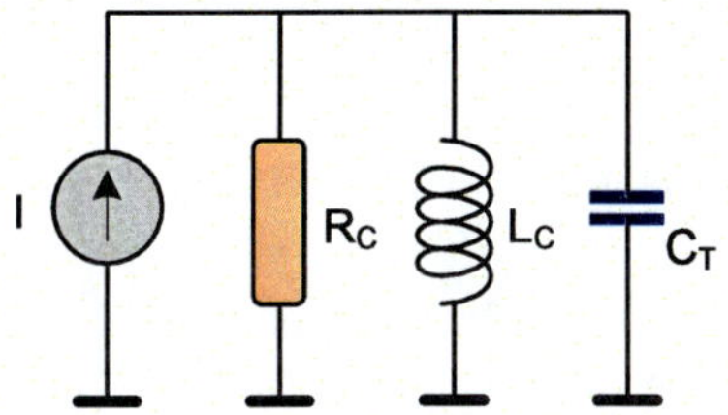

Fig. 14.13 A lossy parallel *LC*-filter

Table 14.5 $(B_T^V)_{RC}$ and $(T_C^I)_{RC}$ for the parallel *LC*-filters

	V_{CT}
V_I	-1
V_{Lc}	1
V_{Gc}	1

(a): $(B_T^V)_{LC}$

	I_{CT}
I	-1
I_{Lc}	1
I_{Gc}	1

(b): $(T_C^I)_{LC}$

The fundamental circuit matrix $(B_T^V)_{LC}$ and the T-matrix $(T_C^I)_{LC}$ associated with this *LC*-filter can be presented as follows (see Table 14.5), where the inductor and the resistor are considered in the cotree branches and the capacitor in the tree branch. ($G_C=1/R_C$).

The signal-flow graph associated with the matrices given in Table 14.5 is shown in Figure 14.14.

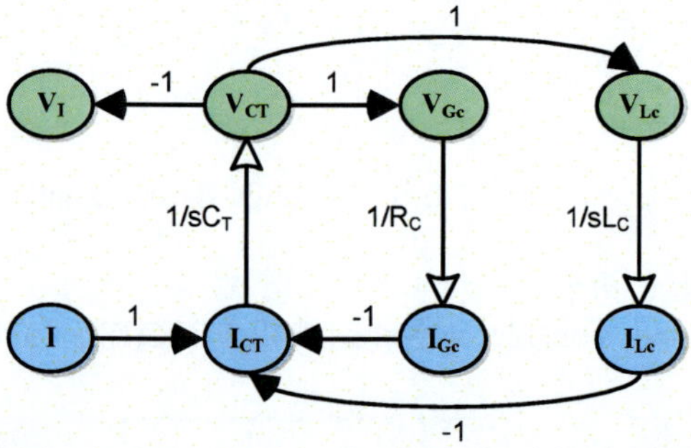

Fig. 14.14 The signal-flow graph associated to the lossy parallel *LC*-filter

It is possible to replace the inductor with parallel losses by a combination of passive elements, as shown in Table 14.6.

Table 14.6 $(B_T^V)_{RC}$ and $(T_C^I)_{RC}$ for the lossy *RC*-filter

	V_{CT}	V_{Cd}
V_I	-1	0
V_{Gc}	1	1
V_{Gd}	1	0

(a): $(B_T^V)_{RC}$

	I_{CT}	I_{Cd}
I	-1	0
I_{Gc}	1	0
I_{Gd}	0	-1

(b): $(T_C^I)_{RC}$

The signal-flow graph of the equivalent *RC*-circuit is shown in the Figure 14.15.

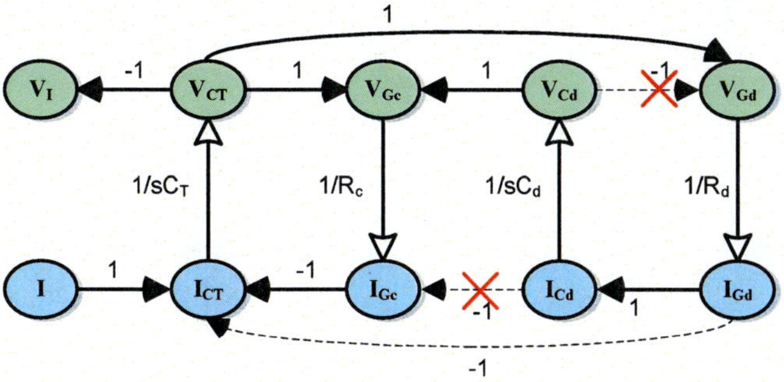

Fig. 14.15 The signal-flow graph of the lossy parallel *LC*-active filter

From the above signal-flow graph we can see that:

$$\frac{1}{Z_{eq}} = \frac{1}{R_C} + \frac{1}{sC_dR_CR_d} \tag{14.6}$$

Thus, the values of the parallel conductance (G_c) and the inductance ($R_cR_dC_d$) can be set orthogonally and the Q factor is equal to:

$$Q = \frac{1}{\omega R_d C_d} \tag{14.7}$$

Using the above proposed procedure, the 'active switches'-based circuit corresponding to the $(B_T^V)_{RC}$ and $(T_C^I)_{RC}$ matrices of Table can be constructed as shown in Figure 14.16.

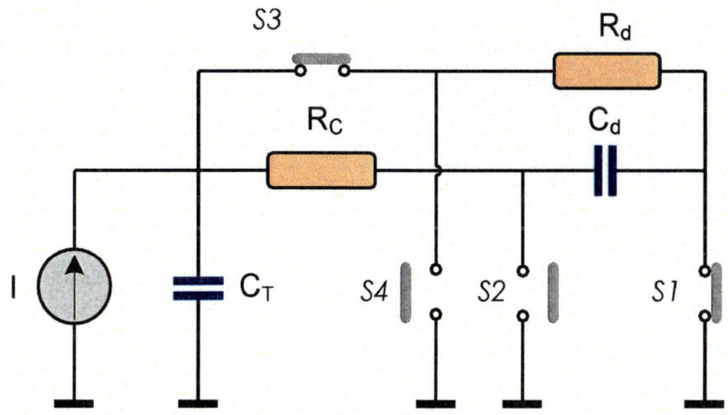

Fig. 14.16 The 'switched' active circuit associated with the signal-flow graph given in Figure 14.15 (The switches are in the position corresponding to the voltage graph)

In Figure 14.17 we give the nullator/norator based circuit that is equivalent to the circuit given in Figure 14.16.

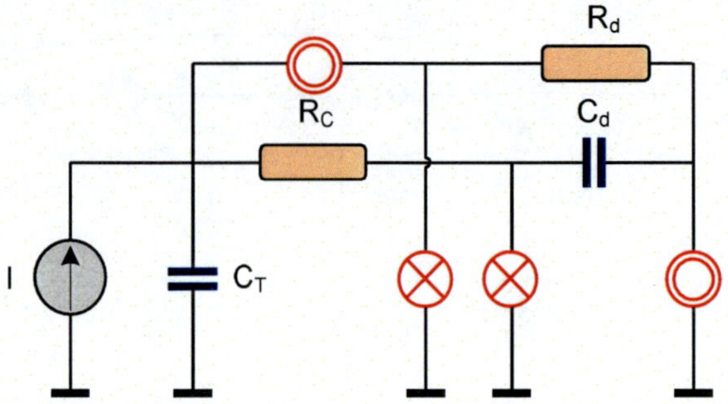

Fig. 14.17 The nullator/norator based equivalent circuit

D- The "simplified" parallel RLC-filter

Similarly, as for the serial *RC*-active circuit, the circuit shown above can be simplified, but at the cost of the *Q* factor. Table 14.7 shows a modification that will result in the elimination of one pair of switches.

Table 14.7 $(B_T^V)_{RC}$ and $(T_C^I)_{RC}$ matrices for the simplified version of the lossy parallel *RC*-filter

	V_{CT}	V_{Cd}
V_I	-1	0
V_{Gc}	1	1
V_{Gd}	1	0

(a): $(B_T^V)_{RC}$

	I_{CT}	I_{Cd}
I	-1	0
I_{Gc}	1	0
I_{Gd}	1	-1

(b): $(T_C^I)_{RC}$

The signal-flow graph associated with the matrices shown in Table 14.7 is given in Figure 14.18.

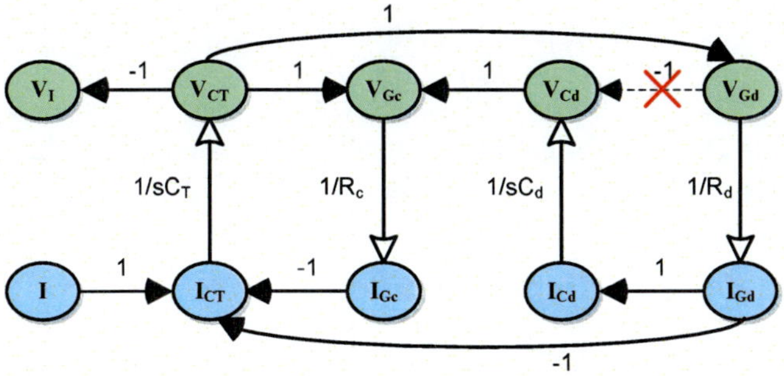

Fig. 14.18 The signal-flow graph of a simplified lossy parallel *RC*-active filter

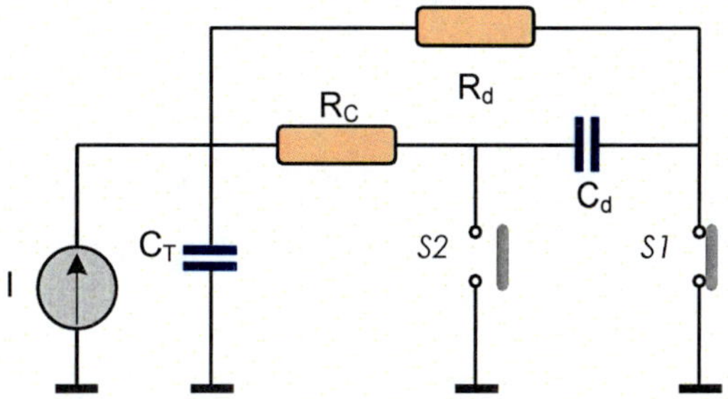

Fig. 14.19 The circuit of a lossy simplified parallel *RC*-active filter with switches. (The switches are in the position corresponding to the voltage graph).

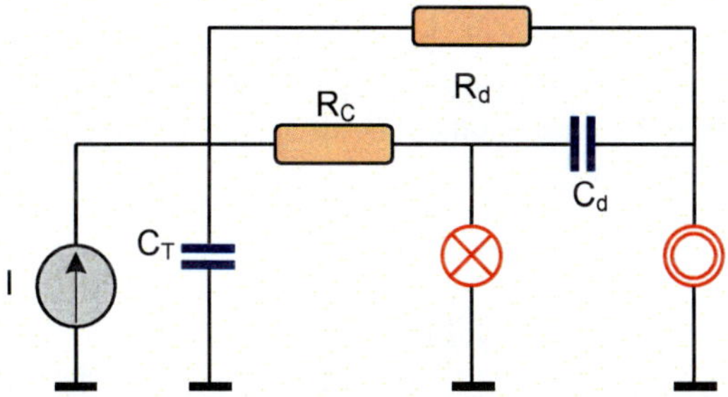

Fig. 14.20 The circuit of a lossy simplified parallel *RC*-active filter with only one nullator and one norator

It is easy to check that in this case we have:

$$\frac{1}{Z_{eq}} = \frac{1}{R_C} + \frac{1}{R_d} + \frac{1}{sC_d R_C R_d} \tag{14.8}$$

So, the values of the parallel conductance $(1/R_C+1/R_d)$ and the inductance $(R_C R_d C_d)$ can be set orthogonally, and the Q factor is equal to:

$$Q = \frac{1}{\omega(R_C + R_d)C_d} \tag{14.9}$$

Of course, the Q factor of this circuit is smaller when compared to the one given by (14.7), but the circuit has a simpler configuration (see Figures 14.19 and 14.20).

14.3 The Practical Implementation

The CMOS negative second generation current conveyor (CCII-) given in [14] can be used as a circuit that carries out the idea of the nullor, as illustrated in Figure 14.21. Hence, this CCII- was used in the following example to construct the equivalent active circuits. It is to be noted that the *AD 844 IC* can also be used for this purpose.

Fig. 14.21 The CCII– nullor equivalency

Example #1

Using the CCII- it is possible to directly realize the above proposed circuits. For example, the series lossy simplified circuit, given in Figure 14.12, will have the form shown in Figure 14.22.

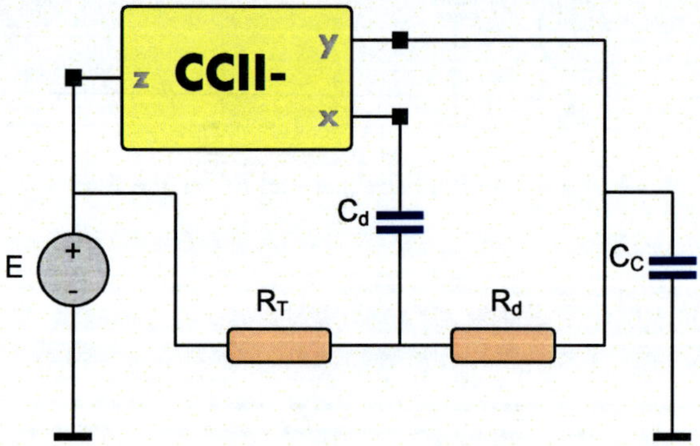

Fig. 14.22 The serial lossy simplified circuit from Figure 14.12

Using expressions (14.4) and (14.5), the proper angular frequency (ω_0) and the corresponding quality factor Q_{ω_0} can be expressed as follows:

$$\omega_0 = \frac{1}{\sqrt{R_T R_d C_d C_C}}$$
$$Q_{\omega_0} = \frac{R_d C_d}{\sqrt{R_d R_T C_d C_C}} \frac{R_T}{R_T + R_d} = \frac{1}{\sqrt{\frac{R_T}{R_d}\frac{C_C}{C_d}}(1+\frac{R_d}{R_T})} = \frac{\sqrt{ab}}{(1+a)} \tag{14.10}$$

where $a = R_d / R_T$, $b = C_d / C_C$.

Accordingly, the expression of ω_0 can be written as follows:

$$\omega_0 = \frac{1}{\sqrt{ab} R_T C_C} \tag{14.11}$$

It is to be noted that Q_{ω_0} depends on the ratio of the values of the circuit's elements and not simply on the elements values.

An examination of (14.10) and (14.11) shows that Q_{ω_0} and ω_0 can be set orthogonally. (Q_{ω_0} can be set via *a* and *b*, while ω_0 can be settled by the product $R_T C_C$).

It is easy to verify that if we would like to reach the maximum value of Q_{ω_0}, than we have to choose coefficient *a* equal to one, i.e. $R_d=R_T$, while coefficient *b* should be as large as possible.

In the following we give an application example and we show through SPICE simulation results, the viability of the proposed approach.

If we choose f_0=50kHz and $Q_{\omega_0} = 5$, than for a=1, $b=(1+a)^2 Q^2_{\omega_0}=100$. In addition, for R_T=5kΩ, we have R_d=5kΩ, $C_C = 1/\omega_0 R_T \sqrt{b} \approx 64 pF$, and $C_d \approx 6.4 nF$.

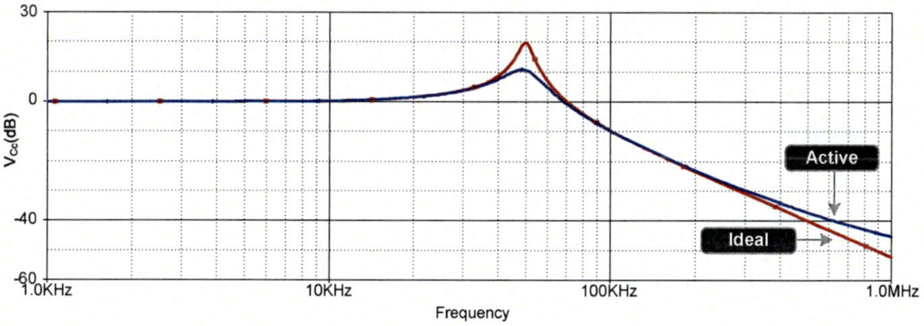

Fig. 14.23 SPICE simulation of the circuit of Figure 14.22

Using the MOS CCII- adopted in [19], SPICE simulations of the circuit given in Figure 14.19 were performed, using the aforementioned components' values. Figure 14.23 shows the corresponding simulations. The deviations in the characteristic from theoretical values are caused by the non-idealities of the CCII-, such as finite output and input impedances.

Example #2

Similarly, the parallel lossy circuit given in Figure 14.20 was realized using a CCII-. Figure 14.24 shows the corresponding simplified circuits.

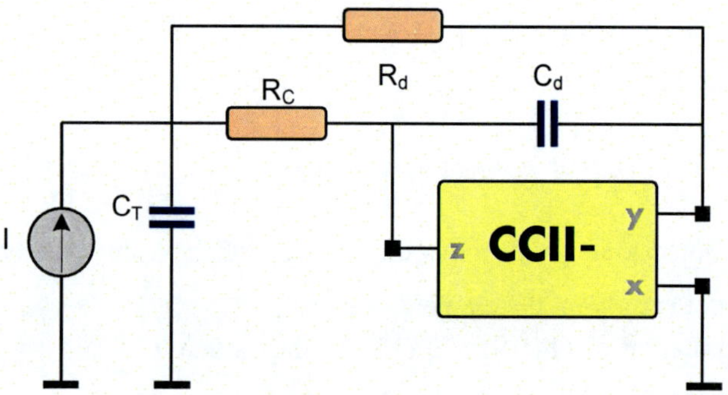

Fig. 14.24 The parallel lossy simplified circuit from Figure 14.20

The corresponding proper angular frequency ω_0 and quality factor $Q_{\omega 0}$ can be deduced from (14.6) and (14.7), thus we obtain:

$$\left|\begin{array}{l} \omega_0 = \dfrac{1}{\sqrt{R_T R_d C_d C_T}} \\ Q_{\omega_0} = \dfrac{\dfrac{R_d R_C}{R_d + R_C}}{\dfrac{1}{\sqrt{R_d R_C C_d C_C}} C_d R_C R_d} = \dfrac{\sqrt{a}}{\sqrt{b}(1+a)} \end{array}\right. \tag{14.12}$$

where $a=R_d/R_C$ and $b=C_d/C_T$.

Introducing, these last values into the expression of ω_0, we obtain:

$$\omega_0 = \frac{1}{\sqrt{ab R_C C_T}} \tag{14.13}$$

It is to be noted that Q_{ω_0} depends on the ratio of the values of the circuit's elements and not simply of the elements values.

An examination of (14.12) and (14.13) shows that Q_{ω_0} and ω_0 can be set orthogonally. Q_{ω_0} can be set by the values of a and b, while ω_0 can be adjusted by the product $R_T C_C$.

It is easy to check that if we would like to reach the maximum value of Q_{ω_0} than we have to choose the coefficient a equal to one, i.e. R_d=R_T while the coefficient b should be as large as possible.

If we choose f_0=50kHz and Q_{ω_0} =0.5, than for a=1, b=$(1/4Q^2_{\omega_0})$=1. In addition, for R_C=50kΩ, than R_d=50kΩ, $C_T = 1/\omega_0 R_C \sqrt{b} \approx 63.66pF$, and $C_d = bC_T = 63.66pF$.

Figure 14.25 gives the SPICE simulation results of the circuit shown in Figure 14.24 that is designed using the CMOS CCII- [16].

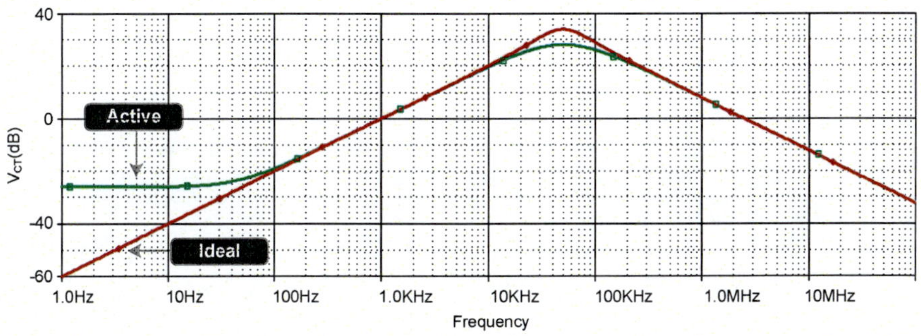

Fig. 14.25 SPICE simulation of the circuit of Figure 14.24

Example #3

As a more complicated application, let's consider the lossy ladder fourth order band-pass filter shown in Figure 14.26 [20]. Such resistively terminated *LC*-ladder networks have been known to posses very low sensitivities to their passive component variations [20]. As a mean of fact, considerable efforts have been spent to attempt to simulate these passive networks with active and passive *RC* elements.

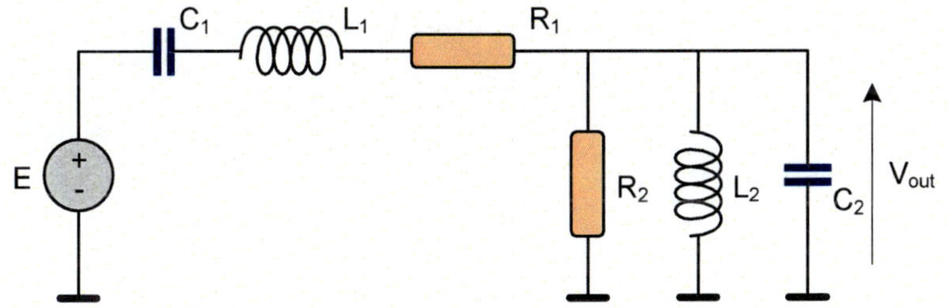

Fig. 14.26 A resistively terminated lossy forth order band-pass *LC*-filter

By using the circuits of the lossy serial and parallel *RC*-active filters, which were given in Figures 14.22 and 14.24, we can construct a circuit which is equivalent to the one shown in Figure 14.26. This circuit is given in Figure 14.27.

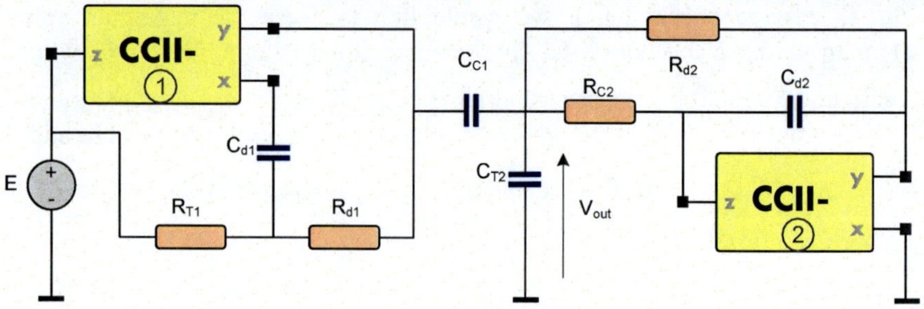

Fig. 14.27 The active resistively terminated lossy band-pass filter

For the SPICE simulations (given in Figure 14.28), the following values were considered:

a.– Circuit of Figure 14.26: f_0=50kHz, Q_1=1.5 and Q_2=5, C_1=C_2=64pF, L_1=L_2= 0.158H, R_1= 5kΩ, R_2=75kΩ.

b.– Circuit of Figure 14.27:

• The parallel resonant circuit: f_0=50kHz, Q_1=1.5, C_{C1}=C_a=64pF, and because a=1, then $b=1/4Q^2_{\omega 0}$=0.111 and $C_{d1}=b\,C_a$=7.104pF≈7pF, $R_{C2}=1/\sqrt{ab}\,\omega_0 C_T$≈150kΩ, R_{d2}=150kΩ.

• The series resonant circuit: f_0=50kHz, Q_2=5, C_{d1}=C_a=6.4nF. Since a=1, then $b=4Q_{\omega 0}^2$=100 and $C_{d2}=bC_a$=64pF, $R_{T1}=1/\sqrt{ab}\,\omega_0 C_C$≈5kΩ, R_{d1}=5kΩ.

The CMOS CCII- [19] was used to design the circuit of Figure 14.27. Figure 14.28 shows the corresponding SPICE simulations.

The deviations in the characteristic from theoretical values are caused by the non-idealities of the CC such as finite output and input impedances.

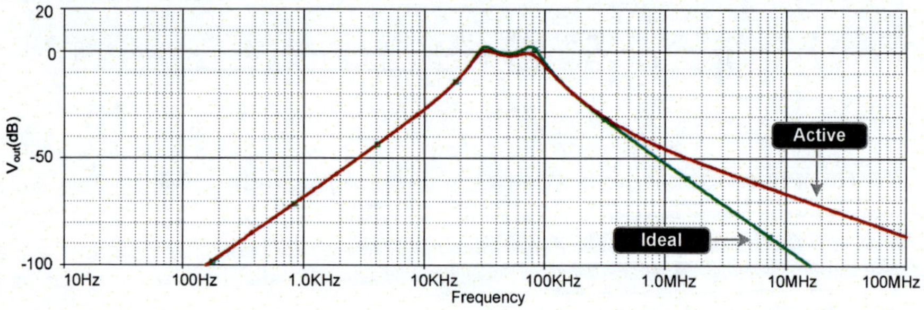

Fig. 14.28 SPICE simulation of the circuit of Figure 14.27 and comparison with the ideal results

14.4 Conclusion

A symbolic framework for designing settable series and parallel *R-L* imittances has been presented. The framework is based on the concept of signal-flow graph technique in the two-graph version. An important feature of this technique is that the circuits with different voltage and current graphs can be described. As a consequence, the method provides a systematic basis for designing setable active inductors, i.e. active inductors which inductance and resistance values can be set individually. It has been shown that a sufficient set of elements to construct such active inductors consists of a number of *RC* passive elements and a single type of active element, the universal active elements. the nullor. These circuits fall into two categories. In the first, a circuit is represented by a simple configuration, at the cost of the lower value of the *Q* factor. In the second, the circuit has more complicated configuration but the corresponding value of the *Q* factor is higher.

SPICE simulation results of a serial, a parallel and a ladder filter were presented to show the viability of the proposed approaches.

References

[1] Chang, C.M., Wang, H.Y., Chien, C.C.: Realization of series impedance functions using one CCII+. The International Journal of Electronics 76, 83–85 (1994)

[2] Cicekoglu, M.O.: Active simulation of grounded inductors with CCII+ and grounded passive elements. The International Journal of Electronics 85, 455–462 (1998)

[3] Wang, H.Y., Lee, C.T.: Realisation of R-L and C-D immittances using single CCIII. Electronics Letters 34, 502–503 (1998)

[4] Kuntman, H., Gulsoy, M., Cicekoglu, O.: Actively simulated grounded lossy inductors using third generation current conveyors. Microelectronics Journal 31, 245–250 (2000)

[5] Wang, H.Y., Lee, C.T.: Systematic synthesis of R-L and C-D immitances using single CCIII. The International Journal of Electronics 87, 293–301 (2000)

[6] Tangsrirat, W., Dumawipata, T., Unhavanich, S., Surakampontorn, W.: Simulation of electronically tunable lossless floating inductor using current-controlled differential current voltage conveyors. In: The International Symposium on Communications and Information Technologies, ISCIT, Sapporo, Japan (2004)

[7] Metin, B., Cicekoglu, O.: A Novel Floating Lossy Inductance Realization Topology with NICs Using Current Conveyors. The IEEE Transactions on Circuits and Systems—II: Express Briefs 53(6), 483–486 (2006)

[8] Kacar, F., Cam, U., Cicekoglu, O., Kuntman, H., Kuntman, A.: Novel OTRA-based grounded parallel immitance simulator topologies. Analog Integrated Circuits and Signal Processing 39, 169–175 (2004)

[9] Cam, U., Kacar, F., Cicekoglu, O., Kuntman, H., Kuntman, A.: Novel grounded parallel immitance simulator topologies employing single OTRA. AEU –International Journal of Electronics and Communications 57(4), 287–290 (2003)

[10] Senani, R.: A novel application of four terminal floating nullors. Proc. of IEEE 35(11), 1544–1546 (1987)

[11] Cam, U., Kacar, F., Cicekoglu, O., Kuntman, H.: Universal series and parallel immitance simulators using four terminal floating nullors. Analog Integrated Circuits and Signal Processing 25, 59–66 (2000)

[12] Kumar, P., Senani, R.: New grounded simulated inductance circuit using a single PFTFN. Analog Integrated Circuits and Signal Processing 62, 105–112 (2010)
[13] Haigh, D.G., Tan, F.Q., Papavassiliou, C.: Systematic synthesis of active-RC circuit building-blocks. Analog Integrated Circuits and Signal Processing 43, 297–315 (2005)
[14] Haigh, D.G., Clarke, T.J.W., Radmore, P.M.: Symbolic framework for linear active circuits based on port equivalence using limit variables. The IEEE Transactions on Circuits and Systems—I 53(9), 2011–2024 (2006)
[15] Saad, R.A., Soliman, A.M.: On the systematic synthesis of CCII-based floating simulators. International Journal on Circuit Theory and Applications 38, 935–967 (2010)
[16] Pierzchala, M., Fakhfakh, M.: Transformation of LC-filters to active RC-circuits via the two graph method. Microelectronics Journal 42(8), 999–1005 (2011)
[17] Pierzchala, M., Fakhfakh, M.: Novel Structures of RC-Active Filters for Tapped Capacitor Resonant Circuits. In: The European Conference on Circuit Theory and Design, ECCTD, Antalya, Turkey (2009)
[18] Pierzchala, M., Fakhfakh, M.: Generation of Active Inductor Circuits. In: The IEEE International Symposium on Circuits and Systems, ISCAS, Paris, France (2010)
[19] Fakhfakh, M., Pierzchala, M., Rodanski, B.: On the design of active inductors with current-controlled voltage sources. Analog Integrated Circuits and Signal Processing 73(1), 89–98 (2012)
[20] Laker, K.R., Ghausi, M.S., Kelly, J.J.: Minimum Sensivity Active (Leapfrog) Ladder Bandpass Filters. The IEEE Transactions on Circuits and Systems CAS-22(8), 670–677 (1975)

Chapter 15
MIDAS: Microwave Inductor Design Automation on Silicon

Luca Aluigi, Federico Alimenti, Luca Roselli, Domenico Pepe, and Domenico Zito

The design of modern radiofrequency integrated circuits on silicon operating at microwave and millimeter-waves requires the integration of several spiral inductors, transformers and transmission lines, which are commonly not available in the process design-kits of the technologies. Moreover, the design and simulation of such devices are not addressed adequately by means of systematic techniques and automation.

In this chapter we present the implementation of an auxiliary CAD tool for Microwave Inductor Design Automation on Silicon, MIDAS. MIDAS is based on VBScripts, exploits commercial simulators and allows us to implement an automatic design flow, including three-dimensional layout editing and electromagnetic simulations of microwave spiral inductors, transformers and transmission lines by means of a systematic approach. A beta version of MIDAS for spiral inductors and transmission lines is available under free Creative Commons license and can also be downloaded from the website: www.midas-project.org. In detail, this chapter reports through several examples characterized by different design constrains how MIDAS allows us to derive a preliminary sizing of the devices on the bases of the design entries (specifications), drawing the layers for the specific process design kit, including vias and underpasses, with or without patterned ground shield, according to a systematic approach for design and simulation, and then launching the electromagnetic simulations. The systematic approach implemented by MIDAS has been validated by means of experimental verifications of several case studies, showing an effective design automation of the microwave inductor and transmission line designs with respect to the traditional RFIC design flow. With the present software suite the complete design time is reduced significantly (typically 1 hour on a PC based on Intel® Pentium® Dual 1.80GHz CPU with 2-GB).

Luca Aluigi · Federico Alimenti · Luca Roselli
Univ. of Perugia, Department of Electronic and Information Engineering, Perugia, Italy

Domenico Pepe · Domenico Zito
Tyndall National Institute, "Lee Maltings", Dyke Parade, Cork, Ireland

Domenico Zito
Univ. College Cork, Dept. of Electrical and Electronic Engineering, Cork, Ireland
e-mail: domenico.zito@tyndall.ie

M. Fakhfakh et al. (Eds.): *Analog/RF & Mixed-Signal Circuit Sys. Design*, LNEE 233, pp. 337–361.
DOI: 10.1007/978-3-642-36329-0_15

This chapter is organized as follows. Section 15.1 reports a short introduction to the open challenges and proposed solution. Section 15.2 summarizes the state-of-the-art design flow adopted commonly by the RFIC design community and based on the most widespread commercial CAD tools including both EM and circuit simulators. Section 15.3 highlights the design automation operated by MIDAS and proposed herein, including script code descriptions and limitations. Section 15.4 reports the results achieved for some representative cases of study. Finally, the conclusions are drawn in Section 15.5.

15.1 Introduction

Thanks to the latest advances in terms of cut-off frequency of active devices in silicon microelectronic technologies, radiofrequency integrated circuit (RFIC) design is extending to millimeter-wave frequency range [1]. Exacerbated by high frequency losses into the silicon substrate, RFIC design is dramatically dependent on the quality factor (Q) of passive components (mostly inductors), and by their prediction accuracy [2]. In particular, the formulas developed for lower frequencies [3] are prone to large deviations from actual performance in the microwave and millimeter-wave frequency ranges. Consequently such formulas are insufficient in order to carry out accurate designs [4, 5]. For this reason, layout editing, electromagnetic (EM) simulations (e.g. three dimensional, 3D) and design fine-tuning are essential to predict accurately and improve the inductor performance. However these crucial design steps lead to extend considerably the time and efforts required for the entire design of circuits and systems.

EM design of inductors lacks of systematic techniques, leading to significant design challenges. The main limitations are due to the computing time and hardware resources required by EM simulation tools, which make the design fine tuning very time consuming. For instance, some tools [6, 7] provide equivalent circuit extracted from the layout, circumventing the need of EM simulations. Another time-consuming design step is the inductor layout drawing within the design and simulation environments. Typically the layout drawing lacks of automation, contributing to increase the design efforts and time. Some tool focused on specific purposes has been proposed in this regard, such as CYCLONE [8] aiming at the automatic design and layout of RF LC oscillators.

In response to the needs of systematic and automatic design approaches, in this chapter we propose an auxiliary software suite for Microwave Inductor Design Automation on Silicon, namely MIDAS, based on scripts and commercial electromagnetic (EM) simulators. In particular, the tool is based on VBScript (Visual Basic Scripting), a "lightweight" interpreted script language developed by Microsoft as a subset of Visual Basic programming language. VBScript can be executed in a wide variety of Microsoft host environments as well as in third party environments, allowing the embedding of the scripts in other programs, such as several commercial EM simulators. In particular, MIDAS allows the design automation of microwave spiral inductors, transformers and transmission lines within commercial CAD environments. A beta version of MIDAS for spiral inductors, with or without patterned ground shield (PGS), and transmission lines is available under free Creative Commons license and can also be downloaded [9]. In detail, on the basis of the design entry, i.e. specifications, such as inductance value, quality factor and operating frequency, MIDAS supports the following automatic design steps i) preliminary device sizing, ii) automatic generation of the device in the 3D EM simulator according to a systematic approach for design and simulation, iii) launching of the EM simulation and iv) layout drawing for GDS-II importing into the Cadence-Virtuoso design environment.

15.2 State of the Art for RFIC Design

In the last few decades, microelectronic design automation engineering provided several accurate automatic CAD tools for design analysis and synthesis both for front-end and back-end design phases, which include advanced techniques for accurate parasitic extractions, sensitivity analyses, routing, optimization, etc. [10-13]. Significant advances have been made also by electromagnetic design engineering, providing CAD tools capable of solving the typical EM problems of interest in IC design with adequate accuracy [14]. Microwave integrated circuits design is characterized by a limited number of active devices (with respect to mixed-signal and digital circuit design) and the EM design challenges are typically predominant.

As for the EM design, typically limited to passive microwave devices for silicon technologies, we can distinguish between two primary classes of simulation tools on the basis of the level of abstraction: schematic and full-wave simulators [15]. The former (i.e. schematic simulator) offers a limited number of predefined devices such as transmission lines, junctions, etc. Such devices are described by parametric models and therefore schematic simulators are very fast, but affected by the accuracy of model equations that often have a quite narrow range of validity. Outside the validity range, the results may experiment large inaccuracy, thus resulting unreliable. For this reasons, schematic models of passive devices, if available, can be used to determine an initial device sizing as the base for next EM design steps. The latter (i.e. full-wave simulator) allow us to overcome the limitations of schematic models, by solving numerically the Maxwell's equations. However this is carried out at the expenses of a high computational complexity, i.e. larger hardware resources and simulation time. Several techniques have been developed by exploiting numerical approximations and theoretical assumptions. The most widespread commercial full-wave simulators are based on Method of Moments (MoM) and Finite Element Method (FEM) (both operating in the frequency domain) and Finite-Difference Time Domain (FDTD) (operating in the time domain). Meshing in MoM is limited to conductors only, which are typically reduced to bidimensional domains (2D), i.e. surfaces, whereas meshing in FEM and FDTD is typically applied to the entire three-dimensional (3D) domain. In general, 2D EM simulations are characterized by shorter simulation time, lower computational efforts, but also lower accuracy with respect to 3D EM simulations. EM simulations, when conditioned properly, ensure accurate and reliable results in good agreement with measurements.

Figure 15.1 illustrates the most widespread design flow used for RFICs on silicon. This chapter is focused on the EM design phase highlighted in red, which will be addressed by means of the MIDAS tool.

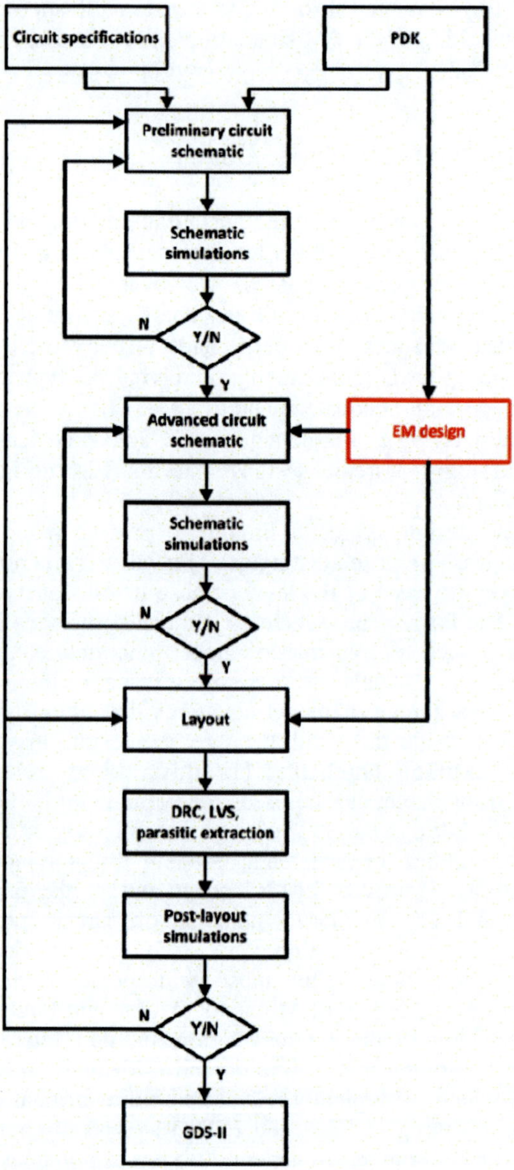

Fig. 15.1 Typical RFIC design flow. PDK stands for process design kit, DRC for design rule check and LVS for layout versus schematic. Note that the device resulted from EM design is excluded from parasitic extraction.

Assuming that the first-guess circuit design is satisfied by basic a resistance-inductance (RL) series model of inductors, the first EM design step will be the preliminary sizing of the inductor (see Fig. 15.2).

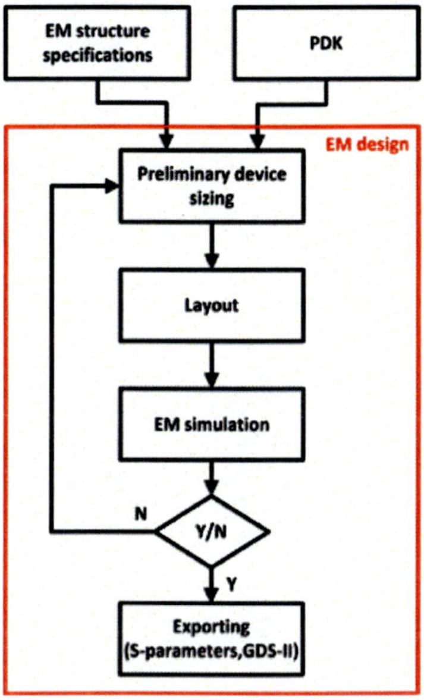

Fig. 15.2 Typical EM design flow on silicon

The preliminary inductor sizing may be based on approximate closed-form analytical equations, which can provide an initial design space exploration [16]. More accurate models based on π-model equivalent circuits could be used [17-25]. Such models proposed in the literature may be derived by numerical techniques (fitting procedures) or physical models. A deep discussion on such models is beyond the scopes of this paper; however, a summary of effective models is reported in Table 15.1. Moreover, it is worth saying that in spite of such models could be exploited to provide more accurate sizing of the preliminary inductor design, it is a common approach to skip such task and focus the design efforts directly on full-wave EM simulations.

Therefore, assuming that 3D EM simulator is the appropriate design tool for the next step, the designer must define therein the typical process cross section, draw the device geometry, set the boundary conditions, excitations and desired outputs, and finally launch the simulation. When the device meets the specifications, typically after a significant number of iterations and fine tuning of the geometry, the next design step is the layout drawing and S-parameters exporting for the subsequent importing into the IC design environment and the completion of the

RFIC design flow including parasitic extractions and post-layout simulations (see Fig. 15.1). It is worth mentioning that, in principle, the completion of the RFIC design flow could require additional fine tuning of the EM design, leading to additional refinement cycles (i.e. design, measurements and redesign). More importantly, note that these typical design phases involving separately EM and IC design environments lack of automation (i.e. integrated software suite) so that it is entirely supported by means of designers' efforts.

Table 15.1 Equivalent circuit models of integrated inductor on silicon

Ref	Skin and proximity effects	Capacitive coupling	Inductive coupling	Distributed effects	Frequency independent
[4]	yes	yes	no	no	yes
[17]	yes	yes	yes	yes	yes
[18]	yes	yes	yes	yes	yes
[19]	yes	yes	yes	no	no
[20]	yes	yes	no	no	yes
[21]	yes	yes	no	yes	yes
[22]	yes	yes	no	yes	no
[23]	yes	yes	yes	yes	yes
[24]	yes	yes	yes	yes	yes
[25]	yes	yes	yes	yes	yes

15.3 Proposed Methodology

The proposed tool for the Microwave Inductor Design Automation on Silicon (MIDAS) allows the designers to generate automatically the layout of the inductor in the EM design environment, set and launch simulations. Moreover, MIDAS provides also the support for the initial design of the inductor by evaluating geometric options on the basis of the performance requirements in terms of L and Q, and provides also the approximate equivalent π circuit of the inductor.

More in detail MIDAS is a software suite consisting of four tools described hereinafter, each independent and designed in order to support a specific design step. The first version of the tool takes into account two possible geometries for inductors: i) octagonal symmetric and ii) square. The tools *Geometric Calculator* and *LQR Calculator* allow the user to determine the geometric values, which suit with the design constraints. *EM Structure Simulator* is useful to generate the layout of the inductor. *Equivalent Circuit Extractor* extracts the equivalent π model, based on the Y parameters computed by the full-wave simulator.

15.3.1 Geometric Calculator

Geometric Calculator offers to the users the possibility to identify the geometric values of the preliminary device, which may potentially provide the desired inductance. This is currently limited to octagonal and square spirals. The prediction formulas used do not account for the technology process (the inductors are supposed to be in air, i.e. $\varepsilon_r \approx 1$). Considering the losses in the silicon substrate, it is evident that these prediction formulas provide only low-frequency approximations for very preliminary device sizing.

The output data set of this tool is an Excel spreadsheet. Therefore Microsoft Excel must be available in the computer in which MIDAS is running. The core of *Geometric Calculator* is a series of conditional statements *if ... then* performing nested sweeps of the four characteristic geometric parameters of spirals, such as number of turns (N), width (w), spacing (s) and outer diameter (d_{out}), as reported in Table 15.2. The output data are the set of such parameters (N, w, s *and* d_{out}), which may provide the desired inductance specified by the user (i.e. design entry) calculated by *Geometric Calculator* according to equation (1). For sake of clarity, the start, stop and step values in Table 15.2 can be easily modified in future versions.

Table 15.2 Sweep parameters in Geometric Calculator

Parameter	Start	Stop	Step
Number of turns (N)	1	3	0.25
Width (w)	1 μm	20 μm	1 μm
Spacing (s)	1 μm	10 μm	1 μm
Outer diameter (d_{out})	30 μm	200 μm	1 μm

On the basis of the settings in Table 15.2, *Geometric Calculator* can cover the range of inductances from about *0.1* to *5 nH* according to the following approximated formula [26]:

$$L_s = \frac{\mu_0}{2} c_1 N^2 d_{avg} f(\rho) \tag{15.1}$$

where

$$f(\rho) = \ln\left(\frac{c_2}{\rho}\right) + c_3\rho + c_4\rho^2 \tag{15.2}$$

$$\rho = \frac{d_{out} - d_{in}}{d_{out} + d_{in}} \tag{15.3}$$

$$d_{in} = d_{out} - 2Nw - 2(N-1)s \tag{15.4}$$

$$d_{avg} = 0.5(d_{in} + d_{out}) \tag{15.5}$$

The coefficients c_1, c_2, c_3 and c_4 given in Table 15.3 allow the formula to be adapted to both octagonal and square spiral inductors.

Table 15.3 Values of the coefficients

Layout	c_1	c_2	c_3	c_4
Square	1.27	2.07	0.18	0.13
Octagonal	1.07	2.29	0	0.19

For instance, we ran *Geometric Calculator* for a required inductance value of *1.2 nH*. The output spreadsheet reports *239* different geometries for octagonal inductor and *764* for square inductor. The running time is about 15 seconds on a pc based on Intel® Pentium® Dual 1.80 GHz CPU with 2-GB RAM. One of the solutions for square spiral is given by $N=2.25$, $s=5\ \mu m$, $w=10\ \mu m$ and $d_{out}=154\ \mu m$. Assuming such a geometry for the hypothetical preliminary design, the outer diameter could be rounded to $150\ \mu m$. As it will be shown in the following sections, the design solution above has been implemented on silicon and characterized experimentally (see Section 15.4.1).

15.3.2 LQR Calculator

The *LQR Calculator* allows us to achieve a preliminary and rough evaluation of the value of inductance, series resistance and quality factor of an integrated inductor on silicon.

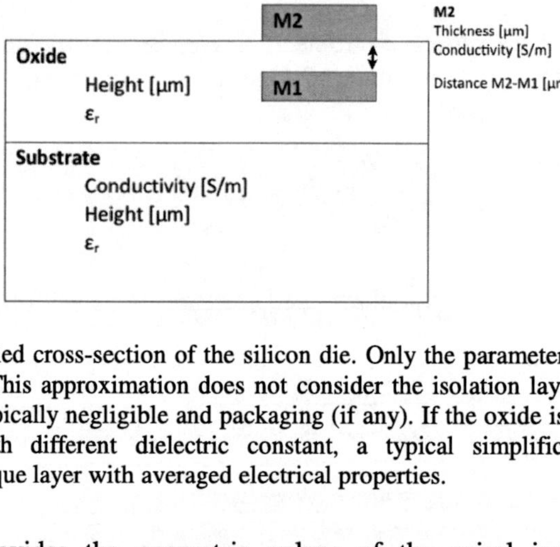

Fig. 15.3 Simplified cross-section of the silicon die. Only the parameters used by the tool are highlighted. This approximation does not consider the isolation layer on top since its contribution is typically negligible and packaging (if any). If the oxide is made of different oxide layers with different dielectric constant, a typical simplification consists of considering a unique layer with averaged electrical properties.

The user provides the geometric values of the spiral inductor evaluated previously, the type of shape (octagonal or square), the frequency of operation and some process-related parameters. In particular, *LQR Calculator* requires the insertion of the geometric and electrical properties of the metal layers used for the implementation of spirals, the oxide layers and substrate. The notation used for

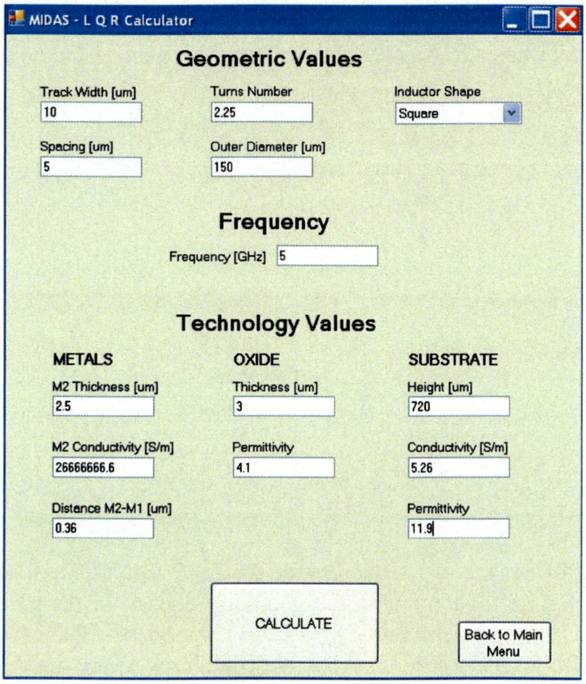

Fig. 15.4 Input window of the *LQR Calculator*

the geometry of the process cross section is reported in Fig. 15.3, whereas the input window is shown in Fig. 15.4.

The output data are the results of the calculations derived for equations (15.1), and the following (15.6) and (15.11) [26-28]:

$$R_s = \sqrt{R^2{}_{dc} + R^2{}_{ac}} \tag{15.6}$$

where

$$R_{dc} = \frac{l}{w t_{TM2} \sigma_{TM2}} \tag{15.7}$$

$$R_{ac} = 1.2 \frac{l}{2\sigma_{TM2}\delta\left(w + t_{TM2}\right)} \tag{15.8}$$

where t_{M2} and σ_{M2} are the thickness and conductivity of the top metal layer, δ is the skin depth and l is the length of the spiral given by:

$$l = N_{SIDE} d_{avg} N \tan\left(\frac{\pi}{N_{SIDE}}\right) \tag{15.9}$$

$$\delta = \frac{1}{\sqrt{\sigma_{TM2}\mu_{TM2}\pi f}} \tag{15.10}$$

where N_{SIDE} is *4* for square inductors and *8* for octagonal inductors. Using the relations (15.1) and (15.6), we can evaluate the quality factor by the following equations [28]:

$$Q = \frac{\omega L_s}{R_s} \frac{R_p}{R_p + \left[\left(\frac{\omega L_s}{R_s}\right)^2 + 1\right] R_s} \left(1 - \frac{R_s^2 C_p}{L_s} - \omega^2 L_s C_p\right) \tag{15.11}$$

where

$$R_p = \frac{1}{\omega^2 C_{ox}^2 R_{Sub}} + \frac{R_{Sub}(C_{ox} + C_{Sub})^2}{C_{ox}^2} \tag{15.12}$$

$$C_p = C_{ox} \frac{1 + \omega^2 (C_{ox} + C_{Sub}) C_{Sub} R_{Sub}^2}{1 + \omega^2 (C_{ox} + C_{Sub}) R_{Sub}^2} \tag{15.13}$$

Figure 15.4 shows the input window of the *LQR Calculator* tool. The particular case of the square inductor with the geometry chosen in the previous subsection ($N=2.25$, $s=5\mu m$, $w=10\mu m$ and $d_{out}=150\mu m$) is reported. The inductor is designed in *0.35μm* BiCMOS technology by Austriamicrosystems (AMS) for operating frequencies in the range *5-6 GHz*. The simulation time is approximately one second on a PC based on Intel® Pentium® Dual 1.80GHz CPU with 2-GB RAM. The *LQR* provides the following preliminary values: $L=1.08nH$, $Q=8.63$ and $R=3.91\Omega$.

15.3.3 EM Structure Simulator

The *EM Structure Simulator* tool requires that the user specifies the shape, i.e. square or octagonal, of the spiral inductor and if it is with or without patterned ground shield (PGS). In other terms, there are four possible combinations (square with and without PGS, octagonal with and without PGS). For each of those selections, MIDAS launches a VBScript routine which guides the designer through the setup and execution of the EM simulation by means of High Frequency Structure Simulator (HFSS) by ANSYS. HFSS is a 3D full-wave EM simulator offering different simulation domains. In particular, we will consider HFSS based on FEM and operating in frequency domain [29]. HFSS generates appropriate meshing (typically derived to be consistent and accurate at the highest frequency of the sweep interval, i.e. stop frequency) and solves numerically the EM problem for each single point of the frequency sweep. It is worth mentioning that the script can easily been adapted to other full-wave simulators commercially available.

Therefore, *EM Structure Simulator* requires specifying the number of dielectric layers of the technology process. The electrical and geometrical parameters (conductivity, permittivity and thickness) can be added one by one (starting from the bottom, i.e. from the substrate). The next step consists of specifying geometrical and electrical parameters (conductivity, thickness and position on the vertical axis) of the metal layers for spirals and underpasses (typically by using the topmost metal layers) and the metal layer used for implementing the ground ring. Then the user has to provide the characteristic geometric values of the spiral inductor.

It is worth mentioning that a one to one correspondence between process geometry and structure described within the EM simulator is not possible in practical cases, since it would lead to smaller unit cell size and then a dramatic increase of the number of mesh cells, leading to a huge computational complexity, typically unfeasible on common workstations. Therefore, a reduction (simplification) of the geometrical details is required for the description within the EM structure simulator. MIDAS takes into account the most common work hypotheses adopted by the scientific community and supported by our common good practice [30] as highlighted in the following sections.

Once all these data are given, the automatic layout generation can start. This is the point in which MIDAS provides a significant contribution to the overall design automation. In fact, the designer could spend a considerable amount of time (e.g. several hours) in drawing manually the complete structure and setting up the simulation environment. In MIDAS this operation is automatic and takes only few minutes to be completed. This is the most significant innovation of the design approach introduced by MIDAS. The tool draws the entire cross section of the layer stack and adds an additional layer of air on top having a thickness (t_a) proportional to the frequency operation (typically $t_a=\lambda_0/10$, where λ_0 is the wavelength at the central frequency f_0). The box horizontal sizes (x and y) are proportional to the outer diameter of the spiral drawn ($x_{side} = y_{side} = 2 \cdot d_{out}$). These general settings are highlighted in Fig. 15.5.

Surrounding the inductor there is a ground ring made out of the lowest metal level available in the technology, which acts as the path for the return current in the structure. The ring is also connected to the Si substrate using a number of vias (contacts). The ground ring has a square shape with characteristic size $d_{ring}=d_{out}+2s$ in case of octagonal spirals, whereas $d_{ring}=d_{out}+4w+2s$ in case of square spirals. The PGS is made of a number of strips (typically polysilicon) proportional to the dimensions of the ground ring: $N_{strips}=(d_{ring} / 2 \cdot s)-3$. The width of each strip and the spacing between them is equal to s. It is worth saying that typically the PGS is realized by using much finer geometries (strip width is maybe be even 1 μm or smaller), but the proposed geometrical simplifications represent a good tradeoff between description accuracy and computational complexity (i.e. simulation time).

The spirals are made of the topmost metal layer, where the underpasses are made of the underneath metal layer. The bottom metal layer is used for the realization of the ground ring, both for spirals with and without PGS. Contacts[1] are added from the ground ring to the substrate or from the ground ring to the PGS

[1] In the current systematic settings, the geometry of contacts has been simplified in order to reduce the mesh requirements and an average resistivity has been considered in order to provide the same overall resistance of the real case.

and from the PSG to the substrate (see Fig. 15.5(c)), which is assumed to be the first technology layer entered by the user, i.e. *Layer0*. The other layers are then numbered progressively. Note that there is not an upper limit to the number of layers to be drawn by MIDAS. The user enters this number.

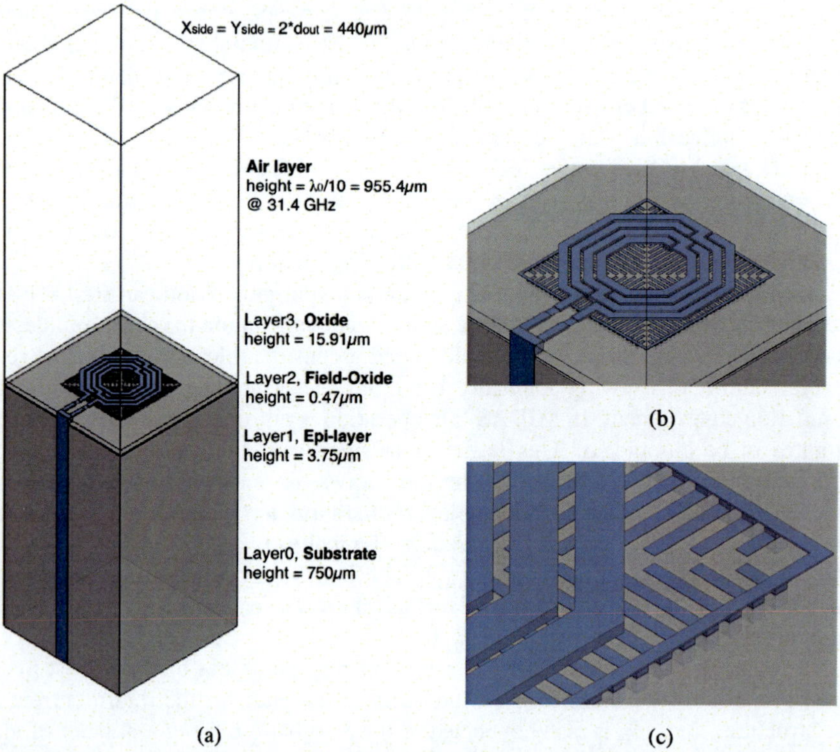

Fig. 15.5 General EM settings in MIDAS. Example of an octagonal spiral inductor with PGS in a commercial 0.25μm BiCMOS process. The inductor has the following characteristic parameters: N=4, s=4 μm, w=12 μm and d_{out}=220 μm. As can be seen, the vertical (axis z) dimension of a typical (Bi)CMOS process is dominated by the height of the substrate, which is the bottom layer (tagged as Layer0 by MIDAS).

Once the steps above have been completed, the designer must set up the excitation ports before the EM simulation has place. HFSS allows the use of two kinds of ports, lumped port and wave port. In case of lumped port the excitation is applied to a specific point of the device as a voltage or current, as reported in Fig. 15.6(a). In case of wave port instead, the excitation is a quasi-TEM wave supported by microstrip line. This is applied to a proper lateral area of the simulation box, which includes the metal layer and the reference ground plane, as illustrated in Fig. 15.6(b).

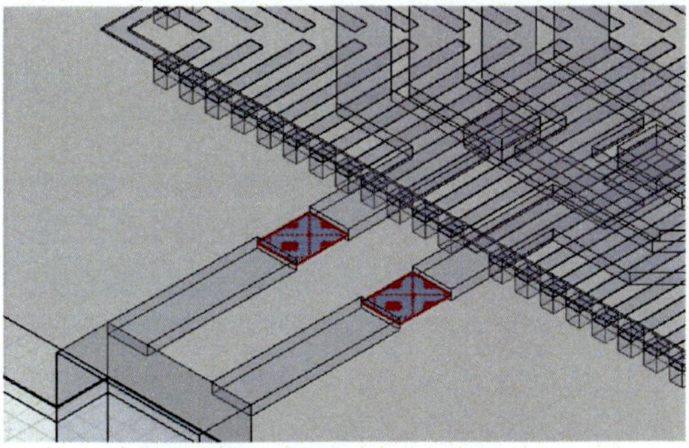

(a)

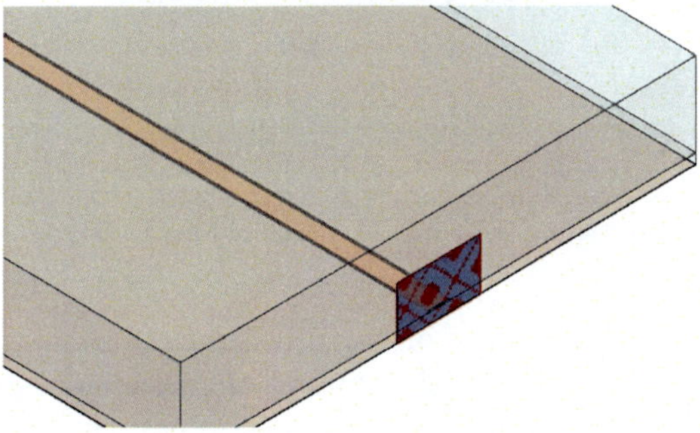

(b)

Fig. 15.6 (a) Lumped ports; (b) wave port

The voltage is a scalar quantity whereas the wave is a vector. Therefore, there are substantial differences between lumped and wave excitations. Wave ports are typically preferred in case of uniform regions, whereas lumped ports are typically preferred in case of non-uniform geometry or material discontinuities, as well as in our case.

Before launching the simulation, the tool requires to specify the central frequency f_0, and the start, stop and step values in order to identify the frequency sweep interval. Once the simulation has been completed, the tool presents the resulting graphs for L and Q according to the following expressions [17]:

$$L = \frac{\mathrm{Im}\left\{ 1/Y_{11} \right\}}{\omega} \tag{15.14}$$

$$Q = \frac{\mathrm{Im}\left\{ 1/Y_{11} \right\}}{\mathrm{Re}\left\{ 1/Y_{11} \right\}} \tag{15.15}$$

It is worth mentioning that the user could also generate the GDS-II file to be imported into the Cadence-Virtuoso layout editor with error-free DRC by means of a layer-mapping file. In the current version MIDAS is able to generate error-free DRC layout only for a specific technology, as reported in [31].

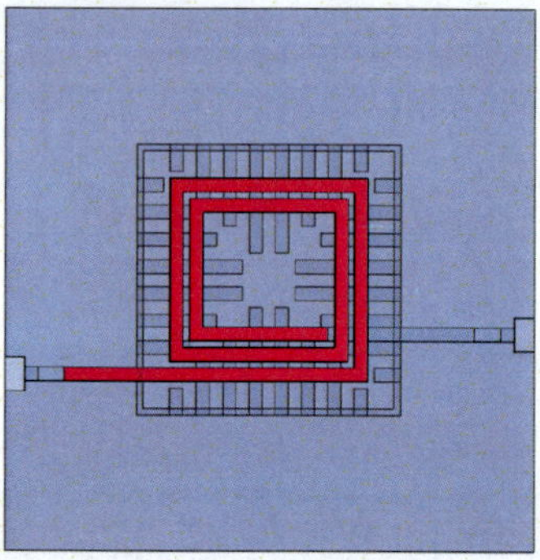

Fig. 15.7 HFSS layout resulting from the EM Structure Simulation tool

Figure 15.7 shows the output of the EM Structure Simulator tool resulting directly from the layout in the HFSS environment. Therein we generated a square inductor with PGS by using the geometric values chosen in the previous subsections ($N=2.25$, $s=5\ \mu m$, $w=10\ \mu m$ and $d_{out}=150\ \mu m$) for the implementation in a 0.35 μm BiCMOS process by Austriamicrosystems (AMS). As result, the simulation provided the values of $L=1.2\ nH$ and $Q=8.4$ at $5\ GHz$. The simulation time was about two minutes (excluding the time required for the HFSS simulation) on a pc based on Intel® Pentium® Dual 1.80 GHz CPU with 2-GB RAM.

15.3.4 *Equivalent Circuit Extractor*

Once the full-wave simulation has been performed, the designer can automatically extract the π circuit model of the inductor (see Fig. 15.8(b)) by means of the Equivalent Circuit Extractor tool. This tool provides the outputs on the basis of the results of the EM simulation results in terms of two-port Y parameters. It is required that Microsoft Excel is installed on the computer where MIDAS is running, since the output of this tool is an Excel spreadsheet containing the extracted circuit parameters. The extraction procedure performed by MIDAS makes use of the method described in [4].

Figure 15.9 shows the input window of the tool for two different examples of inductors (square in Fig. 15.9(a) and octagonal symmetric in Fig. 15.9(b)). Figure 15.9(a) is related to the case of study already introduced in the previous

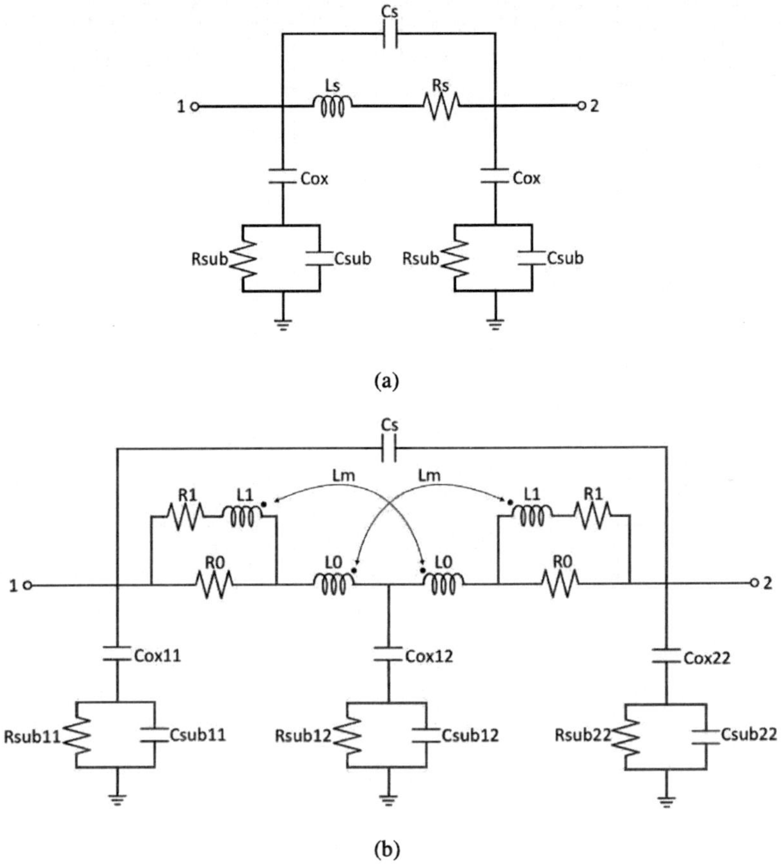

Fig. 15.8 Typical π models of spiral inductors: (a) asymmetric spiral, (b) symmetric spiral

subsections, regarding the EM simulation of a square inductor (d_{out}=*150 μm*, *w*=*10 μm*, *s*=*5 μm*, *N*=*2.25*) in a *0.35 μm* BiCMOS technology. The result was an inductance of *1.2 nH*. Equivalent Circuit Extractor provides a π equivalent circuit model. The comparison in terms of L and Q between the results of π model and EM simulation results is reported in Fig. 15.10.

Figure 15.9(b) instead is related to a case of study presented in [31], regarding the EM simulation of an octagonal symmetric inductor (d_{out}=*48 μm*, *w*=*2 μm*, *s*=*2 μm*, *N*=*2*) in a *0.25 μm* BiCMOS technology. The result was an inductance of *0.27 nH*. The comparison in terms of L and Q between the π model and the EM simulation results is reported in Fig. 15.11. Note that the model is accurate not only at the central frequency of *31.4 GHz*, but over the entire K_a band.

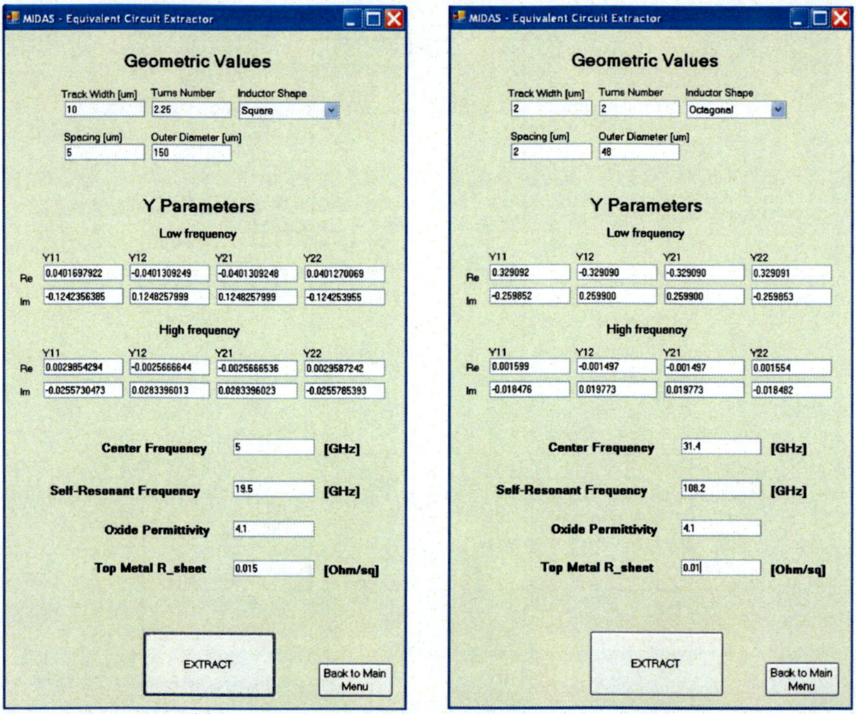

(a) (b)

Fig. 15.9 Examples of input masks of the Equivalent Circuit Extractor tool. (a) square inductor at 5 GHz, (b) octagonal symmetric inductor at 31.4 GHz.

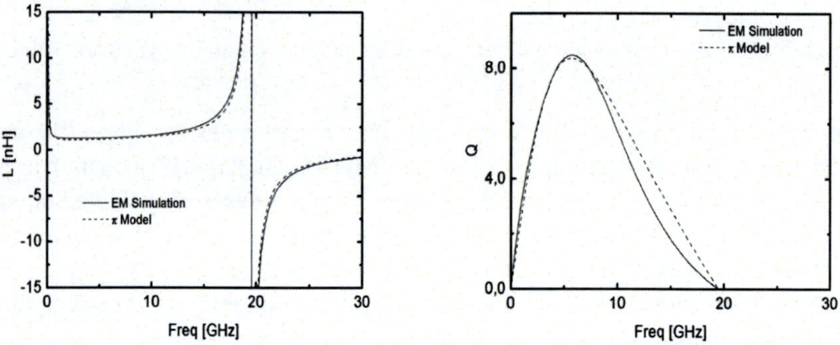

(a) (b)

Fig. 15.10 (a) L and (b) Q provided by MIDAS for the square inductor (dout=150 μm, w=10 μm, s=5 μm, N=2.25) in a 0.35 μm BiCMOS technology. Comparison between the π model and EM simulations.

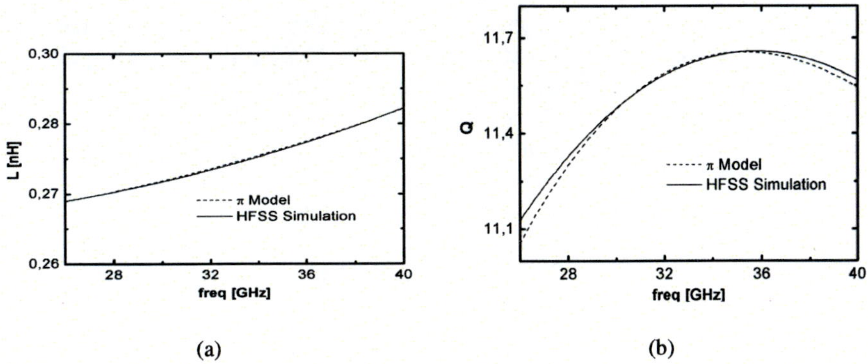

Fig. 15.11 (a) L and (b) Q provided by MIDAS for the inductor presented in [31]. Comparison between the π model and EM simulation results.

15.4 Experimental Results

In order to validate MIDAS, we report in this section the results of three relevant cases of study for different process design kits. For each case of study, the simulation results achieved by following the EM design flow implemented by MIDAS will be compared with the experimental results on test-chips.

15.4.1 Case of Study 1: Square Spiral Inductor in 0.35μm BiCMOS Process by AMS

As a first case of study, MIDAS has been applied to the design of square spiral inductors in *0.35 μm* SiGe CMOS technology by AMS. The die micrograph is reported in Fig. 15.12.

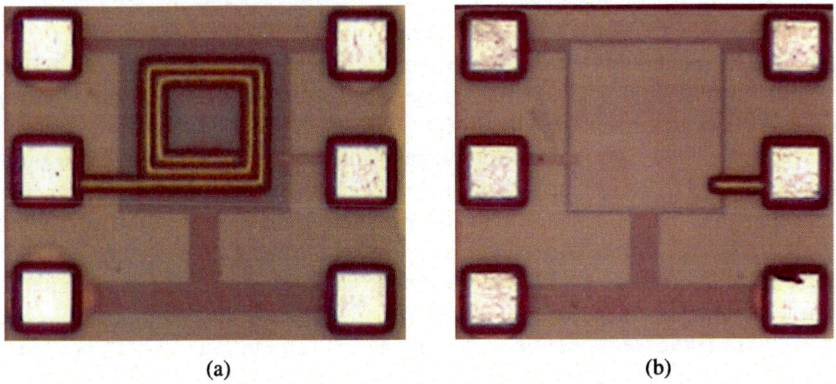

Fig. 15.12 (a) Die micrograph of the square spiral inductors. (b) Test-structure for de-embedding.

The characteristic inductor sizes are: $N=2.25$, $s=5\ \mu m$, $w=10\ \mu m$ and $d_{out}=150\ \mu m$. The inductor has been designed over a polysilicon PGS. The implementation on silicon includes the test structure (see Fig. 15.12(b)), such as pads and additional contact metals from pad to inductor, which are not included in the EM design flow within MIDAS. These additional test structures do not represent any limitation since the measurement results of the inductor used into the RFIC design flow will be derived by using the pad open de-embedding technique [32].

Figures 15.13 show the measurement and simulation results for *L* and *Q*. Note that the measurements and simulations provide a very close result for *L*. As for the *Q*, the peak is very close ($Q=7.8$ measured and $Q=8.5$ simulated), whereas the peak frequency exhibits a deviation.

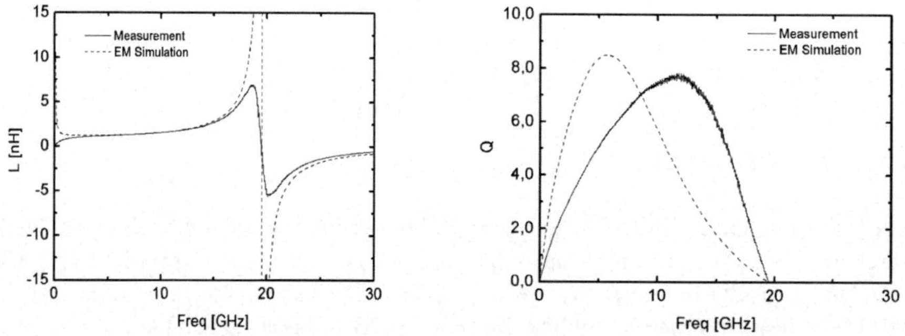

(a) (b)

Fig. 15.13 (a) L and (b) Q for the square inductor (measurement and the simulation results)

15.4.2 Case of Study 2: Square Spiral Inductor in 65nm CMOS Process by STM

As a second case of study we report the experimental results obtained by M. Kraemer et al. in [5]. The measured inductor (about 0.12nH) is realized in a *65 nm* bulk CMOS technology by STMicroelectronics (STM). The sizes are: $N=1.25$, $s=2\ \mu m$, $w=3\ \mu m$ and $d_{out}=31\ \mu m$. The parasitic effects of pads and any other part of the structure not included in the EM problem description given by MIDAS have been de-embedded from the measurement results. Figure 15.14 shows the output of the EM Structure Simulator tool resulting directly from the layout in the HFSS environment. In this particular case the structure generated by MIDAS has been slightly modified to match the shape of the input terminal of the inductor with the geometry of the inductor proposed therein [5]. *L* and *Q* achieved by simulations and measurements are reported in Fig. 15.15. The square spiral inductor has been simulated in HFSS adopting the grounding structure as in Fig. 15.6(a), hereinafter refereed as *GND#2*. Moreover, in order to evaluate the effectiveness of such setting of the EM simulator, we report also the EM simulation results achieved by using a different grounding structure, namely *GND#1*, such as reported therein [5]. *GND#1* is composed by a metallic ground plate made of the bottom metal layer available in the technology process. The ground plate has *80 μm* square hole centred below the inductor. *GND#2* consists of a vertical PEC column connecting each port to the bottom (automatically set by HFSS as a PEC) of the substrate box.

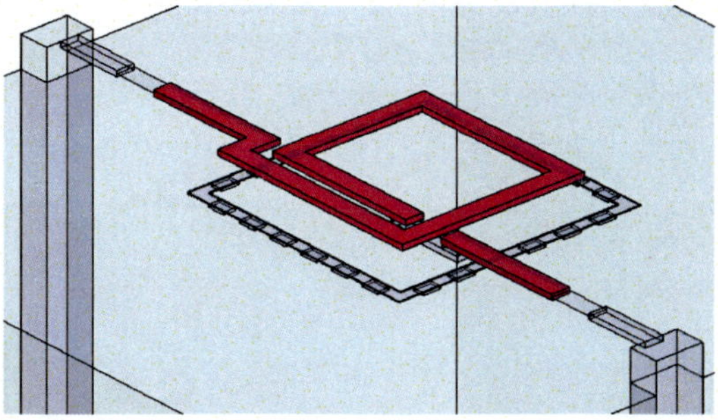

Fig. 15.14 Square spiral inductor (without PGS) simulated in HFSS within MIDAS design flow (i.e. GND#1). The sizes are: N=1.25, s=2 μm, w=3 μm and d_{out}=31 μm.

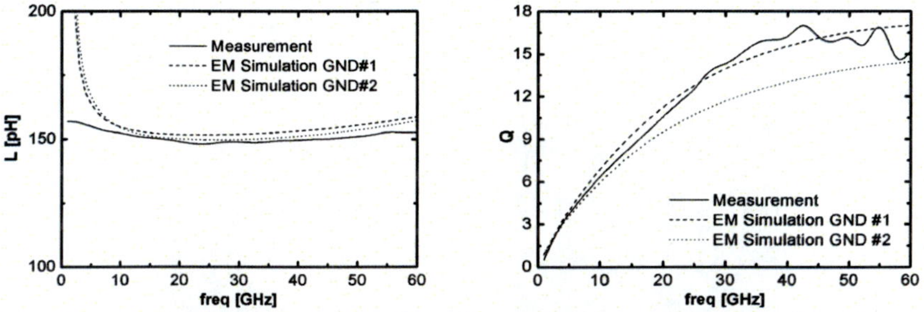

(a) (b)

Fig. 15.15 (a) L and (b) Q from measurements and EM simulations for the inductor in [5]

Both the simulated structures present a good agreement with the measurements. The simulation results for *GND#1*, result in a better matching with respect to the measured *Q*. This is in agreement with the fact that *GND#1* is very close to the grounding structure really implemented in the test chip, and that an accurate description of the grounding geometry (i.e. of the ground current return paths) is very important in these simulations.

15.4.3 Case of Study 3: Microstrip Line in 90nm CMOS Process by STM

As a third case of study, the EM Structure Simulator tool setup has been applied to simulate a silicon microstrip line fabricated on the *90 nm* bulk CMOS technology by STM. The screen shots of the EM structure in HFSS environment are shown in Fig. 15.16. Figure 15.17(a) shows the screen shoot in Cadence design environment, whereas Fig. 15.17(b) reports the picture of the device realized on silicon.

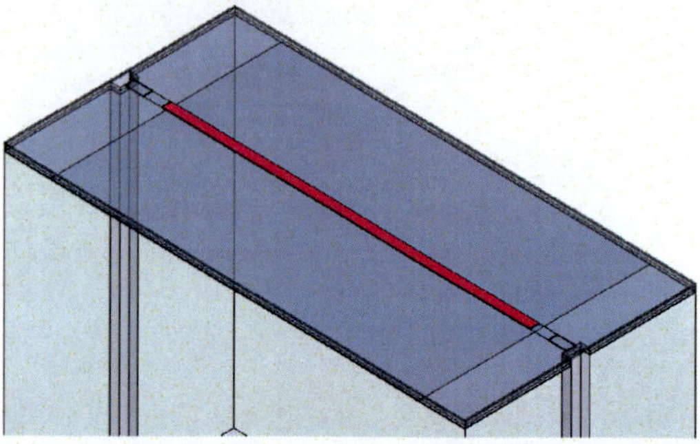

Fig. 15.16 Microstrip line on silicon simulated in HFSS within MIDAS design flow. For sake of clarity, due to metal density rules in the technology process, the ground plane is made of a very fine grid of metal strips so that the uniform ground plane in the EM Structure Simulator takes into account averaged electrical properties.

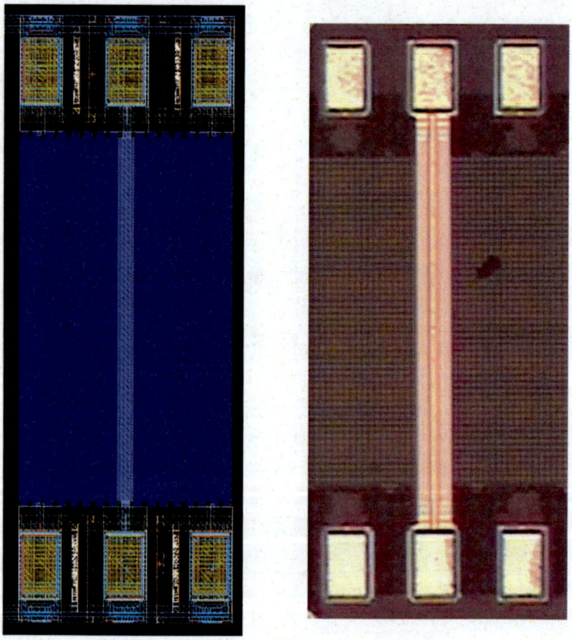

(a) (b)

Fig. 15.17 Microstrip line on silicon: (a) layout in Cadence environment and (b) die photograph

Note that a microstrip line on silicon substrate can be considered as a particular case of square spiral inductor with *N=0.25*. The sizes of the structure are summarized in Table 15.4. The microstrip line is made of the top metal layer and the ground plane is made of the bottom metal layer.

Table 15.4 Geometric values of the microstrip line

	Lenght [μm]	Width
Line	400	10
Ground plane	400	250

Figure 15.18 shows a comparison between the measurement and the simulation results of the *S* parameters. Note the good matching between the simulations and measurements.

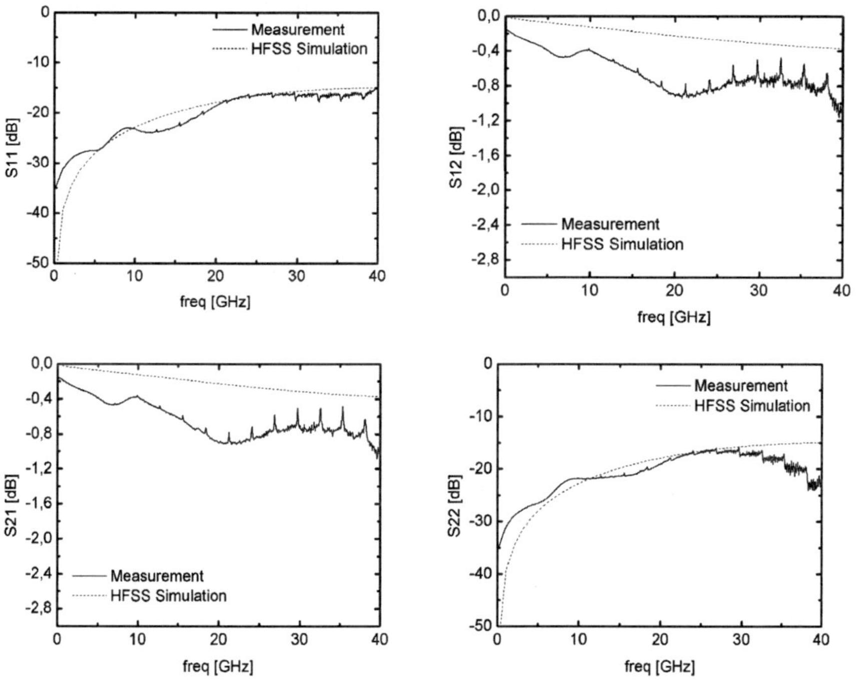

Fig. 15.18 S parameters of the microstrip line. Comparison between the simulation and measurement results.

15.5 Conclusion

Full-wave electromagnetic simulations provide a high prediction accuracy for the design of modern radiofrequency integrated circuits on silicon. Despite the electromagnetic simulation allows us to implement cost-effective design flows with respect to prototyping cycles including verification on test-chip, it still remains quite time consuming due to the computational complexity and the lack of automation.

This chapter has dealt with the lack of systematic design and simulation approaches, and automation. A methodology to automate the design of microwave inductor on silicon has been proposed by introducing an auxiliary CAD tool, namely MIDAS, which reduces drastically the design time of inductor on silicon. Based on full-wave EM simulations, MIDAS allows us to speed up the design phase by introducing automatic steps in the electromagnetic design flow.MIDAS is based on four different tools, which assist the designer from the design entry to the *GDS-II* exporting by means of automatic design procedures, contributing significantly to speed up the entire design flow of radiofrequency integrated circuits on silicon in the microwave and millimeter-waves frequency range. The tool setting have been presented, discussed and applied to different cases of study. The effectiveness of the approach proposed within MIDAS has been confirmed by the experimental results on test-chips.

A beta version of the tool is available to the radiofrequency integrated circuits design community. Possible extensions of the *EM Design Simulator* regard the addition of new inductor shapes. *Equivalent Circuit Extractor* could be enhanced by introducing additional automatic plots of the results obtained by the EM simulation.

Finally, it is worth mentioning the overall extension to the automatic design of transformers.

MIDAS is a unique innovative contribution to the microwave inductor design automation on silicon, which is particularly helpful for the entire scientific and industrial community, as well as in research and education.

Appendix A

Consider the π model shown in Fig. 15.8a, the following parameters (not reported in the previous sections) can be derived as follows.

A.1 Series Capacitance

C_S can be derived by using the following expression [33]:

$$C_s = Nw^2 \frac{\varepsilon_{ox}}{t_{TM2-TM1}} \tag{15.16}$$

where ε_{ox} is the oxide permittivity and $t_{TM2\text{-}TM1}$ is the oxide thickness between the two top metal layers.

A.2 Oxide Capacitance

The parasitic capacitance between the spiral metal and the silicon substrate is estimated by the following formula [33]:

$$C_{ox} = \frac{1}{2} lw \frac{\varepsilon_{ox}}{t_{ox}} \tag{15.17}$$

where t_{ox} is the thickness of the oxide between the inductor and the substrate.

A.3 Substrate Resistance and Capacitance

The substrate resistance and capacitance are obtained by [33]

$$R_{Sub} = 2 \frac{h_{Si}}{lw\sigma_{Si}} \tag{15.18}$$

$$C_{Sub} = \frac{1}{2} lw \frac{\varepsilon_0 \varepsilon_{rSi}}{h_{Si}} \tag{15.19}$$

where σ_{Si}, h_{Si} and ε_{rSi} are the substrate conductivity, height and dielectric constant respectively.

References

[1] Zito, D.: On exploiting new circuit topologies for the next-generation wireless transceivers at the μ- and mm-waves. In: IEEE Int. Conf. NEWCAS-TAISA, pp. 380–383 (2008)

[2] Alimenti, F., Stopponi, G., et al.: Numerical FDTD Modeling of Silicon Integrated Spiral Inductors. Microw. J. 46(8), 68–87 (2003)

[3] Niknejad, A.M., Meyer, R.G.: Analysis, design, and optimization of spiral inductors and transformers for Si RF ICs. IEEE J. of Solid-State Circuits 33(10), 1470–1481 (1998)

[4] Dickson, T.O., La Croix, M.A.: 30-100GHz inductors and transformers for millimeter-wave (Bi)CMOS integrated circuits. IEEE Trans. on Microw. Theory and Tech. 53(1) (2005)

[5] Kraemer, M., Dragomirescu, D., Plana, L.: Accurate electromagnetic simulation and measurement of millimeter-wave inductors in bulk CMOS technology. Top. Meet. on Silicon Monolithic Integr. Circuits in RF Syst. (SiRF) (2010)

[6] `http://www.helic.com/velocerf` (accessed July 19, 201)

[7] Koutsoyannopoulos, Y.K., Papananos, Y.: Systematic analysis and modeling of integrated inductors and transformers in RF IC design. IEEE Trans. on Circuits and Systems II: Analog and Digital Signal Processing 47(8), s:699–s:713 (2000)

[8] De Ranter, C.R.C., Van der Plas, G., Steyaert, M.S.J., Gielen, G.G.E., Sansen, W.M.C.: CYCLONE: automated design and layout of RF LC-oscillators. IEEE Trans. on Computer-Aided Design of Integrated Circuits and Systems 21(10), 1161–1170 (2002)

[9] `http://www.midas-project.org` (accessed July 17, 2012)

[10] http://www.cadence.com (accessed July 17, 2012)
[11] http://bwrc.eecs.berkeley.edu/classes/icbook/spice (accessed July 17, 2012)
[12] http://www.mentor.com (accessed July 17, 2012)
[13] http://www.synopsys.com (accessed July 17, 2012)
[14] Cheng, Q.S., Bandler, J.W., et al.: The State of the Art of Microwave CAD: EM-Based Optimization and Modeling. Int. J. of RF and Microw. Computer-Aided Eng. 20(5) (2010)
[15] Radonic, V., Crnojevic-Bengin, V., Zivanov, L.: Comparison of Commercially Available Full-Wave EM Simulation Tools for Microwave Passive Devices. In: Int. Conf. on Computer as a Tool (EUROCON), vol. 2, pp. 1699–1702 (2005)
[16] Nieuwoudt, M., McCorquodale, M.S., et al.: Accurate Analytical Spiral Inductor Modeling Techniques for Efficient Design Space Exploration. IEEE Electron. Device Lett. 27(12), 998–1001 (2006)
[17] Zito, D., Pepe, D., Neri, B.: Wide-Band Frequency-Independent Equivalent Circuit Model for Integrated Spiral Inductors in (Bi)CMOS Technology. In: IEEE Int. Conf. on Electronics Circuits and Syst. (ICECS), pp. 478–481 (2006)
[18] Ahn, Y.G., Kim, S.K., et al.: Efficient Scalable Modeling of Double-Equivalent Circuit for On-Chip Spiral Inductors. IEEE Trans. on Microw. Theory and Tech. 57(10), 2289–2300 (2009)
[19] Talwalkar, N.A., Yue, C.P., Wong, S.S.: Analysis and synthesis of on-chip spiral inductors. IEEE Trans. on Electron. Devices 52(2), 176–182 (2005)
[20] Rotella, F., Bhattacharya, B.K., Blaschke, V.: A broadband lumped element analytic model incorporating skin effect and substrate loss for inductors and inductor like components for silicon technology performance assessment and RFIC design. IEEE Trans. on Electron. Devices 52(7), 1429–1441 (2005)
[21] Gao, W., Yu, Z.: Scalable compact circuit model and synthesis for RF CMOS spiral inductors. IEEE Trans. on Microwave Theory and Techniques 54(3), 1055–1064 (2006)
[22] Chen, J., Liou, J.J.: Improved and physics-based model for symmetrical spiral inductors. IEEE Trans. on Microw. Theory and Tech. 53(6), 1300–1309 (2006)
[23] Lai, I.C.H., Fujishima, M.: A new on-chip substrate-coupled inductor model implemented with scalable expressions. IEEE J. of Solid-State Circuits 41(11), 2491–2499 (2006)
[24] Huo, X., Chan, P.C.H., et al.: A physical model for on-chip spiral inductors with accurate substrate modeling. IEEE Trans. on Electron. Devices 53(12), 176–182 (2006)
[25] Wang, C., Liao, H., et al.: A wideband predictive double-pi equivalent-circuit model for on-chip spiral inductors. IEEE Trans. on Electron. Devices 56(4) (2009)
[26] Mohan, S.S., Hershenson, M.M., et al.: Simple accurate expressions for planar spiral inductances. IEEE J. of Solid-State Circuits 34(10), 1419–1424 (1999)
[27] Choi, Y.S., Yoon, J.B.: Experimental analysis of the effect of metal thickness on the quality factor in integrated spiral inductors for RF ICs. IEEE Electron. Device Lett. 25(29), 76–79 (2004)
[28] Yue, C.P., Wong, S.S.: Design strategy of on-chip inductors for highly integrated RF systems. In: Proc. of Design Automation Conf., pp. 982–987 (1999)
[29] http://www.ansoft.com/products/hf/hfss/ (accessed July 17, 2012)
[30] Mounet, C., Siligaris, A.: Employing 65nm CMOS for WPAN Applications Partner Companies (2010), http://www.ansoft.com/firstpass/pdf/Employing_65nm_CMOS_for_WPAN_Applications.pdf (accessed July 17, 2012)

[31] Aluigi, L., Alimenti, F., Roselli, L.: Automatic Design and 3D Electromagnetic Simulation of Sub-nH Spiral Inductors. In: Proc. of Prog. in Electromagn. Res. Symp. (PIERS), pp. 1719–1722 (2011)

[32] Koolen, M.C.A.M., Geelen, J.A.M., et al.: An improved de-embedding technique for on-wafer high-frequency characterization. IEEE Proc. of Bipolar Circuits and Technol. Meet., 188–191 (1991)

[33] Voinigescu, S.P.: HF Integrated Circuit. Short Course at the Univ. of Pisa, Pisa, Italy (2008)

Chapter 16
LC-VCO Design Challenges in the Nano-Era

Pedro Pereira, Helena Fino, Mourad Fakhfakh, Fernando Coito, and Mário Ventim-Neves

Abstract. The progressive scaling of CMOS technology towards nanometre sizes has made the implementation of highly integrated systems for the wireless communication systems possible. Additionally, higher speed, lower power consumption and area reduction has been reached. Due to the high-density integration needs, as well as to low cost fabrication, RF applications, such as the *LC*-voltage controlled oscillator (*LC-VCO*), are usually implemented in CMOS technology. The complexity of designing *LC-VCO*s has lead to the development of several design methodologies. This chapter introduces an optimization based methodology for the design of *LC-VCO*s, where its efficiency is granted by the use of analytical models to characterize the active and passive elements' behaviour.

16.1 Introduction

CMOS technology has been responsible for the rapid growth of communication systems. The demands for new services or functionalities have motivated designers competitiveness in order to provide equipments with ever higher performance and lower cost. In fact, and assuming the prediction in related bibliography, during the last two decades, the goal for this market has been to reduce both the power consumption and price of mobile phones by 30% every year [1].

Due to the high-density integration needs as well as to low cost fabrication, fully integrated *LC-VCO*s applied to *RF* applications, such as clock generation, frequency synthesizers or timing-recovery circuits are usually implemented in CMOS technology. Yet, to fulfil the market demands, very stringent design specifications in terms of phase-noise, power consumption or area, among others, must

---

Pedro Pereira · Helena Fino · Fernando Coito · Mário Ventim-Neves
Faculdade de Ciências e Tecnologia, FCT, Universidade Nova de Lisboa, Portugal
e-mail: hfino@fct.unl.pt

Mourad Fakhfakh
University of Sfax, Tunisia

M. Fakhfakh et al. (Eds.): *Analog/RF & Mixed-Signal Circuit Sys. Design*, LNEE 233, pp. 363–379.
DOI: 10.1007/978-3-642-36329-0_16

be attained, leading to the necessity of finding design solutions where technology is pushed towards its limits. Furthermore, as device sizes approach the nano scale, new phenomena affecting the robustness of the solutions must be accounted for. Nonetheless, the technology scaling has lead to a decrease in the supply voltage, thus making the analogue design more challenging, since neither a wide range of linearity nor full output voltage swing are easily guaranteed [2].

To cope with the complexity of the design, optimization based methodologies must be adopted. For the efficiency of the design process, analytical models for both passive and active devices must be considered. The characterization of the integrated inductor has already been addressed in previous publications [3]-[4]. As far as the active devices are concerned, the accurate process-dependent compact transistor model EKV, is considered [5].

In this chapter an optimization based methodology for the design of nano-CMOS *LC-VCO*s, represented in Figure 16.1, is proposed. In this methodology an hierarchical approach is adopted where the optimization of the *LC* tank is performed and then results are integrated into the overall *VCO* optimization. From the optimization point of view, this methodology poses several challenges.

The *LC-VCO* design methodology proposed in this chapter has two major advantages. In one hand the use of accurate compact device models, makes the determination of the design parameters very rapid. On the other hand as the model elements are exclusively based on technological parameters, the adaptability of the design process to new technologies is extremely easy. Yet, additional challenges from the fact that the optimization algorithm must be capable of dealing with both continuous variables and discrete variables such as the inductor's geometrical parameters and the varactor's number of fingers.

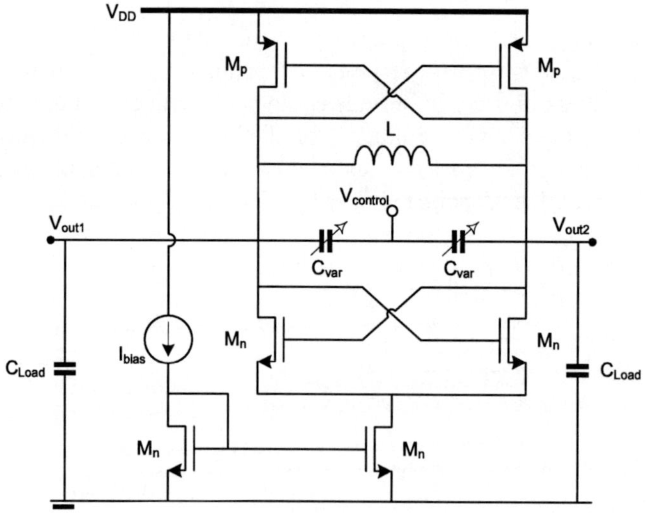

Fig. 16.1 Cross coupled *LC-VCO* topology

For the dimensioning of *LC-VCOs*, the accuracy of the solution relies on the accuracy of models of each element. Even though designers may use very accurate models, for predicting the *VCO* performance, there will always be a certain error due to some parasitic effects that appear in the manufacturing process [6]. For that reason a *VCO* is always tuneable, and varactors must be used as in Figure 16.1. As the core of the design methodology relies on the use of compact models for the *VCO* main blocks, special attention is paid to the models adopted.

In Section 2 this chapter presents a description of the EKV CMOS transistor model, as well as a brief depiction of the parameters extraction procedure. In Section 3 the CMOS varactor analytical model is introduced. The adopted inductor model and its design challenges are offered in section 4. Section 5 and 6 are dedicated to present the proposed *LC-VCO* design approach and optimization working examples, respectively. Finally, conclusions are offered.

16.2 EKV Transistor Model

The EKV model has been developed to facilitate the compact modelling and simulation of low voltage devices for application in low power semiconductor technologies. The main advantage of using the EKV model in the characterization of the MOS transistor behaviour relies on the fact that a single accurate expression, valid from weak to strong inversion and from linear to saturation region, is used [7]. This characteristic makes the EKV model suitable for analytical design and simulation of analogue circuits, allowing a deep insight into the device behaviour. The set of equations that makes the EKV model are listed below [8]-[10]:

$$\begin{aligned} V_g &= V_{Gate} - V_{Bulk} \\ V_s &= V_{Source} - V_{Bulk} \\ V_d &= V_{Drain} - V_{Bulk} \end{aligned} \tag{16.1}$$

where V_g, V_s, V_d are the transistor gate, source and drain voltage referred to bulk.

$$I_{ds} = I_s\left(i_f - i_r\right) \tag{16.2}$$

with I_s, i_f, i_r as the specific, forward and reverse current respectively, and given by

$$i_{f(r)} = \left[\ln\left(1 + \exp\left[\frac{V_P - V_{s(d)}}{2U_T}\right]\right)\right]^2 \tag{16.3}$$

$$I_s = 2nU_T^2 \frac{\beta_0}{1 + \Theta V_P'} \tag{16.4}$$

where U_T is the thermal voltage, β_0 is a transconductance parameter, Θ is the mobility reduction coefficient, V_P the pinch-off voltage and n is the slope factor.

$$V_P' = \frac{1}{2}\left(V_P + \sqrt{V_P^2 + 2U_T^2}\right) \tag{16.5}$$

$$n = 1 + \frac{\gamma}{2\sqrt{V_P + \phi + 4U_T}} \tag{16.6}$$

$$V_P = V_g' - \phi - \gamma'\left(\sqrt{V_g' + \left(\frac{\gamma'}{2}\right)^2} - \frac{\gamma'}{2}\right) \tag{16.7}$$

where V_g' is the effective gate voltage, γ is the body effect parameter, γ' is the correct body effect parameter taking into account the device geometry, and ϕ the bulk Fermi potential.

$$V_g' = V_g - V_{to} + \phi + \gamma\sqrt{\phi} \tag{16.8}$$

$$\gamma' = \gamma - \frac{\varepsilon_o\,\varepsilon_{si}}{C_{ox}}\left[\frac{LETA}{L_{eff}}\sqrt{\phi + V_d} + \left(\frac{LETA}{L_{eff}} - \frac{3\,WETA}{W_{eff}}\right)\sqrt{\phi + V_s}\right] \tag{16.9}$$

where V_{to} is the threshold voltage for $V_{SB} = 0$ V, ε_o is the permittivity of free space, ε_{si} is the permittivity of silicon, C_{ox} is the gate-oxide capacitance per unit area, L_{eff} and W_{eff} are the transistor effective length and width respectively, and finally, *LETA* and *WETA* are the short channel and narrow width effect coefficients.

The next section will focus on the EKV DC parameters extraction process, giving a clear idea on how to perform it.

- **Parameters Extraction**

Although, the analysis and results presented in this section consider an NMOS transistor, a similar methodology is perfectly valid for PMOS transistors.

The very first step on extracting the EKV DC model parameters, is to determine the specific current, I_s, as this information is crucial to correctly determine all regions of operation, as well as to obtain the pinch-off voltage characteristic, $V_P = f(V_g)$ [11]. Considering a MOS transistor operating in strong inversion, the reverse current, I_r, can be neglected and the current I_{ds} may be given by:

$$I_{ds} = \frac{I_S}{4U_T^2}(V_P - V_s)^2 \tag{16.10}$$

thus,

$$\frac{\partial\sqrt{I_{ds}}}{\partial V_s} = -\frac{\sqrt{I_S}}{2U_T} = slope \tag{16.11}$$

which means that the specific current can be obtained from the slope of $\sqrt{I_{ds}}(V_s)$. In Figure 16.2(a) a typical test circuit for simulating the transistor behaviour regarding I_{ds}, is represented. The current I_{ds} as a function of the source voltage, for different gate voltage values, is plotted in Figure 16.2(b). It is clearly identifiable that a similar slope is obtained for all the curves represented in Figure 16.2(b), validating (16.11).

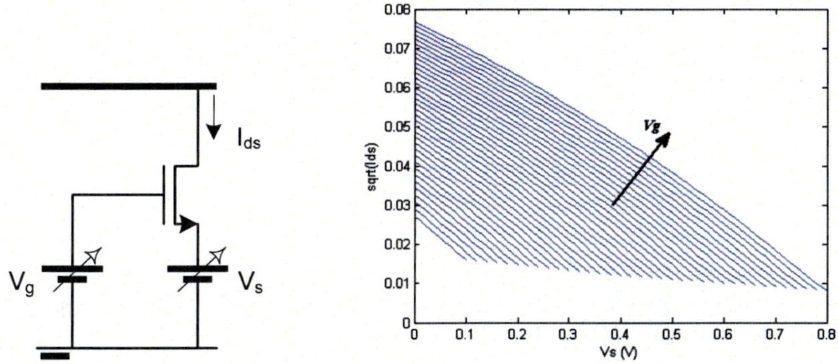

(a) (b)

Fig. 16.2 (a). Circuit for specific current extraction, (b). $\sqrt{I_{ds}}$ as function of V_s

The pinch-off voltage, V_P, is determined by measuring the source voltage in saturation mode, when the transistor current, I_{ds}, is in the order of half of specific current, I_s [9]. Hence, in those conditions, the transistor is biased in moderate inversion region. The use of a simple circuit, similar to the one in Figure 16.2(a), to simulate the transistor performance, does not guarantee a constant drain-source voltage when sweeping V_g. Therefore a small error arises from the channel length modulation, which affects mostly the short-channel devices. The circuit of Figure 16.3(a) overcomes this issue, where a constant V_{ds} voltage is imposed by means of an Op-Amp, and its value is regulated by the resistor R, and the current source I_R [9]. The pinch-off voltage obtained by simulation is represented in Figure 16.3(b).

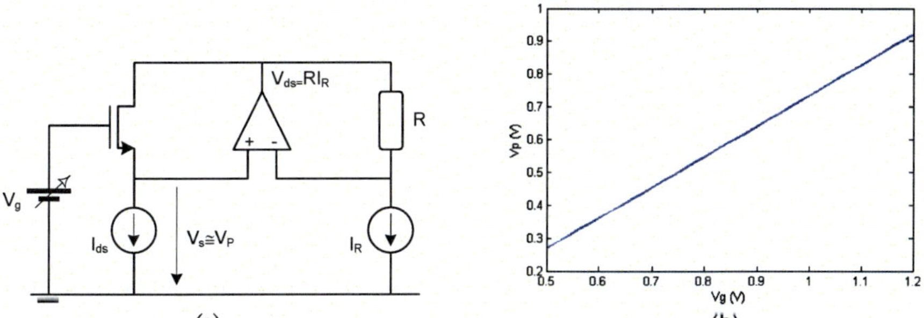

(a) (b)

Fig. 16.3 Circuit for pinch-off voltage (V_P), extraction (a), and V_P, as function of V_g (b)

The first three EKV parameters, V_{to}, γ and ϕ, are determined from the previous results. The value of the threshold voltage, V_{to}, is the value of V_g when V_P equals zero, which is easily obtained by fitting the simulated curve in Figure 16.3(b). The values of γ and ϕ, are extracted by fitting (7)-(8) to the measured characteristic.

The next set of parameters, β_0, Θ, U_T, are determined by fitting (4)-(6) to the simulated characteristics of the specific current and pinch-off voltage. In our approach, we have considered the thermal voltage, U_T, as a constant given by, U_T=*BT*/*q*, where *B* is the Boltzman constant, *T* is the absolute temperature (K) and *q* is the electron charge.

Finally, to determine the parameters *LETA* and *WETA*, two different strategies are needed. In one hand, the parameter *LETA* is obtained through the simulation of $V_P = f(V_g)$ for wide/short (W_{max}, L_{min}) transistors. On the other hand, the parameter *WETA* is obtained through the simulation of $V_P = f(V_g)$ for narrow/long transistors [10]-[11]. In both situations, the simulated results together with the EKV model equations are used for the curve fitting, making possible the extraction of the parameters *LETA* and *WETA*.

As a parameters extraction working example, in Table 16.1, the EKV model parameters obtained for UMC130 technology, and a NMOS transistor with length *L*=0.39μm and width *W*=50*L*, are presented. In Figure 16.4, the error between the transistor current, I_{ds}, obtained with the EKV model and simulations through HSPICE, is plotted. As it is possible to observe, the error range is very satisfactory.

Table 16.1 EKV model parameters

V_{to}	γ	β_0	Θ	U_T	ϕ	W_{eff}	L_{eff}	*LETA*	*WETA*
$261.5e^{-3}$	$175.0e^{-3}$	$31.24e^{-3}$	1.394	$25.9e^{-36}$	$778.1e^{-3}$	$1.94e^{-5}$	$3.67e^{-7}$	$6.85e^{-3}$	$1.15e^{-6}$

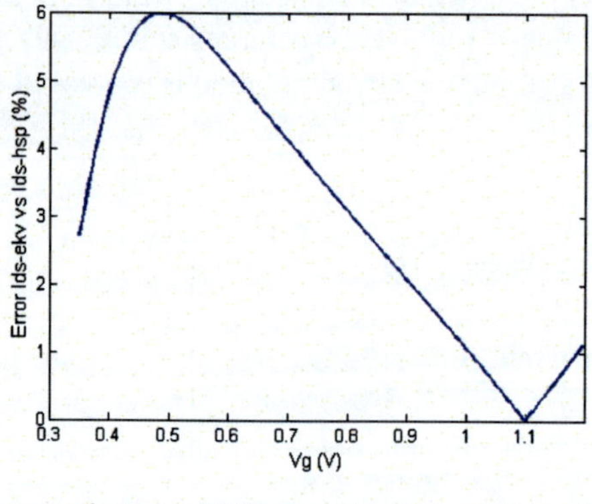

Fig. 16.4 Transistor I_{ds} error between EKV model and simulation (HSPICE) results

16.3 CMOS Varactor Analytical Model

In an *LC-VCO*, the tank circuit is of major importance since it is responsible for producing the required oscillatory signal. Moreover, the varactor is the element which gives the oscillator the capability of being tuneable. In this section a mathematical model for the CMOS varactor characterization is presented. This model has two main advantages. Firstly, the analytical model for the varactor capacitances is based on process and technological parameters, avoiding the undesired empirical/fitting factors. Secondly, the analytical transistor model, which is needed to determine the transistor current, is the well know EKV model, guaranteeing the accuracy of the results for low-voltage circuit design.

The EKV transistor model is suitable to perform the automatic CV-characterization of varactors, due to reduced number of parameters as well as the continuity of the model. According to [12], the intrinsic capacitances of a varactor are obtained through the relative variation of the nodes charge against the node voltage, obtained by:

$$C_{xy} = \pm \partial Q_x / \partial V_y \quad \text{with} \quad x, y = G, D, S, B \tag{16.12}$$

In a varactor, the total capacitance is usually referred to as the gate capacitance, since the drain, source and bulk are connected to a fixed voltage, which allows neglecting the drain/source – bulk capacitance. Additionally, overlap and fringing capacitances – extrinsic capacitances, $C_{extrinsic}$ – must be accounted for. The varactor total capacitance can be obtained through

$$C_{total} = C_{GB} + C_{GD} + C_{GS} + C_{S(D)B} + C_{extrinsic} \tag{16.13}$$

In [12] and [13] simplified expressions to determine each of the intrinsic capacitances are proposed. The varactor intrinsic capacitances are obtained through the following set of equations:

$$C_{GS} = 2/3C_{ox}\left[1-\left(I_{rev}^2 + I_{rev} + 0.5I_{for}\right)/\left(I_{rev} + I_{for}\right)^2\right] \tag{16.14}$$

$$C_{GD} = 2/3C_{ox}\left[1-\left(I_{for}^2 + I_{for} + 0.5I_{rev}\right)/\left(I_{rev} + I_{for}\right)^2\right] \tag{16.15}$$

$$C_{GB} = C_{ox}\left[(n_q - 1)/n_q\right]\left[1 - C_{GS}/C_{ox} - C_{GD}/C_{ox}\right] \tag{16.16}$$

$$C_{SB} = (n_q - 1)C_{GS} \tag{16.17}$$

$$C_{DB} = (n_q - 1)C_{GD} \tag{16.18}$$

where I_{rev} and I_{for} are the normalised reverse and forward current, respectively; n_q is the slope factor, γ is the body effect parameter, and V_P is the pinch-off voltage, obtained by

$$I_{rev} = \sqrt{0.25 + i_r} \tag{16.19}$$

$$I_{rev} = \sqrt{0.25 + i_r} \tag{16.20}$$

$$n_q = 1 + \frac{\gamma}{2\sqrt{V_P + \phi + 10^{-6}}} \tag{16.21}$$

In nano-CMOS technologies, besides the intrinsic capacitances, the extrinsic (parasitics) capacitances may be a major player in the varactor total capacitance. In the extrinsic region the capacitance is bias dependent, and thus essentially influenced by the gate voltage [14]. The extrinsic capacitance may be determined by:

$$C_{extrinsic} \approx 2\left(C_{ov}(V_g) + C_{if}(V_g) + C_{of}\right) \tag{16.22}$$

In (16.22) $C_{ov}(V_g)$ is the parallel plate capacitance associated with the electric field in the gate-to-drain/source overlap region; $C_{if}(V_g)$ is the inner fringing capacitance associated with the inner electric field emerging from metallurgical junction source/drain to the underside of the poly-gate. C_{of} is the outer fringing capacitance, independent of the gate voltage, related to the electric field emerging from the sidewall of the poly-gate, ending at the source/drain region [14]. The gate overlap capacitance is here defined as in (16.23), where L_{ov} is the effective diffusion length.

$$C_{ov}(V_g) = C_{ox} \cdot L_{ov}(V_g) \tag{16.23}$$

Concerning the fringing capacitances, the model proposed in [15] is adopted. For the inner fringing capacitance the equation proposed for nano technologies, does not account for the bias dependence. Since this capacitance is strongly influenced by the gate voltage, a more accurate expression is presented in [14], where:

$$C_{if} = C_{if,\max} \exp\left[-\left(\frac{V_g - V_{fb} - \phi_f/2}{3\phi_f/2}\right)^2\right] \tag{16.24}$$

The outer fringing capacitance is bias voltage independent, and may be calculated by:

$$C_{of} = \frac{2 \cdot \varepsilon_{ox}}{\pi} \ln\left(1 + \frac{T_{poly}}{T_{ox}}\right) \tag{16.25}$$

Figure 16.5 shows the characteristic of an inversion MOS varactor (I-MOS) – a transistor with $B=D=S$, working only in the strong, moderate and weak region – obtained with the proposed varactor analytical model, for different tuning voltages, namely 0.2V (I), 0.4V (II) and 0.6V (III), against simulations obtain with

HSPICE (+) software. The relative error for the mean capacitance was less than 4% in all cases. For those examples, the UMC130 CMOS technology, a transistor width of 10 μm and length of 0.8 μm, with a supply voltage of 1.2 V, was considered. The results in Figure 16.5 show the accuracy of the varactor model thus guaranteeing its adequacy to be integrated in an optimization based design tool.

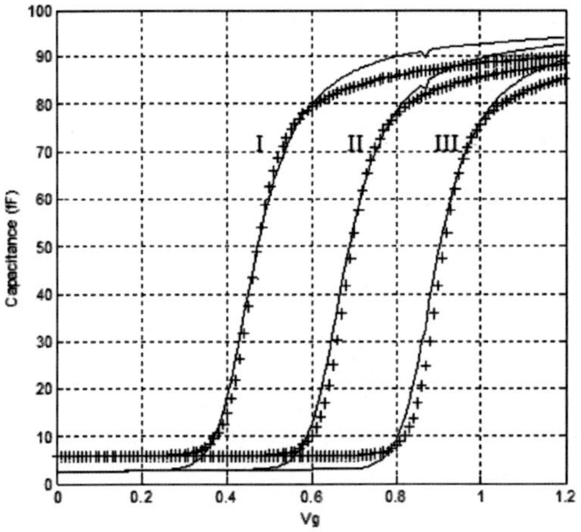

Fig. 16.5 Tuning characteristics for an I-MOS varactor

16.4 Spiral Inductor Model

The efficient design of on-chip planar inductors, Figure 16.6(a), is still a challenging task, since it relies on the availability of accurate models. When modelling an inductor, there are three major sources of losses that must be accounted for. Namely, the series resistance of the inductor, which depends on geometric and technological parameters, such as the inductor length, eddy currents and the skin effect at high frequencies, the capacitive coupling between metal and substrate, and the power losses due to eddy currents in the substrate.

The double π-model has been proposed with the aim of overcoming the lack of accuracy of results obtained with the single π model for frequencies above 1 GHz. This limitation stems from the fact that the analytical expressions for the evaluation of the π-model lumped elements do not take into account some high frequency effects, such as skin and proximity effects [16].

The double π-model is represented in Figure 16.6(b). This model uses a wide range of equations for evaluating the inductor model lumped element values. In spite of the complexity of the model, this equivalent circuit model considerably reduces computation time, when compared to electromagnetic simulation, and supports optimization design. In the double π-model, it is not possible to characterize the inductor quality factor (Q) with a single equation, as for the π model. Here, circuit analysis is needed, such as Kirchhoff's laws and electromagnetic induction theory, for obtaining the value of the input impedance of the one port circuit, and then Q can be calculated.

For the evaluation of the double π-model, the set of equations used can be divided in three blocks; DC inductor parameters, Substrate *Network* and *Ladder* Circuit elements, respectively. Some of those elements such as L_0 and L_p, depend on the inductance value at low frequencies (DC analysis), L_{dc}, where parasitics do not affect the inductor behaviour. For the estimation of the spiral inductance, L_{dc}, several approximate formulas can be found in the literature, such as the Modified Wheeler equation or the Greenhouse Approximation of Grover [17]. Regarding the model capacitors, C_s accounts for all overlap capacitances whereas C_{ox} and C_c account for the parasitic capacitances between the metal-substrate and metal-to-metal. The values for these capacitances may be evaluated with the expressions proposed in [18]. Concerning the substrate elements, the Ohmic losses in the conductive silicon substrate as well as capacitive effects are encapsulated in R_{sub} and C_{sub}. Another resistor, R_{sc} represents the losses due to the electric lines through the conductive substrate. The values for these elements are obtained with the set of equations proposed in [16]. Finally, and considering the *Ladder* circuit elements, in [16] and [18] is offered the range of equations necessary for the computation of the extra elements behaviour.

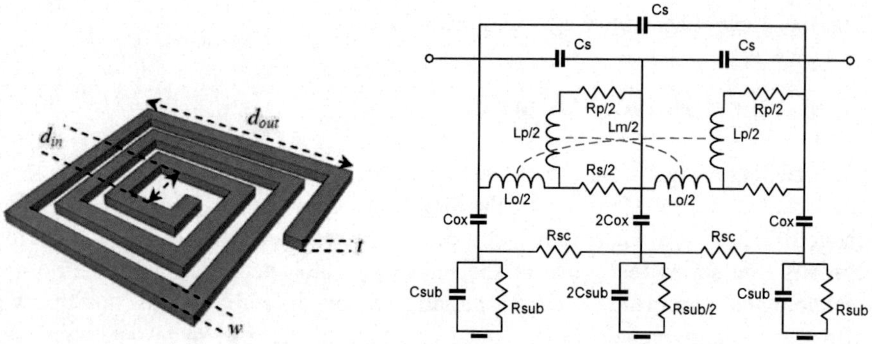

(a) (b)

Fig. 16.6 Layout of a square planar inductor (a), Inductor double π model (b)

• Model Evaluation

As previously pointed out, a single expression for obtaining the inductance and quality factor, cannot be derived. Furthermore, a fully symbolic characterization is also not easily generated, so a semi-symbolic approach is considered. Also a nested-equation approximation is used where the intermediate variables considered are [3]:

$$\overline{Z}_0 = 0.5 j\omega L_s \quad (16.26)$$

$$\overline{Z}_1 = 0.5 j\omega L_p + 0.5 R_p \quad (16.27)$$

$$\overline{Z}_2 = 0.5R_s \tag{16.28}$$

$$\overline{Z}_M = 0.5j\omega L_m \tag{16.29}$$

$$\overline{Z}_{Cs} = 1/j\omega C_{Cs} \tag{16.30}$$

$$\overline{Z}_{Cc} = 2/j\omega C_{Cc} \tag{16.31}$$

$$\overline{Z}_{Cox} = 2/j\omega C_{Cox} \tag{16.32}$$

$$\overline{Z}_{Cox\_mid} = 1/j\omega 2C_{Cox} \tag{16.33}$$

$$\overline{Z}_{Rsc} = 0.5R_{sc} \tag{16.34}$$

$$\overline{Z}_{sub} = \left(1/j\omega C_{sub}\right) // R_{sub} \tag{16.35}$$

$$\overline{Z}_{sub\_mid} = \left(1/j\omega 2C_{sub}\right) // \left(R_{sub}/2\right) \tag{16.36}$$

$$alfa = 1 - \overline{Z}_M^2 / \left(\overline{Z}_0 + \overline{Z}_1\right) \tag{16.37}$$

yielding the circuit admittance matrix represented beneath

$$\begin{bmatrix}
\frac{alfa}{\overline{Z}_0}+\frac{1}{\overline{Z}_{Cs}}+\frac{1}{\overline{Z}_{Cc}}+\frac{1}{\overline{Z}_{Cox}} & -\frac{alfa}{\overline{Z}_0} & -\frac{1}{\overline{Z}_{Cs}} & \frac{alfa\cdot\overline{Z}_M}{\overline{Z}_0\cdot\overline{Z}_1} & -\frac{1}{\overline{Z}_{Cox}} & 0 & 0 \\
-\frac{alfa}{\overline{Z}_0} & \frac{alfa}{\overline{Z}_0}+\frac{alfa}{\overline{Z}_1}+\frac{1}{\overline{Z}_2} & -\frac{alfa}{\overline{Z}_1}-\frac{alfa\cdot\overline{Z}_M}{\overline{Z}_0\cdot\overline{Z}_1}-\frac{1}{\overline{Z}_2} & 2\frac{alfa\cdot\overline{Z}_M}{\overline{Z}_0\cdot\overline{Z}_1} & 0 & 0 & 0 \\
-\frac{1}{\overline{Z}_{Cc}} & -\frac{alfa}{\overline{Z}_1}-\frac{alfa\cdot\overline{Z}_M}{\overline{Z}_0\cdot\overline{Z}_1}-\frac{1}{\overline{Z}_2} & \frac{alfa}{\overline{Z}_1}+2\frac{alfa\cdot\overline{Z}_M}{\overline{Z}_0\cdot\overline{Z}_1}+\frac{1}{\overline{Z}_2}+2\frac{1}{\overline{Z}_{Cc}}+\frac{alfa}{\overline{Z}_0}+\frac{1}{\overline{Z}_{Cox_mid}} & -\frac{alfa\cdot\overline{Z}_M}{\overline{Z}_0\cdot\overline{Z}_1}-\frac{alfa}{\overline{Z}_0} & 0 & -\frac{1}{\overline{Z}_{Cox_mid}} & 0 \\
\frac{alfa\cdot\overline{Z}_M}{\overline{Z}_0\cdot\overline{Z}_1} & 2\frac{alfa\cdot\overline{Z}_M}{\overline{Z}_0\cdot\overline{Z}_1} & -\frac{alfa\cdot\overline{Z}_M}{\overline{Z}_0\cdot\overline{Z}_1}-\frac{alfa}{\overline{Z}_0} & \frac{alfa}{\overline{Z}_0}+\frac{alfa}{\overline{Z}_1}+\frac{1}{\overline{Z}_2} & 0 & 0 & 0 \\
-\frac{1}{\overline{Z}_{Cox}} & 0 & 0 & 0 & \frac{1}{\overline{Z}_{Cox}}+\frac{1}{\overline{Z}_{sub}}+\frac{1}{\overline{Z}_{Rsc}} & -\frac{1}{\overline{Z}_{Rsc}} & 0 \\
0 & 0 & -\frac{1}{\overline{Z}_{Cox_mid}} & 0 & -\frac{1}{\overline{Z}_{Rsc}} & 2\frac{1}{\overline{Z}_{Rsc}}+\frac{1}{\overline{Z}_{Cox_mid}}+\frac{1}{\overline{Z}_{sub_mid}} & -\frac{1}{\overline{Z}_{Rsc}} \\
0 & 0 & 0 & 0 & 0 & -\frac{1}{\overline{Z}_{Rsc}} & \frac{1}{\overline{Z}_{Cox}}+\frac{1}{\overline{Z}_{sub}}+\frac{1}{\overline{Z}_{Rsc}}
\end{bmatrix}$$

Reducing the circuit to its one port equivalent, with the right-side terminal grounded, the input impedance is calculated, Z_{in}, and then the inductance value, L, and the quality factor, Q, are obtained by:

$$L = \Im m\left(\overline{Z}_{in}\right)/2\pi f_o \tag{16.38}$$

where f_0 is the oscillation frequency, and

$$Q = \Im m\left(\overline{Z}_{in}\right)/\Re e\left(\overline{Z}_{in}\right) \tag{16.39}$$

In Figure 16.7 the performance of a four turn inductor obtained through the double π-model and results from [19] are illustrated. Regarding the qualitative behaviour of both curves, a good correspondence was achieved.

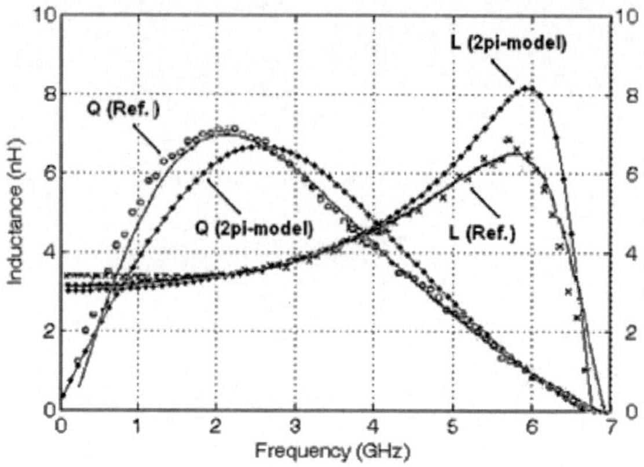

Fig. 16.7 Inductance and Quality Factor plot for a 4-turn inductor

16.5 LC-VCO Design Approach

When designers have in hands the task of designing a *VCO*, one of the main challenges is to obtain results between simulation and on-chip measurement as close as possible. Most of the *LC-VCO* designs have as main criteria to achieve both minimum phase noise and minimum power consumption for a certain oscillation frequency. In a very simplistic point of view, for a specified frequency, the design problem to be worked out is the sizing of all the elements of the *LC-VCO*. However some trade-offs regarding the *LC-VCO* specifications, make the design more and more challenging, becoming a perfect candidate for optimization based design. For instance, if low power consumption is desired, a low bias current must be delivered to the circuit. Yet, parasitic effects will have a major role in circuit behaviour, yielding to the degradation of phase noise. On the other hand, if low phase noise is required, high output voltage swing is desired. To achieve this goal, either the power consumption increases or the *VCO* tuning range shrinks, due to a higher inductance.

In this Section an optimization-based *LC-VCO* design methodology, for the topology represented in Figure 16.1, is proposed. The optimization-based design flow for the proposed methodology is illustrated in Figure 16.8 [20]. Having in mind the oscillator topology represented in Figure 16.1, for a given set of specifications, the design process starts with the design of the active elements, by fixing the DC bias current, since it is the main dominant contributor to the phase noise. The following step aims to optimize the *LC* tank design. This process starts with the optimization of the inductor layout for a fixed inductance, and then the varactor is determined, aiming to maximize the tank quality factor. Finally, the envisaged circuit phase noise and power consumption are validated by means of a figure of merit, *FoM*.

In previous sections a detailed insight to analytical models of each element of the *LC-VCO* was offered. Finally, a last set of equations, concerning the full circuit characterization is offered. The oscillation frequency, f_0, of the *LC-VCO* is given by:

$$f_0 = \frac{1}{2\pi\sqrt{L_{tank}\, C_{tank}}} \tag{16.40}$$

where C_{tank} takes into account both the varactor capacitance, as well as the active elements output capacitance, and L_{tank} is the inductance of the on-chip inductor. The differential output voltage, $2V_{tank}$, is calculated by:

$$V_{tank} = \frac{4}{\pi}\frac{I_{bias}}{g_{tank}} \tag{16.41}$$

Where

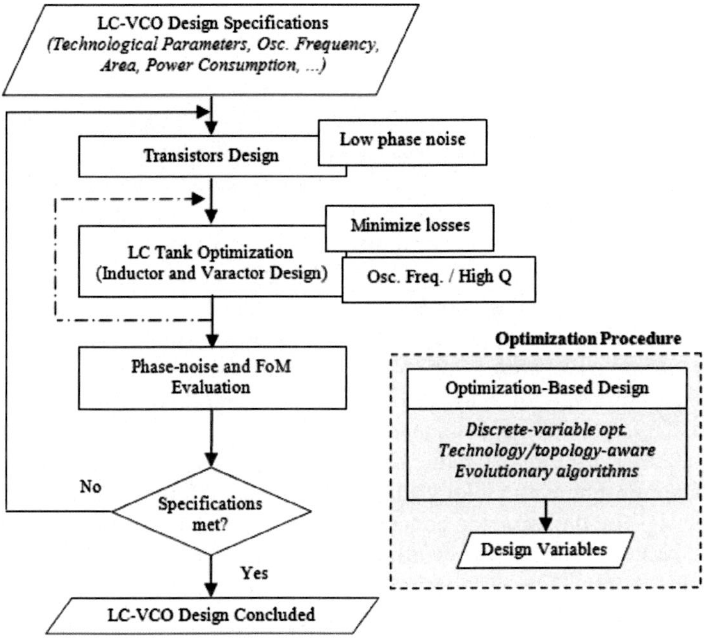

Fig. 16.8 *LC-VCO* design flowchart

$$g_{tank} = g_{ind} + g_{var} + \frac{g_{ds,p}}{2} + \frac{g_{ds,n}}{2} \tag{16.42}$$

where g_{ind}, g_{var}, $g_{ds,p/n}$ are the inductor, varactor and output conductances, respectively.

Phase-noise is an elementary characteristic of a *VCO* that reflects the purity of the oscillation signal in the vicinity (Δ_f) of the oscillation frequency f_0, and is obtained through [21]:

$$L\{\Delta_f\} = 10\log\left[\frac{L_{\text{tank}}^2 (2\pi f_0)^4}{16\pi^2 \Delta_f^2 V_{\text{tank}}^2} \cdot 2K_B T\left[g_{\text{ind}} + g_{\text{var}} + \gamma_{nf}\left(g_{d0,p} + g_{d0,n}\right)\right]\right] \quad (16.43)$$

where g_{d0} is the drain conductance when $V_{DS}=0$, and γ_{nf} is the excess noise factor. Lastly, the figure of merit, *FoM*, is given by:

$$FoM = L\{\Delta_f\} - 20\log\left(\frac{f_0}{\Delta_f}\right) + 10\log\left(P_{\text{dc(mW)}}\right) \quad (16.44)$$

where P_{dc} is the dc power dissipated in the oscillator.

16.6 Design Results

The present work shows an *LC-VCO* obtained with the proposed methodology. It was implemented in Matlab and uses the Genetic Algorithms (GA) toolbox. In this section the design of three *LC-VCO*s for operating frequencies of 1.0, 1.5 and 2.2 GHz in UMC130 technology are addressed. The proposed design methodology approach deals with three major optimization processes: the inductor optimization, the varactor optimization, and the *LC-VCO* performance optimization, regarding phase noise and power consumption.

For the planar inductor design, the methodology adopted aims to achieve a technology/topology-aware solution. Also to be accounted for is the discrete nature of the design variables, as well as, the correlation between them. The design concerns the evaluation of four independent parameters, namely the track width (w), the number of turns (n), inductor shape (N_{side}), and the internal diameter (d_{in}). The inductor design methodology adopted in this work follows [3] and [20].

The design of the varactor concerns the evaluation of three independent parameters, namely the transistor width (W), the transistor length (L) and the number of gate fingers (N_f). The GA optimization algorithm deals with both continuous and discrete variables. The transistor width and length are considered as continuous variables, but the number of gate's fingers is an integer value. Technological and physical parameters, optimization constraints, as well as the optimization approach, are presented in [22].

Regarding the *LC-VCO* performance, the optimization objective functions are the minimization of both the phase noise and the power consumption. The *VCO* envisaged characteristics and parameters range are given in Table 16.2. The results obtained with the proposed methodology as well as those obtained through HSPICE/RF simulations, are presented in Table 16.3, showing a quite good agreement between predicted and simulations results. The dimensions of the tank elements, obtained through the optimization design procedure, are shown in Table 16.4. In Figure 16.9 the *VCO* output signals for an oscillation frequency of 1.5 GHz are presented. The output signal oscillates between 1.11 V and 0.61 V, which represents a tank output swing of 0.50 V.

Table 16.2 *LC-VCO* Characteristics

Center frequency (f_0 / Δ_f)	1.0, 1.5, 2.2 GHz / 1 MHz
Bias current	0.5 mA – 5 mA
Output voltage swing	$V_{DD}/8$ – $V_{DD}/2$
Transistor width	$3L_{min}$ – 1000 μm
Tank inductance - L	1 nH – 10 nH
Tank capacitance - C_{var}	0.5 pF – 10 pF
C_{Load}	1 pF

Table 16.3 Optimization *vs.* Simulation results

	1.0 GHz		**1.5 GHz**		**2.2 GHz**	
	Optim	**Hspice**	**Optim**	**Hspice**	**Optim**	**Hspice**
I_{bias} (mA)	1.00	1.05	1.30	1.36	1.20	1.26
W_p (μm)	217.8	-	283.0	-	261.4	-
W_n (μm)	88.6	-	115.5	-	106.4	-
W_b (μm)	78.6	-	102.2	-	94.3	-
L_{tank} (nH)	7.0	-	2.5	-	3.0	-
C_{tank} (pF)	4.40	4.44	5.90	5.96	0.625	0.634
Tank Q	7.50	-	11.67	-	9.82	-
V_{outAmp}(V)	0.16	0.18	0.22	0.25	0.27	
P_{dc} (mW)	1.20	1.25	1.56	1.60	1.44	1.50
f_0 (GHz)	1.01	0.95	1.50	1.35	2.20	1.95
L{1MHz} (dBc/Hz)	-115.5	-116.2	-131.3	-112.9	-117.7	-118.1
FoM (dBc/Hz)	174.7	174.8	192.9	173.5	183.0	182.1

Table 16.4 Tank inductor and capacitor dimensions

	Planar Inductor				**Varactor**		
	Width (μm)	**Din (μm)**	**Nturns**	**Nside**	**Width (μm)**	**Length (μm)**	**Nfingers**
1.0 GHz (7.0 nH / 4.40 pF)	9.50	137.50	4.5	4	483.0	1.25	198
1.5 GHz (2.5 nH / 5.90 pF)	15.50	180.75	2.5	6	642.0	1.45	198
2.2 GHz (3.0 nH / 0.63 pF)	10.50	192.25	2.5	8	84.7	0.94	52

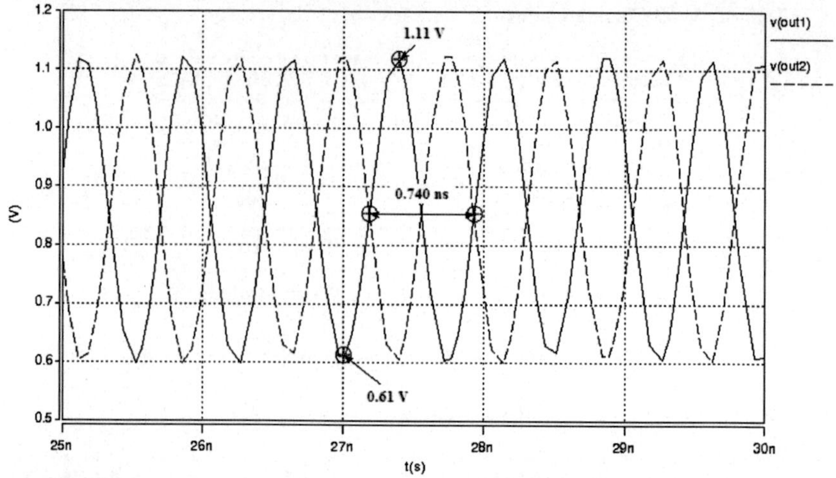

Fig. 16.9 *LC-VCO* output signal (Vout1 and Vout2) @1.5GHz with indication of signal period and output swing

16.7 Conclusions

This chapter introduces an optimization based methodology for the design of *LC-VCOs*. The efficiency of the design process is granted by using analytical models to characterize the active and passive elements' behaviour, which offers an extremely easy adaptability of the design process to new technologies.

The design of *LC-VCO* is supported by a Genetic Algorithms optimization methodology, which is able to deal with both continuous and discrete variables, making possible to satisfy both technological and layout constraints. A set of design examples showing the design of three *VCOs* for different oscillation frequencies were considered. The feasibility of the obtained design solutions is highlighted via comparison with simulated results.

References

[1] Razavi, B.: RF Microelectronics. Prentice Hall (1998)
[2] Masu, K., Ishihara, N., Nakayama, N., Sato, T., Amakawa, S.: Physical design challenges to nano-CMOS circuits. IEICE Electronics Express 6, 703–720 (2009)
[3] Pereira, P., Fino, M.H., Coito, F., Ventim-Neves, M.: RF integrated inductor modeling and its application to optimization-based design. Analog Integrated Circuits and Signal Processing (2011) (in press), doi:10.1007/s10470-011-9682-x
[4] Pereira, P., Fino, M.H., Coito, F., Ventim-Neves, M.: ADISI- An efficient tool for the automatic design of integrated spiral inductors. In: The IEEE International Conference on Electronics Circuits and Systems, Hammamet, Tunisia (2009)
[5] Pereira, P., Fino, M.H.: CMOS delay and power estimation for deep submicrometer technologies using EKV model. In: The International Workshop on Symbolic and Numerical Methods, Modeling and Applications to Circuit Design, Erfurt, Germany (2008)

[6] Gutierrez, I., Meléndez, J., Hernández, E.: Design and characterization of integrated varactors for RF applications. John Wiley & Sons (2007)
[7] Machado, G., Enz, C., Bucher, M.: Estimating key parameters in the EKV MOST model for analogue design and simulation. In: The IEEE International Symposium on Circuits and Systems, Seattle, Washington, USA (1995)
[8] Enz, C., Krummenacher, F., Vittoz, E.: An analytical MOS transistor model valid in all regions of operation and dedicated to low-voltage and low-current applications. Analog Integrated Circuits and Signal Process. 8(1), 83–114 (1995)
[9] Bucher, M., Lallement, C., Enz, C.: An efficient parameter extraction methodology for the EKV MOST model. In: The IEEE International Conference on Microelectronic Test Structures, Trento, Italy (1996)
[10] Bucher, M., Lallement, C., Enz, C., Krummenacher, F.: Accurate MOS modelling for analog circuit simulation using the EKV model. In: The IEEE International Symposium on Circuits and Systems, Trento, Italy (1996)
[11] Stefanovic, D., Kayal, M.: Structured analog CMOS design. Springer (2008)
[12] Bucher, M., Lallement, C., Enz, C., Théodoloz, F., Krummenacher, F.: The EPFL-EKV MOSFET model equations for simulation, Version 2.6 (1998)
[13] Bremer, J., Peikert, T., Mathis, W.: Analytical inversion-mode varactor modeling based on the EKV model and its application to RF VCO design. In: The International Conference on Mixed Design of Integrated Circuits and Systems, Wrocław, Poland (2010)
[14] Prégaldiny, F., Lallement, C., Mathiot, D.: A simple efficient model of parasitic capacitances of deep-submicron LDD MOSFETs. Journal of Solid-State Electronics 46(12), 2191–2198 (2002)
[15] Shrivastava, R., Fitzpatrick, K.: A simple model for the overlap capacitance of a VLSI MOS device. IEEE Transactions on Electron. Devices 29(12), 1870–1875 (1982)
[16] Cao, Y., Groves, R.A., Zamdmer, N.D., Plouchart, J.O., Wachnik, R.A., Xuejue, H., King, T.J., Chenming, H.Y.: Frequency-independent equivalent-circuit model for on-chip spiral inductors. IEEE Journal of Solid-State Circuits 38(3), 419–426 (2003)
[17] Mohan, S., del Mar Hershenson, M., Boyd, S., Lee, T.: Simple accurate expressions for planar spiral inductances. IEEE Journal of Solid-State Circuits 34(10), 1419–1424 (1999)
[18] Wong, S.C., Lee, G.Y., Ma, D.J.: Modeling of interconnect capacitance, delay, and crosstalk in VLSI. IEEE Transactions on Semiconductor Manufacturing 13(1), 108–111 (2000)
[19] Sia, C.B., Yeo, K.S., Goh, W.L., Swe, T.N., Ma, J.G., Do, M.A., Lin, J.S., Chan, L.: A Simple and Scalable Model for Spiral Inductors on Silicon. In: The International Conference on Modeling and Simulation of Microsystems, South Carolina, USA (2001)
[20] Pereira, P., Fino, H., Coito, F.V., Ventim-Neves, M.: Optimization-Based Design of Nano-CMOS LC-VCOs. In: Camarinha-Matos, L.M., Shahamatnia, E., Nunes, G. (eds.) DoCEIS 2012. IFIP AICT, vol. 372, pp. 453–464. Springer, Heidelberg (2012)
[21] Hajimiri, A., Lee, T.: A general theory of phase noise in electrical oscillators. IEEE Journal of Solid-State Circuits 33, 179–194 (1998)
[22] Pereira, P., Fino, M.H., Ventim-Neves, M.: RF Varactor Design Based on Evolutionary Algorithms. In: The International Conference Mixed Design of Integrated Circuits and Systems, Warsaw, Poland (2012)

Author Index